Making Physics

Making Physics

A Biography of
Brookhaven
National Laboratory,
1946–1972

Robert P. Crease

The University of Chicago Press
Chicago and London

The University of Chicago Press, Chicago 60637
The University of Chicago Press, Ltd., London

26 25 24 23 22 21 20 19 18 17   2 3 4 5 6

Excerpts from the series "The History of Brookhaven National Laboratory" are reprinted from the *Long Island Historical Journal*, in which they originally appeared.

Figure 1.4 is reproduced courtesy of the California Institute of Technology. Figure 1.5 is reproduced courtesy of Laura M. Whitman. Figure 9.3 is reproduced courtesy of Time Inc. All other figures are reproduced courtesy of Brookhaven National Laboratory.

ISBN: 978-0-226-12019-5 paper

Library of Congress Cataloging-in-Publication Data

Crease, Robert P.
    Making physics : a biography of Brookhaven National Laboratory,
1946–1972 / Robert P. Crease.
        p.   cm.
    Includes bibliographical references and index.
    ISBN 0-226-12017-1 (alk. paper)
    1. Brookhaven National Laboratory—History.   2. Nuclear physics—
Research—New York (State)—Upton—History.   I. Title.
QC789.2.U62B763   1999
539.7'0720747'25—dc21
                                                98-30327
                                                  CIP

Contents

	List of Illustrations	vii
	Acknowledgments	xi
	Introduction	1
1	A Team of Young General Groveses	7
2	A Reluctant Director, a Remote Site	22
3	National Laboratory	40
4	The "Brookhaven Concept"	47
5	The Pile Project	69
6	Community Relations	93
7	The Accelerator Project	108
8	Reactor Research in the 1950s	152
9	"For the Enlightenment and Benefit of Mankind": Research at the Cosmotron	200
10	Goldhaber's Directorship	257
11	Research at the Alternating Gradient Synchrotron	281
12	The High Flux Beam Reactor	316
13	Crossroad	348
	Appendix 1: Some Key Personnel of Brookhaven National Laboratory and Associated Universities, Inc.	369

Appendix 2: Chairmen of Key Brookhaven National
Laboratory Departments   371

Notes   373

A Note on Sources   411

Bibliography   413

Index   425

Illustrations

1.1	I. I. Rabi	10
1.2	George Collins and Norman Ramsey	11
1.3	George Pegram	13
1.4	Lee A. DuBridge	17
1.5	Mariette Koevesi von Neumann Kuper	18
2.1	Philip Morse	24
2.2	Sites considered for a national laboratory	30
2.3	Plot of proposed lab site at Fort Hancock	34
2.4	Plot plan for the Northeast Laboratory at Camp Upton	36
2.5	The nine AUI universities	38
3.1	Optimistic schedule for completion of major facilities	42
3.2	The three national laboratories, 1947	45
4.1	Entrance to Brookhaven, 1947	48
4.2	Aerial view of lab site, 1947	48
4.3	Organization chart, 1947	49
4.4	Guide map to lab, 1947	59
4.5	Donald D. Van Slyke	64
4.6	Leland Haworth	67
5.1	Lyle B. Borst	70
5.2	Artist's sketch of reactor	71
5.3	Model of the Brookhaven Graphite Research Reactor (BGRR)	73

5.4   First (bottom) layer of graphite   75

5.5   Breaking ground for the BGRR   77

5.6   Tour of the BGRR's construction site   81

5.7   Looking into an exhaust air duct   82

5.8   The Brookhaven Graphite Research Reactor   86

5.9   Aerial view of site   91

7.1   M. Stanley Livingston   110

7.2   Assembling the electrostatic accelerator   112

7.3   Everett Hafner, working on tank   114

7.4   Erection of sixty-inch cyclotron   120

7.5   Proposed 240-inch synchrocyclotron   122

7.6   Surveying base plates for the magnet blocks   130

7.7   Model of Cosmotron magnet   131

7.8   Diagram of Cosmotron magnet   132

7.9   Assembling the vacuum chamber   136

7.10 Final magnet block being assembled   137

7.11 Plot of accelerator history   138

7.12 Setting the vacuum chamber into place   139

7.13 Schematic layout of Cosmotron   140

7.14 BeV Day   142

7.15 CERN scientists visit Brookhaven, August 1952   143

7.16 Four Brookhaven scientists who worked on strong
          focusing   146

8.1   Early Brookhaven logo   154

8.2   West face of BGRR   155

8.3   Donald Hughes at neutron reflector   156

8.4   Fast neutron chopper   161

8.5   Atoms for Peace stamp   162

8.6 Diagram of a neutron cross-section of silver   164

8.7 Gertrude Scharff-Goldhaber   170

8.8 Vance Sailor at his cryogenic refrigerator   176

8.9 Double-axis spectrometer   180

8.10 Lowering first BNCT patient into treatment facility   186

8.11 Space assignments at the BGRR, 1957   190

8.12 HFBR neutron profile   197

8.13 HFBR inventors   198

9.1 View from Cosmotron control room   202

9.2 Request for experimental time at the Cosmotron   203

9.3 Theorists and others who worked at the Cosmotron   204

9.4 Evidence of production of strange particles   209

9.5 Rigging cloud chamber into corn crib   211

9.6 Pion-proton scattering   214

9.7 Gerhart Friedlander working at the Cosmotron   215

9.8 Electron analogue of Alternating Gradient Synchrotron   225

9.9 Cosmotron experimental program   229

9.10 The twenty-inch bubble chamber   240

9.11 Equipment for studying neutrino helicity   250

9.12 AGS injection system   252

10.1 Maurice Goldhaber   260

10.2 Plan for the proposed 200 GeV accelerator   268

10.3 Schematic of proposed fourteen-foot bubble chamber   271

10.4 Diagram of seven-foot bubble chamber   273

10.5 George Cotzias   278

10.6 Lou Dahl   279

10.7	Taking a mouse's blood pressure	279
11.1	Neutrino experiment floor layout	291
11.2	Mel Schwartz with the neutrino detector	293
11.3	CP violation experiment	297
11.4	The first omega minus event	308
11.5	Team that discovered omega minus	309
12.1	Cutaway of HFBR showing building and reactor	319
12.2	Cutaway of reactor vessel showing internal components	320
12.3	Installing the HFBR reactor vessel	322
12.4	The beam tube and irradiation thimble layout	324
12.5	Cabin at the end of a long time-of-flight path	328
12.6	Neutron density map	330
12.7	US-1	333
12.8	HFBR experimental floor	335
12.9	Triple-axis spectrometer	337
13.1	Advertisement for *Brookhaven/Facade*	353
13.2	Moratorium day activity at Brookhaven	355
13.3	First issue of the *Brookhaven Free Press*	356
13.4	Drift tubes in new AGS linear accelerator	362
13.5	ISABELLE	366

Acknowledgments

I conceived this project about eight years ago, while an assistant professor of philosophy at the State University of New York at Stony Brook, and finished the manuscript before Stony Brook became involved in Brookhaven's management (and before the tritium-leak controversy). Brookhaven supported half my time while I finished this project, as its historian, while granting me unrestricted access to and use of its records, for which I am grateful. I am also grateful to Associated Universities, Inc., which gave me unrestricted access to and use of its records during the time period covered by this book. I am indebted, far more than I can say, to the large number of scientists, administrators, and others who shared stories about their lives and work with me, many more than can be mentioned here. Among those who read the entire manuscript were: Robert Adair, Peter Bond, Arland Carsten, Ernest Courant, Stephanie Crease, Gerhart Friedlander, Everett Hafner, Jerome Hudis, Peter Kahn, Charles C. Mann, Lawrence Passell, Mona Rowe, Robert Sachs, Mark Sakitt, Nicholas P. Samios, Julius Spiro, Gerald Tape, and Peter Westwick. I received other helpful comments and assistance from Helen Berman, John Blewett, Martin Blume, Victor Bond, John Castro, Patrick Catt, Robert Conard, Eugene Cronkite, Gordon Danby, Ed Dexter, George Dienes, Paul Foreman, Kathleen Geiger, Maurice Goldhaber, David Gurinsky, Julius Hastings, Eleanor A. Hughes, Doug Humphrey, Andrew Kevey, Thomas Koetzle, Lillian S. Kouchinsky, Leon Lederman, John H. Marburger, Roger Meade, Abraham Pais, Martin Plotkin, Irving Polk, Lewis Porter, Robert Powell, Alex Reben, Vance Sailor, Lyle Schwartz, Mel Schwartz, Dan Slatkin, Lyle Smith, Michael Tannenbaum, Kara Villamil, Joe Weneser, C. N. Yang, and Lynn Yarris. Thanks, as well, to Don Ihde and Donn Welton of the Stony Brook philosophy department, who had the imagination to think that a national laboratory might be persuaded that what it really needed was a philosopher to scour its records, and to the Brookhaven administration, which had the insight to agree.

I silently scoffed, in my first conversation with my prospective editor, Susan E. Abrams, when she told me she was used to going through at least three manuscripts with authors; I thought it was preposterous nowadays for an editor to pay that much attention to a manuscript and assumed the remark was merely a ploy to get me to sign up. To my utter astonishment, I was wrong, and the manuscript benefited greatly from the care and patience she gave it. My agent Richard Balkin, as usual, stood by me when the going got rough. So did my wife, Stephanie Crease, who was supportive during many late nights at work and early mornings teaching. Martin Eger reminded me that the concepts we had been working on in the philosophy of science were useful in interpreting history. Pat Flood taught me how to archive boxes the allowable way. Andy Hull spent a day with me going over two decades of old health physics reports month by month, explaining the source, amount, method of detection, and significance of all the radiation and contamination measured at the lab during that period. Michael Koplow was one of the most careful manuscript editors I ever had the pleasure to work with. If this book is not a page-turner it's because I haven't yet learned how to appropriate all of Charles C. Mann's narrative techniques. Joe Rubino could always be counted on to deliver slides at the last minute. Janet Sillas knew how to find everything. I had many fine afternoons with Roger Stoutenberg, doing things like climbing the BGRR's stack and scouring the woods for old signs and other artifacts. Roger Wunderlich effectively got me going by insisting that I write about Brookhaven's history for the *Long Island Historical Journal*. Patricia Yalden contributed creative and innovative design ideas.

Introduction

This book is about Brookhaven National Laboratory during its first quarter-century. Brookhaven was one of the first three U.S. national labs, and like its siblings was the product of a unique partnership between scientists and the federal government that was forged during and after World War II. Created to build and operate forefront research facilities—mainly reactors and accelerators—that were too big for single universities to afford, the national labs aimed to foster a creative, independent environment while nevertheless being supported financially almost entirely by the federal government. Two of the first three national labs, Argonne National Laboratory and Oak Ridge National Laboratory, were originally constructed on a crash basis during World War II as military institutions in the course of the Manhattan Project to develop the atomic bomb, and were subsequently converted to peacetime purposes. Brookhaven was conceived, constructed, and managed differently. It was the only one of the three that was a civilian institution from its inception, and its program was the most oriented to basic research. In the history of American scientific institutions it was an experiment all its own: an experimental community.

This book is about Brookhaven as a community, and in its subtitle I have called it a biography because communities lend themselves to biographical treatment. One reason is that, like individuals, communities undergo constant struggles for identity and recognition and their actions are interwoven with the political and social environment. National laboratories are "crucibles of uncertainty," wrote a former director of Argonne, whose scientists never know where their research will lead, and whose managers "grapple daily with questions of relevance and survival" (Holl 1997, ix). Another reason to call this book a biography is that the range of sources one has to rely on in describing a community—memos, letters, anecdotes, revealing incidents, etc.—is like what generally appears in a biography. I included, for instance, excerpts from oral interviews in which the subjects reminisced about the bygone atmosphere, because these showed something crucial about the community life. I also included capsule summaries of how key people got to the lab, because it seemed to me important that, long before diversity was a sacred slogan, laboratories far more than other kinds of institutions collected individuals of vastly different temperaments and backgrounds.

Stories of scientific communities are timely for several reasons. One is that, with science facilities ever bigger, the federal stake in science projects ever larger, and the pressures on the government to curtail their independence ever greater, the management of large scientific facilities has become highly contentious. Two recent illustrative events are the 1993 cancellation, by the U.S. Congress, of the Superconducting Supercollider project, and the 1997 cancellation, by the Department of Energy, of the contract under which Associated Universities, Inc. (AUI), had operated Brookhaven for fifty years. One justification cited in each case was poor management. The Department of Energy has grown increasingly negative about academic contractors, and AUI's replacement, by Brookhaven Science Associates, a new institution whose principal members are the Research Foundation of the State University of New York (on behalf of SUNY, Stony Brook) and Battelle Memorial Institute, Inc., marked an important change in what once was a daring and innovative experiment in science policy. Further reductions in the role of academic management of national laboratories are likely in store. It is therefore an opportune time to reexamine the original vision that motivated the founders of the national labs (a vision that was inspired in part by research institutions such as universities), the experiences of trying to realize that vision, and the work that resulted. The founders of the first national labs thought it vital to create institutions that maintained a university-type atmosphere for research even while remaining almost totally dependent on government funding, so that these new institutions could avoid the trade-offs between short- and long-range interests that public corporations and political agencies are constantly forced to make, and allow researchers to pursue long-range plans and embark on risky and speculative projects. The stories of the national labs can help reveal what is living, and what dead, of that original vision.

A second reason why stories of the national labs are timely is that they embody lessons about certain issues that are still problematic for large scientific facilities, involving relations among researchers and between researchers and the community. Various kinds of rivalries arise among researchers, for instance, some of which are productive and others debilitating (what I call performance as opposed to political rivalries). Community relations, too, especially concerning the dangers of pollutants and waste products, and the inevitable gap that arises between the situation as perceived by the scientists, projected by the media, and interpreted by politicians, is more than ever a pressing concern. While right after World War II the national labs' dependence on the federal government insulated them from surrounding communities, today the labs are thrust into contact with them. And while right after the war dependence on Washington was of substantial benefit to the labs, today it is a liability and has given rise to a new political vulnerability. An incident involving tritium leakage from the spent fuel pool of a Brookhaven reactor, which the Department of Energy used in 1997 as the immediate justification for canceling AUI's contract, is emblematic in this respect. Though these events are far too complicated to go into here, their roots, I claim, can be traced back to events at the origins of the national labs.

Finally, I think that looking at stories of life in a scientific community is important for understanding science itself. How science "really" works is hotly debated today. Scholars have compared scientists to entrepreneur-generals waging war (Latour 1987), to lawyers and politicians whose messy negotiating process involves trading off interests (Latour and Woolgar 1979), and to schoolchildren who cannot resolve their own disputes but must be coerced into agreement (Collins and Pinch 1993). The scientific method has been portrayed as like the parrying and thrusting in a duel (Bloor 1976), and like a dialogue between detectives over interpretation of certain evidence (Eger 1997). Even the role of experimentation is controversial. Some say experiment has a "legitimate role in the choice between competing theories and in confirmation of theories or hypotheses" (Franklin 1986, 244), others that "scientists at the research frontier cannot settle their disagreements through better experimentation, more knowledge, more advanced theories, or clearer thinking" (Collins and Pinch 1993, 144–45). In each case we are asked to look at a different type of information to discover what is "really" going on in science. Do we look at the data, the experiments that produced them, the people who planned and executed the experiments, or the programs and policies that funded the experiments? Do we follow the equations or the money?

Each approach says something essential about the activity of science. Yet each, by itself, also makes the meaning of scientific activity something static, reducing it to what can be seen in the light of certain specific images and concepts—and thus threatens to promote a misleading image of science. (Galison [1997] is exceptional in his ability to capture the rich complexity of scientific activity without shoehorning it into a narrow explanatory scheme.) For many features of the dynamic moving event of science (and of other human activities) appear only in stories that unfold over time. Yet many histories depict science as consisting of abstract, technical achievements, with personalities providing at most spice or ornament.[1] To approach science this way is extremely valuable and even indispensable for certain purposes. But to marginalize the human dimension reinforces a highly distorted and misleading view of science as an abstract and blood-less matter of information—and of scientists as cold, uncultured, and dehumanized—in a way that has become a regrettably acceptable stereo-type. Even in the cosmopolitan climate of modern research, scientists are not robotic thinking machines, nor is their world some virtual space from which equations emerge. But it is an overreaction to approach the activity of scientists as do a thriving group of "constructivist" sociologists, as purely a product of social negotiations involving alliances and trade-offs of interests, with the outcome, truth, a mere artifact of discourse. The work of the constructivists has highlighted the role of interpretation in science, and the presence of social meanings implicit in scientific activity. But central to science is the planning, staging, and witnessing of material actions in ex-perimentation, and while this process has social aspects (events do not in-terpret themselves) it makes scientific activity something much different from negotiation of interests (Eger 1997). Central to science, too, is that scientists have lives, not just agendas.

Laboratories are in some sense like theaters: they are special kinds of sites equipped with special kinds of (ever-changing) facilities where differ-ent groups of (unrelated) people come to stage and witness special kinds of actions for the benefit of different communities. A laboratory history, therefore, can follow many different narrative threads: the science itself, or why particular experiments are regarded as crucial to perform; the instru-ments used in such experiments, to whose demands the laboratory has to respond; the individuals who direct, work at, and contribute a vision to the laboratory and its experiments; and the laboratory as a social institution, operating with certain constraints and freedoms amid other social institu-tions. This diversity of narrative threads makes laboratory history difficult and exasperating. Some laboratory histories pursue one or more of these narrative threads and ignore the rest; others are collections of loosely related essays, while a few (notably Holl's excellent history [1997] of Argonne) are in the form of a single narrative, weaving together bits of all these elements. I have adopted the latter approach, trying to balance per-sonality sketches, explanations of scientific concepts and instruments, and characterizations of the political context against a complex background canvas. Specialists will no doubt be baffled and frustrated by my particular mix and object to the shortcuts and to the fact that the space devoted to some events and people is not a measure of their stature in the grand sweep of science history. But this book is not a history of science and does not need to be, given the existence of a number of excellent sources in different fields (e.g., in physics Pais 1986; Brown, Pais, and Pippard 1995; Galison 1997; Krige and Pestre 1997). This book also is not an encyclopedia; I wanted to write something that would be read, not consulted.

My particular focus meant that I had to rely on a wide range of sources, from published articles to memos, letters, and oral interviews. Oral interviews can provide otherwise unobtainable information but, like documents, can also deceive; like documents, they are valuable only if used with care and judgment. My focus runs the risk of parochialism, overplaying Brookhaven's role and underplaying the fact that it was only one institution within the national laboratory system—a set of other, similar communities with which Brookhaven often actively collaborated. This risk is hard to avoid given my focus: all communities are local. I tried to reduce the risk by referring to or mentioning, where appropriate, work at other sites. But my approach also has many advantages, one of which is to illustrate the particular kind of pleasure the scientific life provides in a "university-style" atmosphere. Basic research is a career in which people readily accept night shifts, poor job security, and low pay in exchange for a certain kind of pleasure and freedom that come from planning and executing a research program oneself. This aspect of science is rarely spoken about and usually neglected by philosophers and historians of science. But if a philosophy or history of science does not reveal that and why scientists often love their work, it is not delivering all the goods.

Recently a journalist (a former political reporter) two weeks into work on a major article about the lab stopped in to interview a young, award-winning scientist. The journalist said words to the following effect: "So far I've spent my time in the gee-whiz stage, marveling at the new accelerator with its huge detectors, the brain scan center, the 3-D imaging process, and other things. Now I'm ready to get past gee-whiz to the real story. What are the major rivalries? Who's getting hurt by funding cuts? Which programs are taking the hardest hits?"

The scientist replied, "I've been here five years and I'm still not past the gee-whiz stage. I live for it. I'd leave if I lost it. It's the driving force for hundreds of people around here. If you get past it, you're missing the real story."

The attitude of the journalist—that the "reality" of science consists of the conflicts in its politics, interests, or funding—is all too often shared by contemporary scholars of science, who are prone to dismiss the suggestion that science at the fundamental level might be something more than a business as apology, popularization, or the refusal to accept accountability (e.g., Fuller 1993, 237–51). Intellectual excitement is treated as suspect, a mere emotion disguising a secret agenda, rather than as expressing the consuming commitment of a scientist's whole being. But if you know science only by its politics, interests, funding, or material achievements, you don't know science.

Science is a creative activity that depends on the character, determination, and skill of individuals in multilayered interactions in various kinds of events. Observing the force and sweep of this process yields an almost aesthetic pleasure: the captivating illusions and the neglected breakthroughs on the way to recognition, and the sudden and unexpected reversals of fortune. A good history, even of modern science, can provide all the gratifications of art. Connoisseurs of classical literature know the delights the early Greeks took in detailed stories of the construction and operation of instruments and tools. What reader of Homer can forget, in the *Iliad,* Hephaestus forging Achilles' shield, or, in the *Odyssey,* how the hero builds a boat with which to sail away from Calypso's isle? Ancient authors delighted in drawing such descriptions, ancient audiences enjoyed hearing them, and the tales showed a regard for the *technites* who built and used such equipment, for the knowledge, materials, and processes they involved, and for the double-edged, often dangerous power to interact with and transform nature that they manifested. The stories showed, too, something about the life of the *technites* that would have been missed by any collection of scholarly treatises focused on narrow topics. Perhaps there are still those around who are interested in stories about modern-day *technites* and think that such stories reveal an essential part of the meaning and history of what they do. Such people in particular should enjoy this book.

A Team of Young
General Groveses

NOT MUCH over five feet tall, I. I. Rabi was brilliant, ambitious, and pugnacious; when rendering professional judgments, he was almost entirely lacking in sympathy and compassion. Rabi's students hardly cared; those who came to Columbia University to study physics right after World War II were not looking for sympathy and compassion. By then, Columbia was one of the most competitive and exciting places in the world to study physics. Its faculty were demanding of themselves and their graduate students, and Rabi was the physics department's driving force (fig. 1.1). His career had encompassed many of the extraordinary changes that took place in American science since the turn of the century, and he was the single most influential figure in the events that led to the founding of one of the nation's first large-scale civilian science institutions, Brookhaven National Laboratory.

Rabi always insisted that he was an American and that his was a classic American story—though, he would add, "It is not *Life on the Mississippi*." [1] He was born in the old Austro-Hungarian Empire in 1898 and his mother tongue was Yiddish. He was one of thirteen children in a Hasidic family that immigrated to the United States the year after his birth, and his family spent nine years in a poor Jewish neighborhood on Manhattan's Lower East Side before moving to the Brownsville section of Brooklyn. His parents were devoutly religious, yet in the old talmudic tradition encouraged him to actively question and probe his surroundings. "Did you ask any good questions today?" his mother would inquire when he returned home from school. Learning about Copernicus filled him with "wild excitement," and he saw nature as teeming with marvelous phenomena to be explored. To the horror of his parents, who wanted him to become "a respectful bourgeois Jewish man," he refused a bar mitzvah. When they insisted on holding at least an informal ceremony, his unorthodox topic for a speech was "How the Electric Light Works," and he explained, in Yiddish, the science behind the new electric light bulbs; "platinum conductors in the glass, because of the expansion coefficient—that sort of thing." Presumably, few of the congregation appreciated the quasi-religious character of these interests. "The mystery of the world was with me," he said later. "That's what got me into science . . . I was not terribly interested in God the Administrator."

Rabi majored in chemistry at Cornell, earned a Ph.D. in physics from Columbia in 1927, and in 1929 became the first Jewish member of the physics faculty there. By the late 1930s, he was a powerful figure at Columbia and in the American physics community, known for his dedication and passionate involvement with his field. Once, an interviewer made the mistake of saying that physics was abstract and mathematical. Rabi vehemently objected:

> It has the qualities of art, of literature, of poetry, for the physicist who's immersed in it. Of course it doesn't relate to emotions—emotional factors as between people—but it is intensely moving and exciting when one gets to a new insight, a broader horizon. It can be so moving as to leave one almost speechless, when one has some idea which in a flash reveals a wider understanding. Oh, no, it's very far from cold. In fact, it has never been a cold subject to me at all. It's always full of excitement and symbolisms.[2]

In class, his lectures were informal, yet enthralling. Other professors lectured in a slick, well-organized fashion, with carefully prepared notes and canned jokes, inviting their student audiences to sit back and marvel at the intellectual brilliance of the physics and the virtuosity of the professor. Not Rabi. He would arrive in the classroom, casually pick up a piece of chalk, and ask to be reminded where the previous lecture had ended, usually halfway through some long and difficult derivation. Then he would walk to the blackboard and proceed almost as though it were the first groping search that had ever been made in the subject, talking and explaining the subject to himself in his harsh nasal voice. Sometimes he would find himself in a blind alley, stand back, and scratch his head before plunging in anew to lead students step by halting step through the material as though they were on a voyage of discovery in a new land.

In the laboratory, Rabi had high aspirations. When students approached him with ideas for experiments, he'd ask, "Will it bring you nearer to God?" The students understood: *"Is it fundamental enough so that it will change your outlook on the world?"* Rabi's lab was typical in that it was entirely supported by the university, and many of his assistants worked on their own time. One, Jerrold R. Zacharias, was a full-time assistant professor at Hunter College who put in additional uncompensated hours working for Rabi. As Rabi said, "The whole standard was different . . . It was a privilege—no money at all; it was that sort of thing. Graduate students were not paid. If they needed money they had jobs outside, and some of them did. It was a different world.[3]

In 1940, on the eve of the American entry into World War II, Rabi became assistant director of the Massachusetts Institute of Technology Radiation Laboratory, the MIT Radlab, where microwave radar for military field use was developed. When the Manhattan Project got underway, Rabi was one of the few scientists with enough clout that Brigadier General Leslie Groves, the project director, let him visit its Los Alamos laboratory as he pleased. On one of Rabi's visits, in 1944, the lab threw a party to celebrate his receiving the Nobel Prize that year; on another visit, in July 1945, he witnessed the first man-made nuclear explosion.

The war had a huge impact on American science, especially physics. During the war, average federal support for science shot up from $50 million to $500 million a year. A number of enormous projects—entire laboratories—had been built in record time. The size of the scientific community jumped, thanks to the large number of scientists and engineers trained during the war, and to the number of foreign-born scientists who had sought refuge in the United States. "The Los Alamos generation had come out of World War II disposed to think big—and expensively," science historian Daniel J. Kevles has written. Rabi was a leading member of that generation (Kevles 1978, 367).

In 1945, Rabi returned to Columbia, but was enraged at how its physics department had fared compared to other universities' departments that had contributed to the Manhattan Project. The first federally funded experiment of the project had been performed at Pupin Hall, Columbia's physics building, and much of its early work took place there. The code name "Manhattan Project" (or the work of the "Manhattan Engineer [or Engineering] District" or M.E.D.), arose from the fact that its first office, located not far from Columbia, was in Manhattan. As the project progressed, its work was moved to more remote locations, leaving Columbia behind and forcing many of its star faculty, including Enrico Fermi, to relocate. Project leaders expanded the laboratory of Ernest Lawrence in Berkeley, California (the Berkeley Radiation Laboratory, or Berkeley Radlab, which, though depopulated by the war, benefited from it in the long run), and created new facilities at the University of Chicago (the Metallurgical Laboratory), in Oak Ridge, Tennessee (Clinton Laboratories), and in New Mexico (the Los Alamos Laboratory). After the war, these new facilities became important centers of research—and not only were they far from Columbia, they were luring away some of Columbia's most eminent scientists. When Arthur H. Compton, who directed the Metallurgical Laboratory, secured permanent positions for both Fermi and fellow Columbia faculty member and Nobel laureate Harold Urey in a newly founded Institute for Nuclear Studies (now the Enrico Fermi Institute) at the University of Chicago, Rabi was furious and protested to Vannevar Bush of the Office of Scientific Research and

1.1 I. I. Rabi at left, 1947.

Development, the single most influential scientist in government, that Compton was misusing his position in the Manhattan Project to help his own institution at Columbia's expense. "The east had nothing," Rabi said. "We had the preponderance of students and of researchers, and no tools." In Rabi's eyes, Columbia had been rewarded for its wartime efforts by being stripped of faculty and facilities. Rabi hated administration—he liked to say that he was not the kind who was ever elected class monitor—but insisted on becoming chairman of Columbia's physics department on his return to have leverage in restoring its eminence.[4]

Rabi's chief associate in this quest was his younger colleague Norman Ramsey (fig. 1.2). A highly accomplished scientist (and future Nobel laureate) who had just turned thirty, Ramsey had worked at both the MIT Radlab and Los Alamos. Rabi's opposite in aspect and affect, Ramsey was athletic and stood 6'1" tall. Ever meticulous, in argument he would have a clear, well-developed position that he would press firmly, point by point, and if in the end his opponents had somehow withstood the force of Ramsey's logic and booming voice that never needed amplification, they

1.2 George Collins and Norman Ramsey (left and second from left), with colleagues, 1947.

emerged with a thorough grasp of his thinking. Like Rabi, Ramsey was relentless, but if Rabi had the air of a skilled pugilist, Ramsey was more a genial bulldozer.[5]

As a first step, Rabi and Ramsey decided in fall 1945, Columbia would need a reactor. The first reactors had been built as part of the Manhattan Project, to study nuclear fission and create fissionable material, but they had an enormous significance for nonmilitary scientific research. Before the war, physicists had discovered that neutrons are useful probes for studying nuclear structure and the structure of solids and for creating radioisotopes, but neutrons were available only in small quantities. Following the discovery of fission and the onset of war, fundamental research in nuclear physics was set aside in favor of military applications. After the war, the development of reactors, with their ability to produce neutrons in unheard-of quantities, even *beams* of them, promised to revolutionize nuclear physics, solid-state physics, and any field in which radioisotopes could be used. Rabi, Ramsey, and other physicists who were paying attention realized that access to a reactor was essential to university physics departments that wanted to do work of any importance. At war's end, half a dozen research-sized reactors had been built in the United States, but none was located in the northeast, where much of the American physics community worked.

Initially, Rabi and Ramsey kicked around the idea of having the Columbia physics department build one. But funding, constructing, and operating a reactor was a huge undertaking that would exhaust most of the physics department's resources, eating up support for projects in other areas of physics. Dealing with security and classification restrictions imposed by the government also would be a problem. By the end of 1945, Rabi and Ramsey concluded that Columbia could build a reactor only in collaboration with other local institutions and with federal support.

This conclusion amounted to a first conceptual leap in their thinking. For the project now could not be a matter of building a reactor alone, but would mean having to create a new kind of institution—government funded, but with a university-type atmosphere. Such an atmosphere was considered essential because universities, unlike corporations and political agencies, are not faced with trade-offs between short- and long-range interests, and would allow the lab a vision, idealism, and independence. Government funding would be needed because of the scale of the project and its extensive support structure: special engineering, machining, and monitoring facilities; special equipment and laboratories with which radioactive materials might be studied; and "hot labs" able to process highly radioactive substances. A staff of physicists, chemists, health scientists, and engineers would need to be on call. Moreover, ways would have to be developed to form collaborations and negotiate conflicts between the client institutions sharing the lab, to equitably allocate research resources, and to handle collaborations between the lab staff and scientists from the client institutions. The project would mean, in short, the formation not just of a new type of research plant but of a new kind of partnership between government and science, an experimental community—better, mutually supportive and evolving *groups* of such communities.

There were some wartime precedents for such an institution, and a couple of the Manhattan Project-built labs were indeed en route to conversion into civilian laboratories. Also, it was already clear to many others besides Rabi and Ramsey that maintenance of independent laboratories to serve many university clients was the only way to fully exploit the intellectual resources of contemporary science given the jump in scale of scientific instruments. Still, Rabi and Ramsey were proposing what would become the first large laboratory designed and built from scratch to serve a multiuniversity community in peacetime.

Although Rabi and Ramsey envisioned a multiclient laboratory, they still saw Columbia as its principal sponsor. Precedent existed for such an arrangement: MIT and the University of California, Berkeley, had played such a role for their respective radlabs, and the University of California had also served as the contractor that operated the Los Alamos lab. In an effort to assemble a consortium to back a Columbia-sponsored reactor, Rabi and Ramsey compiled a list of New York City-area institutions that might be interested, and Rabi convinced George B. Pegram, Columbia's dean of

1.3 George Pegram, dean of Columbia University's graduate school, 1937–1949.

graduate faculties and a pioneer of American nuclear physics, to convene a meeting of their representatives (fig. 1.3).

Pegram was the right person to arrange such a meeting. Still spry and athletic despite his seventy years, he had been the principal force behind the buildup of Columbia's physics department; both Rabi and Ramsey were among his recruits. "He was pooh-bah at Columbia," Rabi said. Pegram was also a liaison between top scientists and Washington. In March 1939 (months before the famous letter from Albert Einstein to President Franklin Roosevelt), a phone call from Pegram had first directly connected government officials and physicists concerned about the military possibilities of uranium fission, setting in motion the events leading to the Manhattan Project.[6]

Pegram hosted the meeting to discuss the project in the Trustees' Room at Columbia's Low Library on 16 January 1946. Thirty-five participants represented sixteen institutions, including colleges and universities, hospitals and medical schools, and corporations, like Bell Telephone Laboratories, with major research facilities. Pegram chaired the meeting, Ramsey took notes, and in ninety minutes those at the cordial, upbeat gathering agreed unanimously on the need for a research reactor near New York City. Most of the meeting was devoted to going through a letter, drafted by Ramsey and addressed to General Groves, seeking approval of the facility. The letter was addressed to Groves because, even though World War II was over, the Manhattan District he headed remained in existence and in charge of American nuclear research. Its huge infrastructure was due to be taken over by a new government agency whose form was then being debated by Congress, but meanwhile control of matters related to atomic energy and research was still in Groves's hands.[7]

The letter, sent on 19 January, requested that "a regional research laboratory in the nuclear sciences" be built near New York City. The letter marshaled five arguments in support of such a laboratory; one was that more than one out of five members of the American Physical Society lived within eighty miles of New York City. On 22 January, Groves replied, suggesting a meeting on 8 February of his deputy, Colonel Kenneth D. Nichols, and representatives of the group proposing the laboratory.[8]

Rabi and Pegram brought some heavy ordnance to the meeting with Colonel Nichols: Hugh S. Taylor, a chemist and Princeton's graduate dean, and Henry D. Smyth, chair of Princeton's physics department and author of *Atomic Energy for Military Purposes,* an account of the development of the atomic bomb written at the request of General Groves himself. A bestseller, the Smyth report had sold over a hundred thousand copies in five months (Smyth 1945).

Nichols was a quiet, confident, bespectacled thirty-eight-year-old who had graduated from West Point and earned a Ph.D. in hydraulic engineering. As Groves's right-hand man during the atomic bomb project, he was fully capable of standing up to Rabi, Pegram, Taylor, and Smyth. But there was no confrontation: Nichols was ready to accept the idea of a "regional laboratory."

The idea of permanent government-sponsored laboratories for nuclear research had been raised as early as the onset of World War II (Holl 1997, 7). By the war's end it was clear to leading scientists that efficient postwar research would require the establishment of regional laboratories, and the many who discussed the idea included Arthur Compton and his brother, MIT president Karl Compton; William Watson, chairman of the Yale physics department; and Smyth himself. Nichols had heard Arthur Compton discuss the notion and persuaded Groves to endorse it. Rabi and Ramsey were therefore merely presenting Nichols with an opportunity to realize the idea. From Nichols's perspective, the main unresolved issues were where the laboratory should be built and which institution should be its prime contractor (Needell 1983, 96–97; Greenbaum 1971, 14–15; Holl 1997, 32–39; Galison 1997, 297–303).

These issues were more complex than Rabi and Ramsey wished. Word of the meeting had spread to Boston, where another group of physicists at MIT, led by John C. Slater and Rabi's former assistant Zacharias also had planned to approach Groves about a reactor in the northeast. Hearing of Pegram's actions, they hurriedly passed their own thoughts on to Groves, who now had on his hands two strong-willed and competitive groups of scientists. Nichols told the Columbia-Princeton group that Groves wanted them to work with their Boston colleagues. "Groves said, 'You'll have one place, or none,' " Rabi recalled.[9] Nichols also passed on a warning. The Manhattan District, Groves had said, was in its final days. At the end of the year its functions would be taken over by some yet-to-be-created federal agency, which then would need time to organize itself and consider, let alone approve, such a major project. Given the nature of bureaucracy, the northeastern scientists needed to cut a deal quickly with the War Department through the Manhattan District. If they didn't, they would lose a lot of time.

Adding to the pressure was the knowledge that competitors in Manhattan Project laboratories (especially those in Oak Ridge and Chicago) already had a substantial lead in returning to nuclear physics work. "We may find," said Compton, "that the available resources to support such work are all devoted to these regions if we are too slow in getting our specific proposals before the Manhattan District."[10]

The day after Nichols and the New York–area scientists met, Slater drafted a "Proposal for Establishment of a Northeastern Regional Laboratory for Nuclear Science and Engineering." "Nuclear research has entered an altogether new order of magnitude," it said, whose "scale is too large to be carried out by even the largest university organization." Given MIT's experience with its radlab, Slater suggested that Boston would be an excellent location and MIT a potential sponsor.[11]

On 10 February, the New York-area scientists met for the first time with their Boston-area rivals. In the meantime, Rabi had indulged in a bit of gamesmanship and solicited the support of scientists in institutions from the south and west. Among the sixteen people present on 10 February, therefore, were new faces from Rochester, Johns Hopkins, and Cornell, keeping New York City at the geographical center of the group despite the presence of the Boston contingent. (Industrial laboratories did not participate in further deliberations.) The factions knew they had no alternative but to work together, but it was also clear that site selection would be a divisive issue.

The group agreed to ask the Manhattan District that "a major pile [reactor] be located immediately" in the northeast. The group agreed that the laboratory had to be "accessible in an overnight trip from all major laboratories between Washington and Boston," and near a good institution of higher learning. It also needed power and water supplies, about ten square miles of land, and readily available labor and materials.[12]

On 2 March, the group met again to draw up further plans for what they were calling the "National Nuclear Science Laboratory." The word "national," signifying that the lab would serve not only the region but also the nation as a whole, was a second conceptual leap for the group, though also an idea clearly in the air: Groves's Advisory Committee on Research and Development was preparing a document on national laboratories that would specifically recommend a lab in the northeast (Greenbaum 1971, 16). This, too, was gamesmanship, designed to encourage the government to look favorably on the project. The group also agreed on the text of a letter from Pegram to Groves that would summarize the rationale for the lab and contend that in terms of scientific, technical, research, and educational facilities, "the resources of the northeastern part of the country are greater than those of any other," and that these resources could best be exploited "by the establishment of a major government laboratory in this area." The letter urged the Manhattan District to take "immediate steps" to create such a laboratory, including "a major chain reacting pile." As for administration, the laboratory would be "operated by a single institution as contractor, preferably a single university," with scientific direction "in the hands of a board representing the sponsoring institutions and appropriate government agencies." The authors of the letter asserted, "We see no virtue in waiting until an atomic energy commission is set up. . . . Everything is therefore to be gained by action."[13]

At this point, nine participating universities were involved: Columbia, Cornell, Harvard, Johns Hopkins, MIT, the University of Pennsylvania, Princeton, Rochester, and Yale. On 23 March, representatives met at Columbia and formally established the Initiatory University Group (IUG). The IUG would be composed of eighteen members, one research scientist and one administrator from each university. The IUG would also have a planning committee consisting of one representative from each institution. The IUG's membership included some of the ablest scientists and science administrators in the field—an indication of the high priority sponsoring universities accorded the project and their realization that they needed this lab in order to remain at the forefront of physics research.[14]

When it came to selecting a head for the IUG, the group had an obvious and willing choice in Lee DuBridge, who had directed the MIT Radlab (fig. 1.4).

"We realized," DuBridge recalled, "that even the process of talking about [the new laboratory] would take some money. We had to bring people in and pay their travel expenses and hotel bills, we had to get a secretary to do some of the secretarial work, we had to have telephone calls with various people around the country." [15] Pegram therefore approached Groves and secured supplemental funds to finance the IUG under Columbia's contract with the Manhattan District, which would remain in effect until 1947. Money in hand, the group set out to hire its first employee, a person who

1.4 Lee A. DuBridge, director of the radiation laboratory, Massachusetts Institute of Technology 1940–1945, head of the Initiatory University Group that planned Brookhaven in 1946, and president of the California Institute of Technology, 1946–1969.

would have to do everything: secretarial work, serving as liaison with Groves and top-ranking scientists, investigating Columbia's government contract to see what costs it covered, and setting up the proper machinery to run the IUG accordingly. Within days, Ramsey found Mariette Kuper (fig. 1.5).

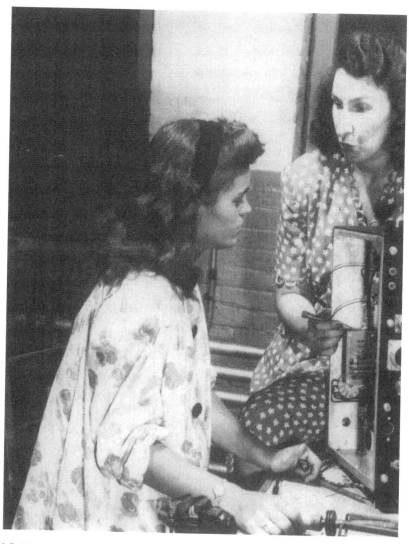

1.5 Mariette Koevesi von Neumann Kuper (right) at the radiation laboratory, Massachusetts Institute of Technology, ca. 1943, training a technician in circuit wiring.

Mariette Koevesi von Neumann Kuper was a determined, self-confident woman whose aristocratic mien turned heads when she entered a room. She turned heads, too, when she swore, as she often did. Overhearing someone sighing romantically over the beauty of new-fallen snow, she would impatiently dismiss the stuff as "vite shit." When a problem erupted and she had to call on someone's help, she liked to tell them, "You have to go pee on a fire," and affectionately refer to her problem solvers as "firemen." The daughter of a Budapest doctor, she had left Hungary in 1930 to settle in Princeton with her first husband, the eminent mathematician John von Neumann. In 1937, she divorced von Neumann and married physicist J. B. H. Kuper (whom she insisted on calling "Desmond," saying she had no reason other than wanting to be married to a man with that name). The couple quickly gained renown, not least for their elaborate parties, launched by her formidable martinis. As a consequence of her marriages, Mariette was on a first-name basis with many of the important American and European physical scientists and mathematicians.

During the war, the Kupers had worked at the MIT Radlab, where Mariette started out assembling radar sets in the Technical Training Program for Female Personnel and Metal Shop Production. Told by skeptics, she once recalled, that she was just a socialite who would quit as soon as she was bored, she quickly became a foreman and then supervisor, hiring and training some two hundred technicians to wire circuits from diagrams. She was confidante, facilitator, and hostess, loved both for her strengths and her idiosyncrasies. She typified the positive spirit of the place, facilitating the fierce concentration as well as the energy spent in relaxation, helping to ensure that people laughed, enjoyed, and were proud of what they were accomplishing. The Kupers became the focus of the radlab's social life, which, like the lab itself, leveled social hierarchies ("We all come from the same bar," Rabi once said when asked about the lab's organizational structure). Among her achievements was to integrate the department; using a combination of sheer force of personality and Hungarian profanity, she persuaded her supervisees to vote "yes" on the question of whether they were willing to work with "negroes." [16]

That was not her only blow for civil rights. For a few years after the war, when MIT Radlab personnel had dispersed to dozens of universities and research institutes, Kuper organized reunions at posh Washington hotels during the spring meetings of the American Physical Society. A few black physicists had been employed at the radlab, and at the first reunion, at Washington's Wardman Park Hotel, one of them, a radioisotope specialist, came with his wife. To the horror of the hotel management, the couple was accepted as professional and social equals by the group and also as dance partners. The next year, the Wardman refused Kuper its facilities, and she moved the event to the Hotel Washington, near the White House. Again, the presence of the black radioisotope specialist finished off the reunion's relationship with that hotel. Eventually, the dances were discontinued, but they deserve mention, in the words of one participant, "as the first unsegregated dances at first class hotels in Washington D.C., setting a precedent for the enlightenment which followed many years later." [17]

Small wonder that several radlab veterans thought first of Mariette Kuper as ideal for the fledgling laboratory organization. DuBridge, en route to a meeting, called her from an airport pay phone. Luckily, she was between jobs and enthusiastic about joining a group that included several of the old MIT Radlab crowd. "See the Dean at Columbia," DuBridge said, then hung up and dashed off for his plane. [18] Kuper started work on 27 March 1946. Her first job was to figure out how to get herself hired, cleared, and paid under the terms of Columbia's Manhattan District contract. She went on to spend twenty-eight years at the lab, first as executive aide to the director, then as staff assistant.

On 30 March, the IUG planning committee divided its responsibilities among six subcommittees. One, led by Smyth, would plan an institutional structure to manage the lab and initiate contract negotiations between it and the government. Another, led by Rabi, would iron out personnel issues. A third, led by Watson, would tackle problems of security and classified research. A fourth, led by Ramsey, would handle site selection. Robert Bacher of Cornell would head the subcommittee to plan the piles (more than one were projected), and Zacharias would direct the subcommittee for "electronuclear machines," an early name for particle accelerators. [19]

Thus by the end of March 1946, the IUG had a head in DuBridge, a rudimentary organization, a temporary funding arrangement, six subcommittees, and two employees (Kuper and Clarke Williams, a Columbia physicist, hired the same day as Kuper to help Bacher coordinate the pile project). The subcommittees had nine months in which to create a laboratory ex nihilo, but their task appeared possible because of the wartime experience of quickly realizing large and complex projects. "[T]here were people there who knew how to do things fast," recalled M. Stanley Livingston, who would head the accelerator project. "We didn't have to have General Groves to push things along—we had our own team of young General Groveses to do the job." [20]

Scientists elsewhere in the country, however, tended to look skeptically at the attempt by northeastern scientists to force a new laboratory into being. With a few exceptions, members of the IUG team had little experience with reactors, and the notion that neophyte easterners virtually overnight could create a lab to compete with the likes of the labs at Berkeley, Chicago, Los Alamos, and Oak Ridge seemed ludicrous. DuBridge, Rabi, and

Zacharias were on a recruiting mission to Berkeley in the late spring of 1946 when they informed Ernest O. Lawrence, head of the Berkeley laboratory and father of the cyclotron, that they were setting up an eastern version of the Berkeley Radlab. Lawrence was unimpressed, and, his visitors on a recruiting mission, perhaps a little defensive. Zacharias recalls Lawrence telling them, "You can't do this in the east, you haven't got what it takes." What was the reply? "My guess," Zacharias said, "is that I looked at Rabi and at DuBridge and they looked at the others and we said, 'Screw you,' essentially, probably more politely." [21]

CHAPTER 2

A Reluctant Director,
a Remote Site

IUG HEAD Lee DuBridge was a mentor to many scientists who had worked at the MIT Radlab, including Rabi and Ramsey. Beginning with very little, DuBridge had made the radlab one of the largest single-purpose scientific plants ever. It had more personnel than Los Alamos, and arguably more of an impact on the progress and outcome of the war, due to its role in transforming microwave radar into an indispensable tool that gave the Allies control of sea and sky. The huge undertaking had brought DuBridge into contact with most other major American physicists. "Getting together with this [IUG] planning committee," he said later, "was just like getting together with a group of old friends." [1] At the end of March 1946, when the IUG planning committee began looking for a director of the new lab, DuBridge was a natural first choice. But though willing to direct the IUG, DuBridge wanted to return to teaching and research and was uninterested in yet another long-term administrative position.

In April, the IUG approached Milton G. White, a Princeton professor who had been a radlab division head, but he, too, declined. Planning committee members revived their efforts to persuade DuBridge, who began to waffle. But the California Institute of Technology invited him to become its president, a much more inviting offer if he were to return to administration. In May, DuBridge accepted the presidency of Caltech and stepped down as IUG chair.

Eyes then turned to another radlab figure, F. Wheeler Loomis of the University of Illinois. Loomis was initially encouraging, but was afraid that the government would not allow a contract sufficiently flexible for the laboratory to be effective and that a sudden change in national policy toward science might cause the project to collapse. [2] A meeting with Nichols did not allay Loomis's fears, and in June he too declined.

In desperation, Zacharias placed a call to his MIT colleague Philip Morse (fig. 2.1). Morse was a theorist, not an experimenter, and like the others wanted to return to research. But his wartime role as head of an underwater sound laboratory in Washington, D.C., had given him administrative experience and extensive government contacts. Morse was not DuBridge's equal in status or experience, but scientists regarded him as a solid member of DuBridge's generation. Government officials also trusted him because he had directed a laboratory that had significantly aided the war effort. Everyone, in short, treated him as an insider. "He was," Zacharias said, "one of the clan." [3]

As a graduate student at Princeton, Morse had studied under Smyth and Karl Compton and met many of the leaders of the next generation of American science, including John von Neumann and his then-wife, Mariette, and Ukrainian-born chemist George Kistiakowsky. One of the first Americans to become well versed in quantum mechanics, Morse and his teacher, Edward Condon, coauthored the first major English-language textbook on the subject while Morse was still a graduate student (Condon and Morse 1929). After following Compton to MIT in 1931, Morse wrote what became a standard text on acoustics (Morse 1936). His expertise led to his appointment as director of the MIT underwater sound laboratory during the war, and to a Medal of Merit for his role in antisubmarine warfare projects. He was just gearing up to write another textbook on theoretical physics with an MIT colleague when Zacharias asked him to direct the new lab.

Morse's answer: No. "To accept," he explained later, "would mean re-entering the political-administrative arena, in the region of greatest political pressure, just when I was beginning to relax from the wartime stress" (Morse 1977, 221). He hadn't kept up with nuclear physics, was not an experimenter but a theorist, and, most important, he loathed the secrecy that would inevitably cloak a laboratory with a reactor. While during the war government officials had trusted Morse enough to give him latitude to manage his own lab's security, he had been infuriated by what he considered the military's petty-minded obsession with security.

Zacharias persisted, saying that the lab desperately needed a well-connected scientist with Washington experience and promising the assistance of top scientists at the nine collaborating universities. Zacharias pointed out that it was imperative to regain the pace of nuclear research, and that a major new lab in the east would be an important step. Moreover, the lab would be well supported thanks to a consensus among influential members of the executive branch, Congress, and the military that nuclear research ought to rank high on the national postwar agenda. Morse wavered, and agreed to go to Washington to meet General Groves.

"This visit," Morse wrote later, "almost made me decide to back out" (Morse 1977, 222). He tried to stay in the background, letting Zacharias do

2.1 Philip Morse, Brookhaven's first director.

the talking and responding only to direct questions. Still, he did not hide his opposition to secrecy or his intention to leave the position as soon as he could; Groves, meanwhile, made it clear that his vision of security was stricter than that of Morse's wartime government associates. Morse and Groves each exhibited some contempt for the other, producing an atmosphere of "muted antagonism." Morse left feeling that Groves would never accept him as lab director and not at all unhappy about it (Morse 1977, 222).

Morse was stunned to learn a few weeks later that Groves, bowing to Morse's reputation among soldiers and scientists alike, had approved his appointment. Morse cheered himself by thinking of the challenge of creating a new organization and of the many friends he would be working with, including Karl Compton, Kistiakowsky, Zacharias, and Mariette von Neumann Kuper. He then treated himself to a monthlong vacation in New Hampshire. Hoping for the best, he wrote a representative of the new lab that he assumed the government would give the contractor "a more or less free hand" to run the lab, but he still worried that he would spend too much time fending off government security restrictions.[4]

Morse was right to worry. His relatively liberal wartime experience with secrecy was anomalous, and the issue would cost him even more time and aggravation than he feared. His departure from the lab would be sooner than he expected, or even hoped.

WHILE ZACHARIAS was courting Morse, the IUG was preparing a scientific program and administrative organization. Scientific aspects of the lab were addressed by the accelerator and reactor subcommittees planning the major facilities, and by a "proposed program" being drawn up by the planning committee. The subcommittee on "electro-nuclear machines" met in April 1946 and made initial plans to build a number of different types of accelerators. To head construction, the subcommittee turned to M. Stanley Livingston, one of Morse's MIT colleagues and an experienced accelerator builder whom Morse had recruited for his U-boat countermeasures project during the war.

Robert Bacher's subcommittee on reactors ran into more difficulties. The subcommittee met on 15 April with representatives of the Metallurgical Laboratory (soon to become Argonne National Laboratory) and Clinton Laboratories (later Clinton and then Oak Ridge National Laboratory) to discuss building a research reactor at their own "national laboratory." A research reactor seeks to maximize production of neutrons rather than power, like a commercial reactor, or fissionable material, like a military reactor. The basic measure of neutron production is *flux;* the greater the flux, the greater the reactor's value as a research tool. Because the first reactor had been built only three and a half years earlier and reactor technology was still poorly understood, nobody knew how much flux could be achieved at a reasonable cost. The highest flux created thus far had been by the plutonium-producing piles at Hanford, Washington. These had involved highly classified technology, extensive water treatment plants, and costs of several hundred million dollars, an expense justified by the war but clearly far more than any research lab would receive from Congress. To start the research program rolling, therefore, the reactor subcommittee decided that the lab's first reactor "should not be of too radical a design"— about the size and flux of the more modest X-10 at Clinton. Moreover, the scientists, with a naïveté encouraged by the success of the Manhattan Project, assumed that they would have little trouble contracting outside their group for assistance in design and production. After gaining experience with this first reactor, they felt the new lab would then be in a position to build another, more advanced reactor with higher flux (Needell 1983, 103–4).

At the meeting, the subcommittee asked Eugene Wigner, director of research at Clinton, to head a design team for the low-flux reactor. He at first agreed, then backed out. Wigner and most other Clinton scientists were more interested in the development of a nuclear power industry and believed that a first step toward realizing this goal was the establishment of a reactor training school. The northeastern scientists, Wigner felt, ought to attend his reactor-training school and then design their own reactor. His idea was not well received. The northeastern scientists felt they had no time to sit through Wigner's courses, but needed a first-class reactor as soon as possible (Needell 1983, 101).

They again tried to court Wigner in May, when the IUG planning committee discussed reactor design while meeting at Clinton, which was now home to much of the then-small pool of qualified reactor engineers. Rabi aggressively tried to recruit some members of this valuable pool with words to the following effect: "You fellows out here in the wilderness, why don't you move the laboratory to civilization?"[5] Although the northeastern scientists did successfully woo some key figures, most notably Lyle Borst, supervisor of research at the X-10, Rabi's patronizing tone backfired and served mainly to irritate the Clinton officials. A disappointed Bacher then told the planning committee that he might be able to get Walter Zinn of Argonne instead of Wigner to head the design group. News of this predatory behavior got back to Groves, and an alarmed Nichols demanded that the northeastern scientists "refrain from competitive bidding for key personnel now located at other laboratories" (Needell 1983, 102).

Despite these setbacks and rebukes, the group remained optimistic about its ability to build a reactor quickly and cheaply; "Hanford flux for a million bucks" was their only half-joking slogan (Needell 1983, 100). The group's optimism is clear in the "Proposed Program for the New National Laboratory of Nuclear Science," a draft document prepared at the end of May that described a central facility that would include two reactors together with a set of electronuclear machines.[6] The two reactors were a "low-power (300–1000 kw [kilowatt]), air-cooled, graphite-and-uranium pile similar to the pile at Clinton Laboratories," projected to cost $2.5 million, and "a high-power, high-flux reactor of unique design, probably using enriched uranium and ordinary water" that would follow after "about two years" and for which no cost figure was provided. Because the low-power reactor would be modeled on Clinton's X-10, "all design, engineering, drafting and construction need consume only about nine to ten months, making allowance for some delay due to winter weather." Speedy completion of the low-power reactor would be essential to the success of the lab, for construction of the other major instruments described in the "proposed program"—the high-flux reactor and the accelerators—would take several years. The document asserted that the low-power pile would be in full operation by 1 July 1947, and that its completion would quicken interest among researchers in the northeastern region in the new laboratory.

Meanwhile, the other four subcommittees were developing an administrative structure for the laboratory. Wartime laboratories that had been sponsored by the National Defense Research Committee were not really useful models because they had focused on applied research and relied on the military's virtually unimpeded wartime ability to command resources, while the new laboratory would promote basic research, have to coordinate researchers with different interests from different universities, and need to compete for funds with other institutions under peacetime conditions.

Basic and applied research are intertwined in practice, but their intellectual missions can be formulated distinctly: basic research aims to discover or understand previously unknown structures or characteristics of nature, whereas applied research is addressed to practical problems. The management of basic research is difficult: generally, management means coordinating resources and people to achieve a particular goal, but how can one coordinate an activity when its end is unforeseen and unforeseeable? To do so involves providing scientists with some amount of freedom to follow their hunches in research programs, and with the facilities and freedom to collaborate and compete with colleagues. Creating and maintaining a healthy scientific culture requires tending to both the intellectual and institutional conditions of basic research and is a key element in making a laboratory effective. The various IUG subcommittees wanted to solve this management problem by having the new, federally funded lab preserve as much as possible the flexible working conditions they had known in their universities, which allowed researchers to change directions rapidly to capitalize on new developments and opportunities. Watson's subcommittee on security and classification, for instance, attempted to reduce the amount of work that would be treated as secret, both to simplify classification procedures and to protect employees and the laboratory from clearance requirements that could interfere with collaboration and collegial interaction. Meanwhile, Rabi's subcommittee on personnel policy decided on a fluid laboratory staff, with short-term and on-leave appointments, hoping to reduce the bureaucratization that can ossify research.[7]

The greatest challenge to flexibility was confronted by Smyth's subcommittee on contract. First, the subcommittee had to decide whether the contracting organization should be a single university, an industrial or nonprofit sponsor, or a new corporation formed by the nine founding universities. The wartime arrangement of lab sponsorship by a single university or corporate sponsor had worked well, but given the unwillingness of the nine university members to accept a secondary status (parity among them was an important issue at early meetings), Smyth's subcommittee decided that the schools would form a new corporation, a kind of holding company, to sponsor the laboratory jointly. This novel arrangement—another major conceptual leap in the lab's development—would promote, Watson claimed, the "democratic and national" character of the enterprise (Watson 1946, 7).[8] It would also influence the structure of subsequent laboratories both in the United States and abroad.

This decision taken, Smyth's subcommittee then had to specify the role each university would play in the new organization. For instance, should each provide financial support? According to DuBridge,

> This was a delicate problem, because it was evident that this laboratory was going to eventually run into many millions of dollars a year [for its] budget, and it was clear that the universities didn't have that kind of money to contribute, that it would have to be financed eventually by the government. And yet it was also felt that at least to keep the universities interested, they ought to have some kind of stake in it financially.[9]

In April 1946, IUG representatives asked each university to pledge funds to help pay expenses not properly chargeable to the government. Although Harvard, the wealthiest institution, initially balked, eventually all nine IUG universities agreed.

With the universities yoked into an acceptable plan for an organization, Smyth's subcommittee approached the Manhattan District with its ideas for a contract. It was here that conflicts erupted over issues so deep they are still fiercely argued over half a century later, involving the extent of government control. The scientists could cite an important wartime precedent. When Groves was organizing Los Alamos, he initially had envisioned a wholly military facility staffed by commissioned officers in a military hierarchy; lab director J. Robert Oppenheimer would have become a lieutenant colonel. But Rabi, Bacher, and several other influential scientists had felt this plan was inimical to the conduct of science and refused to agree. Under pressure to get the project moving, Groves reluctantly had agreed to civilian management for Los Alamos under the aegis of the University of California (Orlans 1967, 6).

The subcommittee on contract proposed a management scheme similar to the Los Alamos model in some respects, except that the sponsor would be a consortium of universities rather than a single one. But in peacetime, the scientists had less leverage, and the fight to keep control of the laboratory out of government hands was much more difficult. Government negotiators for the Manhattan District, seeking to guarantee accountability for such a large project, insisted on federal ownership of every item of importance at the lab, including lab notebooks, and wanted all major decisions, even those involving hiring and salaries, made in Washington. Members of the subcommittee objected vehemently. By August 1946, three different drafts of the contract had been debated and rejected. Despite the 1 January 1947 deadline, self-imposed based on Groves's suggestion, the planning committee felt the battle was worth the risk. Pegram feared that "under the contract proposed by the District, the District officer in charge of salaries and wages, rather than the Corporation's Director, would run the laboratory." [10] Negotiations, and relations between the IUG and the Manhattan District, grew increasingly strained as the summer of 1946 wore on.

While Smyth's subcommittee on contract battled the government, Ramsey's subcommittee on site (which included also Bacher of Cornell, Smyth of Princeton, and Zacharias of MIT) was engaged in internecine warfare. Harvard and MIT naturally wanted the lab near Boston, Yale promoted a New Haven site, Princeton thought highly of the benefits of New Jersey, etc.; only Rochester, bowing to reality, did not propose a site near its campus. But none of the locations met the primary criterion, a sizable amount of unused, isolated land conveniently accessible to all the major research institutions in the northeast. As DuBridge remarked much later, "Accessibility and remoteness were two criteria which were hard to satisfy together." [11] In April 1946, when the site subcommittee held its first meeting, it was suggested, only half facetiously, that the ideal site was ten square miles of land adjacent to Grand Central Terminal (Ramsey 1967, 5).

But the subcommittee had wisely chosen Ramsey as its head. The job tapped his two outstanding talents, handling details and negotiating compromises with strong-willed adversaries. His first move was to get battling subcommittee members to agree that "the site could be within one hour's drive from some major station on the coastwise lines of the Pennsylvania or New Haven Railroad, so scientists traveling by train during the evening would be able to spend one full day at the laboratory and still be at their own university on the days immediately preceding and following." Ramsey then dutifully clocked the time it took to drive to each site from the nearest major station, picking up his first-ever speeding ticket while measuring one site's accessibility (Ramsey 1967, 5).

To get a large site, the subcommittee hoped to take over swamp land or other undesirable property, or a military base that might soon become surplus. With the help of Major Emery L. Van Horn, an engineer and liaison officer with the Manhattan District, subcommittee members drew up a list of seventeen possible sites, most of them close to one or more universities (see fig. 2.2). A site at Lake Zoar in Connecticut was thirty miles northwest

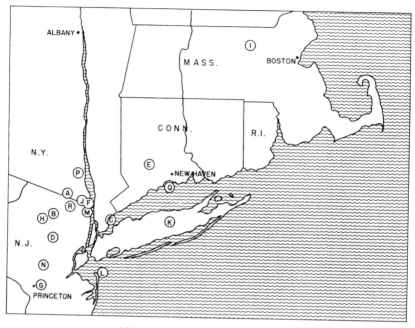

SITES CONSIDERED FOR A NATIONAL LABORATORY

A. BEAR LAKE - in Ramapo Mountains, near Suffern, N.Y. (43 min. from Columbia)
B. BIG PIECE MEADOW, in bend of Passaic River, near Towaco, N.J. (46 min. from Columbia)
C. FORT SLOCUM - on David's Island, off shore of New Rochelle, N.Y. (30 min. from Columbia)
D. GREAT SWAMP - on Passaic River, near Chatham, N.J.
E. LAKE ZOAR - on Housatonic, N.W. of New Haven, Conn. (30 min. from Yale)
F. PALISADES - bordering Palisades Park, near Rockleigh, N.J. (22 min. from Columbia)
G. ROCKY HILL - 5 miles N. of Princeton, N.J. (10 min. from Princeton)
H. UNTERMEYER LAKE - in mountains near Boonton, N.J. (50 min. to Columbia)
I. FORT DEVENS; near AYER, Mass. (60 min. to Harvard)
J. CAMP SHANKS - near Tappan, N.Y. (30 min. from Columbia)
K. CAMP UPTON - near Patchogue, N.Y. (100 min. from Columbia)
L. FORT HANCOCK - on Sandy Hook peninsula, near Red Bank, N.J. (90 min. to Columbia)
M. ALPINE, N.J. on Hudson, River opposite Yonkers (30 min. from Columbia)
N. MILLSTONE RIVER - section near New Brunswick, N.J. (15 min. from Princeton)
P. North of BEAR MOUNTAIN
Q. CONNECTICUT shore
R. SADDLE RIVER country

2.2 Sites considered for a national laboratory.

of Yale, whereas one at Millstone River in New Jersey was fifteen miles north of Princeton. The Boston-area scientists energetically promoted the virtues of Fort Devens, thirty miles west of Cambridge. "It was charming," said Zacharias of Harvard. "People could live in Concord and Lexington; it would have been a wonderful thing." [12] But Bacher of Cornell, worrying about accessibility from his institution and fearing that Harvard and MIT would be dangerously close and powerful neighbors, objected.[13]

Farthest from any university and least favored of all seventeen sites was Camp Upton in Yaphank, Long Island. "At first, everyone took a dim view of going way out on Long Island," Bacher said.[14] Upton if anything was too isolated. After all, the reason IUG members had banded together in the first place was to create a laboratory they could use conveniently. Upton, a full sixty miles from Columbia, was reachable only via local roads or the bumpy Long Island Rail Road; Yale was actually closer as the crow flies, across Long Island Sound. Still, all sites were dutifully explored.

On 15 April, the subcommittee narrowed the list of sites from seventeen to three. The leading contender was Fort Slocum on David's Island in Long Island Sound, about a mile offshore from New Rochelle. On 3 May, the planning committee officially selected Fort Slocum and asked the Manhattan District to acquire the site. David's Island was only about three thousand feet long and fifteen hundred feet wide, not a sizable piece of land. But that criterion had been established primarily because of the need for isolation in case of a reactor accident, and "the water surrounding Fort Slocum effectively replaced such a safety separation," Ramsey said (Ramsey 1967, 5). The Manhattan District now did an about-face, fearing that too many people lived downwind; moreover, the military decided to keep Fort Slocum after all. The news of Slocum's unavailability hit the IUG at about the same time as news that Loomis was unavailable as director. Finally, the IUG members learned that the Argonne laboratory outside of Chicago was successfully en route to becoming the first national laboratory. The result, Ramsey once recalled, was "the all-time low" in morale.

> For a period of several weeks I believe that only Mariette Kuper and I had any hopes that there ever would be a laboratory and neither of us was very hopeful. The entire group felt that the project had lost its initial momentum and was approaching a desperate state such that if a positive development did not soon take place the entire project would be abandoned. We had all been accustomed to the relatively rapid rate of Government decisions which took place during the war. A slowing down of the project for even a month or two appeared almost fatal. (Ramsey 1967, 6)

As it turned out, Slocum would have been a disaster. Although the IUG thought the lab would have accelerators, it foresaw neither the future importance of accelerators nor their eventual size. "In those days nobody dreamed of accelerators four miles in circumference," Zacharias said in 1983.[15]

With Slocum out of the picture, the subcommittee rethought its site criteria. It would be difficult to move rapidly into undeveloped sites, not only because of the time needed but also because building an infrastructure would require congressional authorization. Several army bases suddenly sprang back into contention, including Camp Shanks, on the Hudson River north of Columbia; Fort Hancock, in Sandy Hook, New Jersey; Fort Devens; and Camp Upton. But Camp Shanks was slated to become a public housing project, and the military decided it wanted to keep Fort Devens. Hancock and Upton alone remained real possibilities. Stone & Webster, a Boston construction and engineering firm that had been the principal contractor for the Manhattan Project, was engaged to survey them. Hancock, the leading contender, occupied a sandy spit of land at the mouth of New York harbor and commanded views of the gleaming towers of Manhattan to the northwest and the ocean to the east (fig. 2.3). Unfortunately, the military wanted to keep parts of Hancock and would make only a few undesirable buildings available to the lab.

Glumly, Ramsey and his wife, Elinor, visited Camp Upton, the sole remaining candidate, early in June 1946 (fig. 2.4).[16] It was a bleak sight. It had housed some prisoners of war and barbed wire still ringed a weedy stockade. Canvas tents and deserted cement-block buildings lined the roads, and the barracks were gray, unheated, and shabbily maintained. Elinor insisted they visit Bellport, fifteen miles away on the South Shore, where she had spent summers as a child. The sparkling water and long, empty beaches revived Ramsey's spirits, and when he took several members of the planning committee to Upton on 27 June, he made sure the tour wound up on the Bellport town dock. There, in the words of Kistiakowsky, the planning committee decided to resign itself to the "equalization of disappointment" and proceed with Camp Upton (Ramsey 1967, 6).

Many IUG members worried that Upton's distance from the universities would make commuting difficult, but Bacher put as good a face on it as he could. Upton had plenty of space, he said, the government had sunk money into the site but was willing to dispose of it, the gymnasiums could be converted into laboratories, and living quarters (the barracks) and a rudimentary infrastructure were already on hand. Rabi, too, was irrepressibly realistic, pointing out that Upton was "equally inaccessible to everybody" (fig. 2.5).[17]

On 8 July, the first two lab employees, Mariette Kuper and Clarke Williams, made their initial visit to Camp Upton. When they arrived at the site, they were stunned. Their future business address was a muddy army camp in the middle of rural Long Island, with pitched tents, temporary wood shacks, and drafty barracks with broken windows. Near the barbed wire and watchtowers of the stockade was a golf course with cement tracks for the wheelchair bound. The landscape was monotonous pitch pine and scrub oak. This was not the Mariette Kuper style. On the way back, she and Williams sat in the car staring blankly out their windows in stony silence. A week and a half later, on Morse's first visit, he too found it "a flattening experience." [18]

Ramsey feared that the site's appearance would harm recruitment, which would be difficult enough in any event because the lab was new, there was little housing and no job security, and salaries for junior scientists were low even for that time and occupation, only about $250 to $290 a month. As executive secretary of the group, he was putting together many documents and needed some name for the lab. He guessed that if he used a good one in his documents, it would probably be adopted. His list of candidates included a number of local names like Yaphank, Upton, Suffolk, Long Island, Middle Island, Central Brookhaven, Northeastern, and Brookhaven Laboratory. Ramsey asked Elinor which name would be attractive to the wives of scientists being recruited, and she chose Brookhaven for its connotations of "quiet, shady streams" (Ramsey 1967, 7).

The corporation, too, needed a name. Pegram had used Universities, Inc., as a filler name in early documents. Other IUG members produced more descriptive names (University Institute of Research, Atlantic University Research, University of Advanced Research), acronyms fashioned from the initials of the nine universities (Pyjohmitch Corp., Phytch Corp.), or abbreviations based on the lab's mission (Univas, Inc., for *university association* or Nures Corp., for *nuclear research*) (Brookhaven National Laboratory 1948, 10). In the end, the IUG chose the name Associated Universities, Inc. (AUI), scarcely an improvement on Pegram's placeholder. A well-respected law firm, Milbank, Tweed, Hope, Hadley & McCloy, was retained to draw up incorporation forms. On 8 July, AUI was incorporated in New Jersey (the state with the largest number of prospective sites), and two days later AUI had its first and only meeting as a New Jersey corporation, but back in New York at Columbia's Low Library. The trustees formally voted to ask the government for the Upton site and to press for incorporation in New York as quickly as possible. On 18 July, AUI was granted a charter as an educational institution by the Board of Regents of the State of New York. The IUG then passed out of existence, having done its job by giving birth to AUI.

AUI's trustees met officially as a New York corporation on 30 July. The incorporating trustees adopted a set of bylaws giving AUI, like the IUG before it, a governing board of eighteen members, an administrative officer and a scientific trustee from each of the nine member universities.[19] Again Upton was formally approved as the laboratory site.

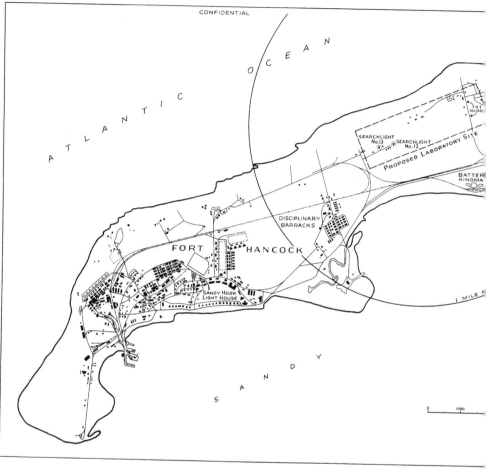

2.3 Stone & Webster plot of proposed lab site at Fort Hancock on Sandy Hook, June 1946. Note the small amount of land (1500′ × 4000′) thought adequate for a national lab.

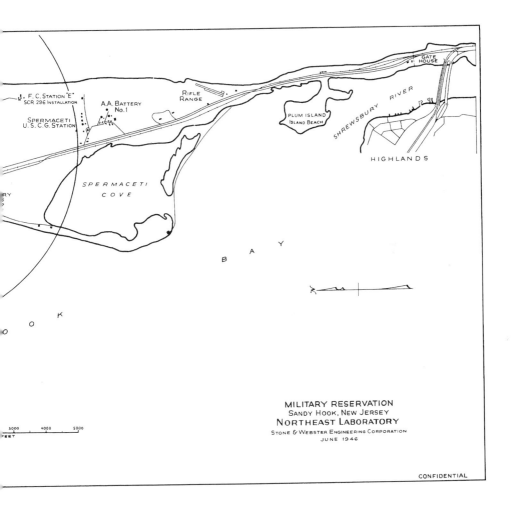

MILITARY RESERVATION
SANDY HOOK, NEW JERSEY
NORTHEAST LABORATORY
STONE & WEBSTER ENGINEERING CORPORATION
JUNE 1946

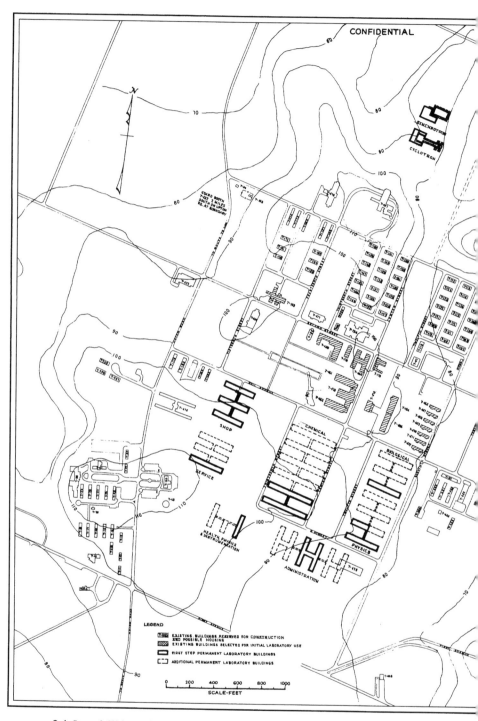

2.4 Stone & Webster plot plan for the "Northeast Laboratory" at Camp Upton, June 1946.

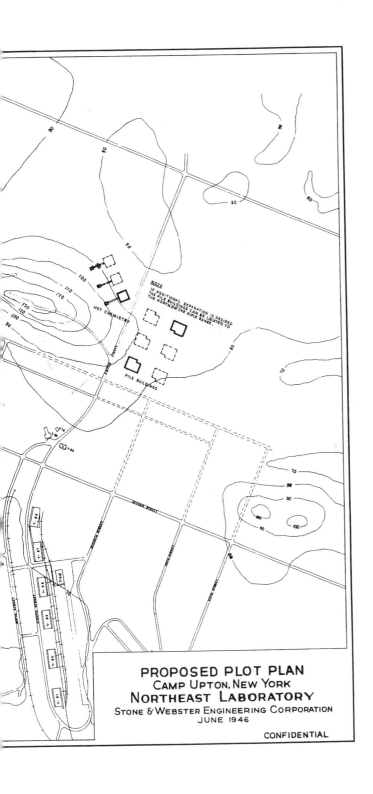

PROPOSED PLOT PLAN
CAMP UPTON, NEW YORK
NORTHEAST LABORATORY
STONE & WEBSTER ENGINEERING CORPORATION
JUNE 1946

CONFIDENTIAL

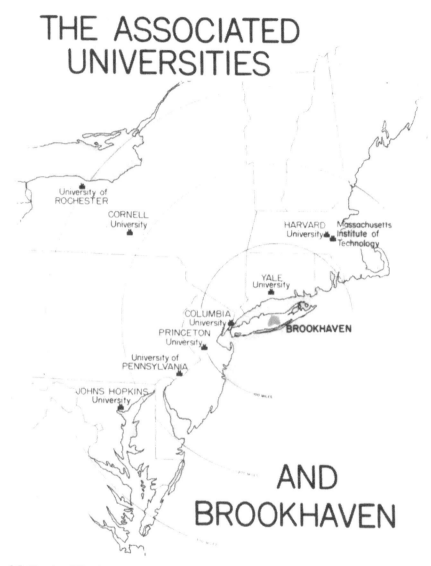

2.5 The nine AUI universities. (From Brookhaven National Laboratory 1948)

Donald Mallory, Camp Upton's post engineer, befriended Kuper, Ramsey, and other AUI associates. In August, Mallory began readying the site for transfer to the Manhattan District, prior to his terminal leave from the army. His orders were to ensure that, before the transfer, the site meet army specifications for its most recent purpose, which was as a convalescent hospital in anticipation of the expected flood of casualties during a land invasion of Japan. Mallory's sympathy for the site's future owners led him to look for loopholes that would allow him to fix up the site rather more than the army may have intended. He discovered a regulation requiring that all buildings at convalescent hospitals be periodically painted. "And I said, 'Fine—we are a convalescent hospital. That means *every* building on the site gets painted!'"[20] And so on 27 August 1946, when Mallory transferred property accountability for the site to the Manhattan District, the entire depressing place at least stood under a fresh coat of white paint.

National Laboratory

AS PLANS for Brookhaven National Laboratory inched forward in the spring and early summer of 1946, Congress had been battling over what kind of institution should handle atomic energy in the future. Two rival scenarios had emerged: the May-Johnson bill would retain military control of atomic energy, while the McMahon bill would put atomic energy under the control of a civilian agency. Sponsors of May-Johnson had expected it to pass handily but did not anticipate the bitter, stiff opposition of scientists, whose wartime exploits had earned them considerable prestige on Capitol Hill. The McMahon bill, and civilian control of atomic energy, prevailed, and the Atomic Energy Act, as it was called, was passed by Congress in late July and signed by President Truman on 1 August 1946. The act created an Atomic Energy Commission (AEC), which would take over the resources of the Manhattan District on 1 January 1947, as well as a congressional Joint Committee on Atomic Energy (JCAE), composed of members of the Senate and House, charged with making "continuing studies" of the activities of the AEC (Smith 1970).

Thus, when Morse returned from his New Hampshire vacation at the end of summer 1946 to begin work as director of the newly named Brookhaven National Laboratory, he and the other AUI members had only four months to get the lab on a firm footing with the Manhattan District before the bureaucratic chaos that would inevitably ensue after transfer of responsibilities to the AEC.

Morse hired J. Georges Peter (pronounced "Pay-terre") of the Harvard Graduate School of Design to transform the army camp into a laboratory. Like many AUI scientists, Peter had MIT Radlab experience, and was looking for an opportunity to plan an entire environment rather than merely design individual buildings. At Brookhaven, Peter got what he wanted— sort of. He had an environment but little money to work with, and his clients had no interest in innovative architecture. Still, he kept a positive outlook, telling a reporter that "there's more vegetation here than you'd expect to find in any [army] camp," and gamely setting about the job of turning barracks into labs and living quarters, and old army huts into summer dormitories (Lang 1947, 36). Many buildings could be used as they stood, especially the cement block buildings on the main street. Against the wishes of the AEC, Morse and Peter decided to keep most of the recreational facilities, including the theater, gymnasium, and swimming pool. To head up maintenance of the physical plant, Morse approached Mallory. Like many nonscientists, Mallory had been captivated by the mystique of the atomic bomb and intrigued by nuclear research. To him, this job sounded much more appealing than going "back to the New York City Board of Water Supply on design work on the Delaware aqueduct." [1]

On 9 September, Morse attended the first meeting of the AUI executive committee. After unanimously adopting Brookhaven National Laboratory as the lab's name, the committee turned to ongoing contract negotiations with the Manhattan District, acting on behalf of the War Department. Since its last meeting in August, the subcommittee on contract had been fighting with the War Department over the fourth contract draft and was still attempting to stave off government attempts to control various aspects of the lab, from security to ownership of documents. Still, the executive committee felt confident enough to schedule the first AUI annual meeting for October, assuming that the final contract with the War Department would be the chief agenda item. [2]

On 27 September, the board of trustees debated a fifth draft and objected to one provision stating that the AEC could force AUI to fire employees who failed to obtain clearance. The trustees authorized their officers to sign the draft but to push the War Department to agree that AUI could bar such employees from access to restricted data rather than dismiss them. [3]

Thus the first annual meeting of the AUI board of trustees, on 24 October 1946, did not review a final contract and instead reorganized the board's structure and elected officers. The board established an executive committee consisting of its nine administrative members, and a scientific advisory committee consisting of its nine scientific members, who would advise the director on the laboratory's research program. As its president, the board selected Edward Reynolds, who as Harvard's administrative vice president had supervised a vast empire of business and financial arrangements. A stereotypical New Englander, Reynolds held himself slightly aloof, seldom smiled, and spoke no more than necessary. "He wasn't a forbidding sort of person—quite approachable," Morse said. "It's simply that it didn't occur to him that he needed to talk very much." [4] Joseph Campbell, Columbia's assistant treasurer, was chosen treasurer. The post of executive vice president went to Eldon C. Shoup, a tough-minded Harvard Business School graduate who, as director of rationing and then regional administrator for the Office of Price Administration, had established and supervised wartime gas- and food-rationing programs for New England. As for the contract still being negotiated with the government, Reynolds confidently told the board "that there were no disputes with the

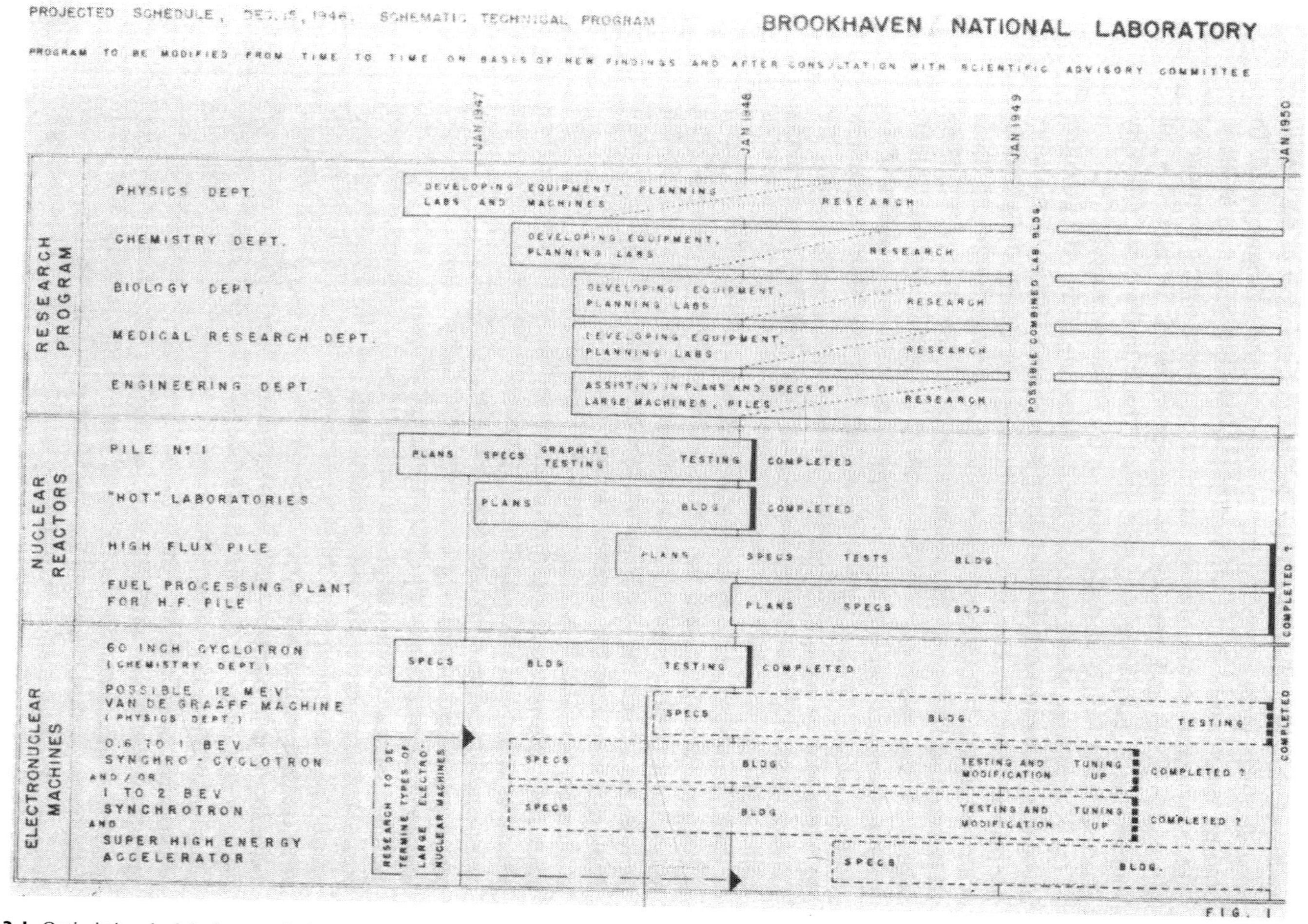
PROJECTED SCHEDULE, DEC. 15, 1946, SCHEMATIC TECHNICAL PROGRAM
BROOKHAVEN NATIONAL LABORATORY
PROGRAM TO BE MODIFIED FROM TIME TO TIME ON BASIS OF NEW FINDINGS AND AFTER CONSULTATION WITH SCIENTIFIC ADVISORY COMMITTEE
JAN 1947
JAN 1948
JAN 1949
JAN 1950
RESEARCH PROGRAM
PHYSICS DEPT.
DEVELOPING EQUIPMENT, PLANNING LABS AND MACHINES
RESEARCH
CHEMISTRY DEPT.
DEVELOPING EQUIPMENT, PLANNING LABS
RESEARCH
BIOLOGY DEPT.
DEVELOPING EQUIPMENT, PLANNING LABS
RESEARCH
MEDICAL RESEARCH DEPT.
DEVELOPING EQUIPMENT, PLANNING LABS
RESEARCH
ENGINEERING DEPT.
ASSISTING IN PLANS AND SPECS OF LARGE MACHINES, PILES
RESEARCH
POSSIBLE COMBINED LAB BLDG.
NUCLEAR REACTORS
PILE Nº 1
PLANS
SPECS
GRAPHITE TESTING
TESTING
COMPLETED
"HOT" LABORATORIES
PLANS
BLDG.
COMPLETED
HIGH FLUX PILE
PLANS
SPECS
TESTS
BLDG.
FUEL PROCESSING PLANT FOR H.F. PILE
PLANS
SPECS
BLDG.
COMPLETED ?
ELECTRONUCLEAR MACHINES
60 INCH CYCLOTRON (CHEMISTRY DEPT.)
SPECS
BLDG.
TESTING
COMPLETED
POSSIBLE 12 MEV VAN DE GRAAFF MACHINE (PHYSICS DEPT.)
SPECS
BLDG.
TESTING
COMPLETED
RESEARCH TO DETERMINE TYPES OF LARGE ELECTRO-NUCLEAR MACHINES
0.6 TO 1 BEV SYNCHRO - CYCLOTRON
AND / OR
1 TO 2 BEV SYNCHROTRON
AND
SUPER HIGH ENERGY ACCELERATOR
SPECS
BLDG.
TESTING AND MODIFICATION
TUNING UP
COMPLETED ?
SPECS
BLDG.
TESTING AND MODIFICATION
TUNING UP
COMPLETED ?
SPECS
BLDG.
FIG. 1

3.1 Optimistic schedule for ...

[War] Department on any points and that the contract was expected to be executed in the immediate future." [5]

On 15 December, the scientific advisory committee issued its initial program report.

> The welfare and preparedness of the United States demand that fundamental research in physical, chemical, biological, medical, and engineering aspects of the atomic sciences be prosecuted with the utmost vigor. . . . The world has still not recovered from the awesome effects produced when nuclear energy was released for military purposes. . . . Everyone now hopes for the benefits to mankind that can come from the harnessing of nuclear forces in peaceful activities. But further advances in the atomic sciences will very largely await the collection of new fundamental scientific information. It is to this end that the research program of Brookhaven Laboratory is primarily directed.[6]

The laboratory would resemble a university to the greatest extent possible, and therefore research would be organized by discipline "rather than around the individual machines." Brookhaven would have five departments (physics, chemistry, biology, medicine, and engineering), served by unique facilities built by two "projects" (reactors and accelerators).

Two days later, and a mere two weeks before termination of the Manhattan District, the AUI executive committee received the initial program report for consideration and approval. Reynolds also produced a contract to which both AUI and government negotiators had agreed, which he signed and passed over to the Manhattan District.[7] Groves attempted to make a few final changes, which were unacceptable to AUI, giving rise to a day of telephone negotiations that finally produced a mutual agreement. Groves sent the contract over to the AEC with his approval on 20 December. A mere eleven days before the deadline the AUI trustees had given themselves, the final contract seemed set.

It wasn't. In the last week of 1946, and of the Manhattan District's existence, Groves approached David E. Lilienthal, who was designated to head the AEC, to ask whether his agency wanted to review the contract, to which, after all, it would be a party. To Groves's dismay, Lilienthal was eager. The possibility of Manhattan District–AEC disagreement over Brookhaven was precisely what Groves had feared and why, through Nichols, he had urged the northeastern scientists to complete the preliminaries before a civilian bureaucracy took over. Late on Christmas Eve, Reynolds received the following telegram from Groves:

LILIENTHAL HAS JUST INFORMED ME THAT THE ATOMIC ENERGY COMMISSION OBJECTS TO MY APPROVING THE CONTRACT WITH ASSOCIATED UNIVERSITIES PRIOR TO ADDITIONAL REVIEW IN DETAIL BY THE COMMISSION AND THAT IT IS NOT POSSIBLE TO COMPLETE SUCH REVIEW BEFORE THE END OF THE YEAR.

THIS MEANS THAT I WILL BE UNABLE TO APPROVE YOUR CONTRACT AS I HAD
INTENDED TO DO. MY RESPONSIBILITIES PASS TO THE COMMISSION ON JANU-
ARY 1ST AND I WOULD NOT FEEL IT PROPER TO GO COUNTER TO ITS WISHES. [8]

Reynolds spent a miserable Christmas. Early on 26 December, he telephoned Groves. Did this delay mean, he asked, that all ongoing recruiting, financial commitments, and transfer of staff and equipment to Upton
should suddenly cease? Groves, helpless, said the matter was out of his
hands and advised Reynolds to contact Lilienthal directly. Reynolds did.
While Lilienthal indicated the AEC needed time to review the contract, he
did assure Reynolds that the work of preparing the lab could continue, and
he followed through on 27 December with a confirming telegram:

THE COMMISSION IS RELYING UPON YOUR GROUP TO PRESS AHEAD WITH THE
DEVELOPMENT OF PLANS AND ARRANGEMENTS, INCLUDING RECRUITMENT,
FOR THE EARLY ESTABLISHMENT OF THE LABORATORY. [9]

Instead of a signed contract, AUI received a two-page letter that confirmed
the terms of the final draft contract for a three-month period. While this
arrangement was not ideal, it was effectively a go-ahead, for the implicit
agreement expressed in the phrase "plans and arrangements" meant that
the AEC would have a difficult time terminating the project or making
major contract revisions.

The special partnership was on the verge of coming into being. Not all
lab scientists were happy with it. On 27 December Livingston, discouraged
and depressed, had a long talk with Mariette Kuper in which he related the
contents of a series of recent conversations. One prominent scientist had
told him that AUI should have held out for a guarantee of a better contract
and that the current one would create "great difficulties in hiring top-notch
personnel"; another told him that Brookhaven had lost the original concept
of a free research laboratory. Most devastatingly, MIT's Slater had said that
he was not sure that the lab would succeed at all, that the planners should
tell the AEC that "under these conditions, we could not operate a laboratory," and that if things do not change soon, he would think seriously of
recalling all MIT personnel on leave to the lab. [10]

Nevertheless, having taken to heart Groves's insistence on speed, Brookhaven's planners had made it in just under the wire. Groves had been right
to worry. Around Christmas 1946, Brookhaven reactor scientist Irving
Kaplan ran into George Weil, head of the AEC's reactor development division. Kaplan recalls:

George was a little surprised and even a little annoyed at how far we were along
on this project, and he asked me, "Who on Earth gave you permission to build
this reactor?" By that time, we were already building away. I realized that if we

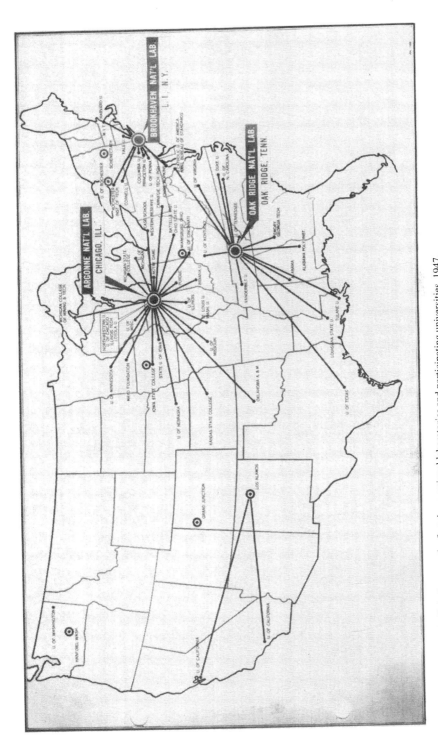

3.2 Diagram showing the links between the first three national laboratories and participating universities, 1947.

had had to wait for permission from those guys, it might have taken years. I remember saying to myself, "Nuts to you, George!" [11]

Moreover, the AEC was virtually paralyzed by a bitter fight over the Senate confirmation of Lilienthal. This delayed several important decisions at the other national labs, such as Argonne's attempt to move to a new site (Holl 1997, 53–55).

On 1 January 1947, responsibility for Camp Upton was transferred from the War Department to the AEC, and Brookhaven National Laboratory officially sprang into being as one of the first three national laboratories. Its two siblings were Argonne National Laboratory (the first to emerge officially as a national laboratory, on 1 July 1946) and Clinton National Laboratories, and for a few months the U.S. national laboratory system was as easy as ABC: ANL, BNL, and CNL, though the last would soon disrupt the pleasant acronymic sequence by changing its name to Oak Ridge National Laboratory (ORNL). But Brookhaven was unlike its two sister labs both scientifically and organizationally. Scientifically, its program would focus more intensely on basic, academic-style research. Though the AEC had not yet assigned specific missions to the labs, and their research overlapped somewhat, the agency viewed Argonne and Oak Ridge as places for applied projects related to atomic energy, and Brookhaven as a basic research lab. (Argonne and Oak Ridge each also had basic research programs, for which they sometimes had to struggle; Holl 1997, 62–63, 87–88, 243–45.) Organizationally, Brookhaven was the only one managed by a consortium of universities. Argonne's contractor was a single institution, the University of Chicago; other nearby universities served in an advisory group. Oak Ridge's contractor was a corporation, first Monsanto, then Union Carbide, and later Martin Marietta, and here, too, various regional universities served as an advisory group. These advisory groups had nothing approaching the kind of influence on Argonne and Oak Ridge that the nine universities of AUI (initially, at least) had on Brookhaven.

But on 1 January 1947, the most striking difference among the three labs was that, while Argonne and Oak Ridge were already functioning laboratories, furnished with facilities and staff, Brookhaven was still an army camp, furnished with barracks and deer. [12]

The "Brookhaven Concept"

ON MONDAY, 6 January 1947, Donald Mallory was the first to appear on-site, now as superintendent of buildings and grounds at a laboratory rather than as post engineer of an army camp. His task that chilly Long Island day was to fire up the old coal-burning heating plant in preparation for the arrival, the following week, of the scientists and staff moving out from their Columbia offices as well as the AUI staff moving from their downtown Manhattan office. The heating plant, which had sat idle for months, belched, protested, and finally roared into life, announcing the resumption of activity at the site with a long plume of black smoke. As Ramsey recalled, the plume provoked a neighbor to threaten a lawsuit, complaining that she had been exposed to radiation, long before any equipment or staff had moved in (Ramsey 1967, 9). The threatened lawsuit served notice, literally from day one, that the laboratory operated within a larger social community with its own expectations and fears.

Philip Morse arrived a week later. He faced a daunting task, for the lab existed only on paper and administrative as well as scientific challenges lay ahead. Administratively, the special partnership between the lab and the AEC existed only in outline and needed fuller specification. The lab still did not have a final contract, for instance, and policies regarding security clearance and classification were up in the air. One major obstacle to enacting the founders' ideals in these issues was the continuity of many personnel, and therefore habits of thinking, from the Manhattan Engineer District to the Atomic Energy Commission. Morse was not temperamentally well suited to the aggressive diplomacy that would be called for in breaking those habits, and his strategy in dealing with problematical Washington officials was that which he had adopted in his first encounter with Groves: fade into the background in meetings, respond only to direct questions, and hope that reason will eventually prevail. In late 1940s Washington, the wisdom of that strategy was questionable, and resolution of many issues would have to wait until Morse was replaced by a more activist director, Leland Haworth, in 1948. On the science side, Morse first faced an extraordinarily difficult task in recruitment. The other national labs, already well under way, were involved in a feverish enough competition for scientific staff with prominent universities. Brookhaven was saddled

4.1 Entrance to Brookhaven, 1947.

4.2 Aerial view of lab site, 1947, before any construction work was underway. Brookhaven still resembled the army camp on whose site it was built.

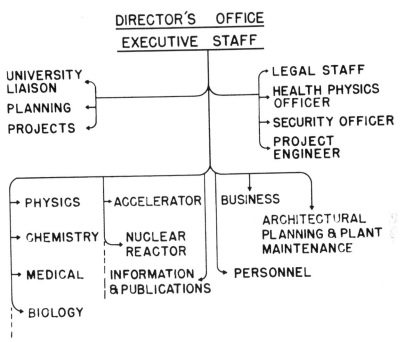

4.3 Organization chart, 1947, showing the early organization into departments and projects.

with additional difficulties in that it didn't even exist yet as an operating lab (with some skeptics even convinced it would fall through) and had no major working facilities and clearly would not have any for several years. Morse would have to rely on a combination of incentives for qualified senior staff and clever hiring of bright junior staff overlooked elsewhere. These obstacles to recruitment, bad enough in physics, were especially difficult in the life sciences. (Much support staff was recruited from the local community through advertisements in newspapers.) Furthermore, Morse would also have to see that the pile and accelerator projects were on a sound footing—but the desires of the lab's founders collided with the AEC's policies and interests here as well (as discussed in chapters 5 and 7). Finally, community relations would prove surprisingly difficult for the new laboratory (chapter 6).

Security Clearance

Few would have been troubled if the AEC and FBI had handled the clearance process in a thorough, responsible manner. National security interests obviously required great caution in deciding who could work in the field of atomic energy. Following provisions of the Atomic Energy Act, the AEC could deny access to restricted data to anyone until the FBI had conducted a thorough investigation into "character, associations, and loyalty" to ensure no threat to "the common defense or security." If the AEC was satisfied, it would grant "Q clearance," an authorization to work with restricted materials.[1] But the Soviet Union's new power in eastern Europe, coupled with the fear that it would somehow acquire the secret of the atomic bomb, contributed to overwrought fears of communist subversion and the beginnings of the red scare. The 1946 elections gave Republicans control of both houses of Congress for the first time since 1928, and the extremely conservative Eightieth Congress put pressure on the AEC, which responded by tightening its supervision of its labs and placing nuclear physicists under greater scrutiny (Holl 1997, chap. 3). "Although not one American atomic scientist was ever shown to have turned traitor, no group was more closely inspected or forced so often to prove their loyalty" (Weart 1988, 121). Even those who had received top-secret clearance during the war could be required to apply for a new clearance, existing clearances could be called into question at any time, and charges were often kept secret from the accused. Investigations could be sloppy and random, hearsay and innuendo were often treated as fact, and efforts to clear up confusion and error met with stubborn resistance. "It was really a very, very serious problem," says reactor physicist David Gurinsky, who was hired by the lab in 1947. "If they took your clearance away, you couldn't make a living. And they didn't care what they did to you—they used it. We were scared."[2]

Consider the case of chemist Gerhart Friedlander, a Jewish refugee from Nazi Germany. Friedlander arrived in the United States with ten dollars in his pocket in 1938, went on to undergraduate and graduate studies at Berkeley, and wound up at Los Alamos. After the war, he worked for two years at G.E.'s research laboratory in Schenectady. He was one of the first members of Brookhaven's chemistry department, arriving in January 1948, but after a few months was surprised and alarmed to learn that his security clearance was being questioned. Later, he would discover that the FBI had been tracking an alleged Soviet agent who had worked in Berkeley and Schenectady under the name of Ignatsi Witczak, and that one of this agent's associates was a man named Friedlander. The FBI had managed to mix up *that* Friedlander with the Brookhaven Friedlander. The consequences did not stop there. The mix-up had a domino effect; the suspicion faced by Friedlander spilled over onto his associates. Chemist Morris Perlman, in whom Brookhaven was also interested, also fell under suspicion; one charge against him was that he was a friend of Friedlander. For months, neither Friedlander nor Perlman could understand exactly why suspicion was falling on them, let alone act effectively to counter it.

Frustrated, Perlman took a job elsewhere, but Friedlander continued his efforts to clear his name. In November 1948, five members of the AEC's Personnel Security Board, chaired by Admiral John Gingrich, questioned him at the AEC's New York offices. Years later, it is difficult to read the transcript of that session, which occasionally verges on the absurd, and feel the full weight of seriousness the proceedings carried at the time for each participant.[3]

> The Chairman: Did you know a man up there [in Schenectady] named Ignatsi Witczak?
> Dr. Friedlander: Witczak?
> The Chairman: Yes, W I T C Z A K.
> Dr. Friedlander: I never heard of the man . . . Is it possible there is some case of mistaken identity about this gentleman in Albany I'm supposed to know? There was another Friedlander in Schenectady and I've got mixed up with him before. I don't know him at all and he's not related to me.

Talk continued for several minutes about Friedlander's associates in Berkeley. Friedlander was asked again about Witczak, and again he mentioned the existence of another Friedlander at Schenectady and denied yet again knowing a man named Witczak. More discussion of Friedlander's former associates followed. He was asked again about Witczak, again denied knowing such a person, and again suggested that the AEC look into the possibility of a mix-up. Friedlander was asked to state "for the record" his position with respect to Nazism, and the Jewish refugee obligingly complied. He was opposed to it. Then:

> The Chairman: You don't know Ignatsi Samuel Witczak?
> Dr. Friedlander: No.

Friedlander was then twice asked if he had been married by a Presbyterian pastor, and twice replied no. A few minutes later:

> The Chairman: Now, you and your wife were married [in] 1941, February?
> Dr. Friedlander: That's right.
> The Chairman: Who married you?
> Dr. Friedlander: A judge in the Oakland County Court House.
> The Chairman: Are you sure?
> Dr. Friedlander: Yes, I'm absolutely sure. That is one of the things one remembers.
> The Chairman: Have you ever attended a Presbyterian Church out there on College Avenue?
> Dr. Friedlander: I have not.

Mr. Hammack: You're not a Presbyterian, are you?

Dr. Friedlander: No, I'm not a Presbyterian and I've never been one.

Mr. Hammack: You see, that's why we have these hearings, just in order to correct the record. Yes, and I can tell you, as you probably well know, that in Russia they wouldn't be having these hearings.

Dr. Friedlander: No, I realize that and I have no desires of being in Russia, I can assure you.

Friedlander asked if his friend Perlman could be cleared as well. The admiral said Perlman would have to wait. Friedlander reminded the admiral that Perlman had already waited six months. The admiral advised Friedlander not to go off "half-cocked." Friedlander then exploded.

Dr. Friedlander: I meant that it would have been very helpful, for instance, if Perlman at that time had been called in and interviewed like this, say, and been asked this question and he would have undoubtedly said he doesn't know Mr. Witczak—at least, I think he doesn't—and then you could have immediately investigated this further and this could have been cleared up. Now, in the meantime, he has spent six months worrying what in the world was going on with his clearance and he has taken another job for the time being and is still waiting to hear what goes on. This is pretty hard on a person's morale, you see, just sitting around not knowing what the score is.

The admiral resumed questioning Friedlander about other matters, asking him again if he had been married in a Presbyterian church.

The Chairman: Is there anything you'd like to say, Doctor, before we conclude?

Dr. Friedlander: I just want to say that I hope this gets cleared up with some dispatch.

The Chairman: Well, in the meantime, let me tell you this. . . . We look at these things pretty objectively and pretty sensibly. We're not on any witch hunt or anything of that sort. . . . You wouldn't think of investigating a scientific matter without going into it thoroughly. . . . Well, you sleep well at night and you go ahead and do your work and you forget about it. We'll tell you when we get this thing done as we still have to check those points.

Four months later, Friedlander received a letter from the AEC informing him that his "continued employment on work of a classified nature would not endanger the common defense and security." [4]

The lab's efforts to hire British physicist Ronald Gurney, coauthor of a famous paper that explained radioactivity utilizing quantum mechanics for the first time, had a less happy ending. Morse thought Gurney one of Brookhaven's best senior physicist prospects, and in the summer of 1946 he offered a scientific position to Gurney and a job in the information division to his wife, Natalie. He was cleared but she was not, for no apparent reason other than lack of U.S. citizenship. Morse was convinced that the misunderstanding blocking her clearance would soon be resolved, and he continued to write encouraging letters to the Gurneys. A year passed with no clearance and no explanation. The Gurneys declined other job opportunities, then lost their lease and had to move. Morse tried to persuade AUI trustees to take up the case with the AEC, but they refused, saying the case was not worth a fight. Finally, in October 1947, Morse was forced to withdraw his job offer. Gurney lived in the United States on assorted short-term jobs for a few years, barred from further work in the kind of research he had helped initiate. High blood pressure and a heart condition, aggravated by problems caused by Natalie's failure to get security clearance, led to his premature death in 1950, "a victim," Morse maintained, "of the treatment he had received, a sacrifice to our paranoiac fear of losing 'the secret'" (Morse 1977, 229).[5]

The AEC initially required that all senior staff members at Brookhaven have Q clearance. Although AUI trustees objected, they feared a fight over clearance would be counterproductive. The AEC was increasingly sensitive to charges it was not doing enough to detect communist sympathizers. Operating as Brookhaven did under AEC contract, the executive committee members decided that the lab was "not the place in which to make an issue of personal liberties out of security clearance procedures."[6]

Morse was unable to move them, and the issue was one source of conflict between the lab and the AEC whose resolution would have to await a new director. Later, a security policy was worked out involving a P clearance as an alternative to Q clearance, though a Q would still be required for permanent staff. P clearance allowed temporary employees who had had a routine FBI file check to work in unclassified areas, a system "unique for major Laboratories wholly financed with Atomic Energy Commission funds."[7] The system was occasionally cumbersome; for one thing, a separate set of bathroom facilities had to be maintained for P- and Q-cleared males and females. (Toilets for the uncleared, at Brookhaven and elsewhere, were known as "P clearances.") And as the red scare grew, the system was not easy to maintain. In late August of 1949, the explosion of Joe I, as the first Soviet atomic bomb was nicknamed, inaugurated a new wave of anxiety and the era of the "red menace." Klaus Fuchs was arrested for passing atomic secrets to the Soviets in 1950, and Julius and Ethel Rosenberg were arrested the following year. In February 1950, Senator Joseph McCarthy made his famous announcement that he had a list of State Department employees loyal to the Soviet Union, and later that year accused seven nationally known scientists of being communist sympathizers, two of whom worked at Brookhaven. Also in 1950, an increasingly suspicious Congress required new P employees to sign a certification of noncommunism.[8] But despite these developments, the lab had many fewer troubles with the new system than the old.

Classification of Research

Initially the AEC's atomic energy labs had to classify every report or paper, with declassification only on request after submission of a final report, which would be closely scrutinized in Washington.[9] Morse, with Rabi and most of the trustees, wanted the lab to perform only unclassified work that could be discussed publicly. This was a key element of the partnership envisioned by the lab's founders, who wanted to maximize interaction among scientists at the lab, and with scientists outside the lab so that ordinary university researchers could use the lab without excess bureaucracy. "It seemed to me," Morse recalled, "that the future of the popularity of the laboratory was going to depend a great deal on that problem."[10] Indeed, other national labs were severely hampered by this problem (Holl 1997, 158–59).

Morse hoped that if he kept quiet, Brookhaven might be able to take charge of its own security, as his navy group had done during the war. In mid-October 1946, he attended a meeting of directors of major labs supported by the Manhattan District, at which the policy of routine classification of all research was restated. He was appalled but waited until the district representatives left. Then he told the other lab representatives, as he later recorded in his notes, that given its special status, "the rules of the game for Brookhaven were so different from the other laboratories that I did not intend to stick to the rules as commented upon at that meeting." The other laboratory directors were surprised, but Morse did not elaborate, "feeling that it was hardly worthwhile crystallizing feelings at this point."[11]

Morse was disturbed that he received little support for his position from the heads of some other AEC labs, and was particularly angered by the lack of help from Ernest Lawrence, head of the Berkeley lab, who seemed perfectly comfortable working under the AEC rules (Morse 1977, 231). Also, Morse had to deal with William Kelley, who had managed the Manhattan Engineer District's New York–area office on Madison Avenue and simply stayed on to manage the New York office of the AEC. Morse was thus facing a person whose approach to security and classification was forged in wartime, and Kelley's actions sometimes came to be regarded even by outsiders as extreme; when Dwight D. Eisenhower visited Brookhaven in 1948, Kelley's insistence that he first obtain clearance annoyed even the general. Fights with Kelley and the AEC over the clearance issue, Morse later estimated, took between a quarter and a third of his time during the first months of his directorship.[12] But here Morse was able to make some progress, and worked out a set of security regulations to get most but not all of what he wanted. Brookhaven, save the reactor and the library building where classified material was kept, would be open to unclassified individuals. Most research would be unclassified—a set of categories of unclassified research was agreed to—but some papers on nuclear research were to be withheld from publication until declassified by the AEC.[13]

Contract

Contract negotiations with the AEC, which were handled on Brookhaven's end by a special contract committee headed by AUI's tough executive vice president Shoup, were also proving unexpectedly difficult. The contract was regarded as especially important, for it would spell out the terms of the special partnership between the lab and the AEC. The lab had opened with a temporary contract that extended the terms of the previous Manhattan District contract for three months, and the agreement had to be extended several times more as negotiations continued. One major sticking point involved the division of responsibilities between AUI and the AEC. AUI trustees worried over the AEC's intention to claim ownership of everything on- and even off-site related to the contract, and to control contracting and subcontracting provisions. Knowing that the AEC demanded one-fourth of the operating time of the Columbia cyclotron, they also wanted to ensure that Brookhaven would be free to pursue its own research agenda rather than the government's. In October 1947, ten months after the lab had opened and with yet another draft contract in their hands, the trustees were protesting "too much management and control," "too much expensive red tape," and "restrictive" patent provisions. They demanded "the right to refuse classified research," and rejected the government's intent to own "all records and every scrap of paper." The AEC draft, they felt, revealed "an attempt to manage the details of the laboratory, showing lack of confidence in the contractor's own abilities."[14]

In November 1947, just before the fourth temporary contract ran out, the special contract committee and the AEC met again. Beforehand, vice president Shoup had sent a memo to AUI president Edward Reynolds. The contract, Shoup wrote, should create "an equal partnership between the government and the contractor"; instead, it promotes "a master-servant relationship" because the government controls the purse strings. But this "simply would not work with free educational institutions." Shoup then outlined a plan for a long-term special partnership between the AEC, AUI, and the laboratory.

> The plan provides for a large, centrally located laboratory . . . [whose] . . . *financial* support will be provided by the *government,* while the resources of scientific knowledge, ideas and initiative will derive from all universities, private and public laboratories and industry. The function of *program* and *policy control* is a joint responsibility of government and the contractor. The function of *management* is entrusted to the contractor. The contractor in this case assumes the responsibilities of management, not for financial profit, nor for special benefit of any one or all nine universities, but as a trustee on behalf of all participants in nuclear research and education in this region.[15]

In the face of the trustees' united front, the AEC relented, and on 23 December Reynolds signed a contract essentially endorsing the AUI position. "They now clearly have confidence in the ability of this Corporation to build, organize, and operate a laboratory," one of the trustees said with relief at the next trustees meeting.[16]

This contract between Brookhaven and the AEC, though based on wartime precedents, was enormously influential in contracts with other labs in the U.S. and abroad. Article 4 was of particular significance, for it embodied an important concept later known as administrative contracting. As opposed to the usual fixed-price and cost-plus-fixed-fee type contracts, an administrative contract enacts a "no-profit, no-fee," long-term collaborative relationship between the two parties. Almost immediately after the contract was signed, the AEC decided to use passages like article 4 in its other contracts involving laboratories and other large research and projects (Furman 1990, 355). "It is the desire of the Commission," the AUI contract reads, "to procure for the Government managerial skill and responsibility which will permit flexibility in administrative controls and freedom from detailed supervision."[17] AUI would be an independent, nonprofit contractor, exempt from many of the usual restrictions written in other government contracts, an idea soon known as government owned, contractor operated (or "Goco") contracting. Gocos were described in the AEC's booklet on contract policy and operations as follows:

> Because of the greater difficulty in describing goals in basic research, and because responsible scientists plan their own basic research projects, such programs are not budgeted or controlled in the detail found in applied research planning. Dollar estimates and dollar limitations are coupled with classes of work such as "neutron physics" and "chemistry of the rare earths," leaving to the laboratory director and his scientific staff the choice of specific investigations and methods of attack, within the totals fixed under each major budget program.[18]

While administrative-type contracting had been used before, this was the first time the contractor was an organization consisting of a consortium of universities operating a project on such a large scale.

Still, many of the novel ideas embedded in the contract were insufficiently spelled out and thus subject to conflicting interpretations by the two parties. By 1949, too, the AEC had become concerned over delays and cost overruns in certain of the lab's projects.[19] Other factors, some Cold War–related, were also making the AEC increasingly sensitive about the management of the national labs (Holl 1997, chap. 3). Thus as negotiations to extend the contract with AUI approached, tensions persisted. In April 1949, Shoup complained that the AEC was increasingly tightening its control of Brookhaven and curtailing AUI's independence in managing the lab. Some provisions in the AEC's contract manual, if applied literally, "would be a serious impediment to efficient operation." The AEC's budget requests had required increasingly detailed presentations that are "time-consuming and of little apparent value." Finally, the AEC had demanded budgets for future years on extremely short notice. Other trustees, however, were afraid that protests might backfire, and counseled approaching the AEC before attempting to renegotiate the contract, which would expire at the end of 1950, to discuss the benefits and costs of the lab's current operating procedures.[20]

AEC general manager Carroll Wilson welcomed the suggestion, and the AUI appointed a subcommittee headed by Reynolds to evaluate the administration of the lab and its relations with the AEC. The subcommittee reported in September 1949; it complained about the AEC's tendency "to impose uniform procedures of a type better adapted to governmental organizations or Atomic Energy Commission production plants than to a laboratory like Brookhaven." The report then outlined what it called "the Brookhaven concept" consisting of "a scientific program or series of scientific objectives in combination with a unique plan by which the necessary resources are mobilized for their achievement." To operate its unique facilities, the report said, can only be brought about by a "contract relationship between the Government and the universities . . . which will join the resources of both in a constructive partnership favorable to the successful progress of fundamental scientific research." Because the partnership between the federal government and AUI to run Brookhaven was unprecedented, the AUI report continued, "it is bound to have profound effects in Government, in education, in research and, ultimately, in our economic life." The report emphasized that the success of the partnership depended on mutual recognition of the fact that "the contracting parties are of equal importance and equally indispensable," and that the continuing success of the contracting system that had been worked out would depend on whether this equality was protected and encouraged.[21]

Initially, however, the AEC reacted with suspicion, and in the late spring of 1950, contract negotiations began in an atmosphere of distrust. The

AEC's area manager Emery Van Horn (who had been the MD's area manager) told Shoup that "the Commission was by no means satisfied with the management provided by the corporation to date," and hinted darkly that the AEC might be considering another contractor.[22] AUI responded with, among other things, a corporate restructure that put an end to the criticism, and a contract was finally agreed to.

Culture and Housing

Brookhaven's founders were well aware that they were creating a novel type of institution that was neither university, business corporation, nor government agency, but combined elements of all three. They were also aware of the potential problems; among the most obvious of these was that of striking the right balance between fundamental, programmatic, and applied research for the lab.[23] Other potential problems loomed in the lack of a liberal arts environment and university culture in general; this, some feared, might create a tendency toward detached, remote, or insular thinking and behavior on the part of the staff. The problem was exacerbated by the lab's location in what at the time was a cultural hinterland. ("Car pools and cesspools" was Rabi's favorite phrase describing life in the new surroundings.) An employee recreation association (BERA) was created to help manufacture cultural activities, and clubs for painting, singing, dancing, ceramics, bird watching, chess, etc. were created and a number of small orchestras and combos formed. Brookhaven, too, brought Suffolk County a substantial cultural diversity. In the 1950s, it was almost certainly the only place on Long Island where soccer was regularly played, thanks to the large numbers of European employees. By the end of the decade, the AEC would disapprove Brookhaven's request to build a golf course on the lab site on the grounds that there was now "at the laboratory sufficiently well balanced cultural, athletic, and social activities to meet the needs of the employees."[24]

The trustees periodically raised the question of whether formal courses ought to be offered at the lab, which was, after all, operated under a charter granted under the education laws of the state of New York to a group called Associated Universities, and perhaps even degrees granted. The lack of students could be both a plus and a minus; a plus in that absence of teaching duties left more time for research, a minus in that it deprived researchers of assistants. The consensus was that formal training would take staff members too far afield from the research that was the lab's main function. However, other kinds of teaching programs were given from time to time, as at other national labs (Holl 1997, 104–5, 135–37). For a few years in the

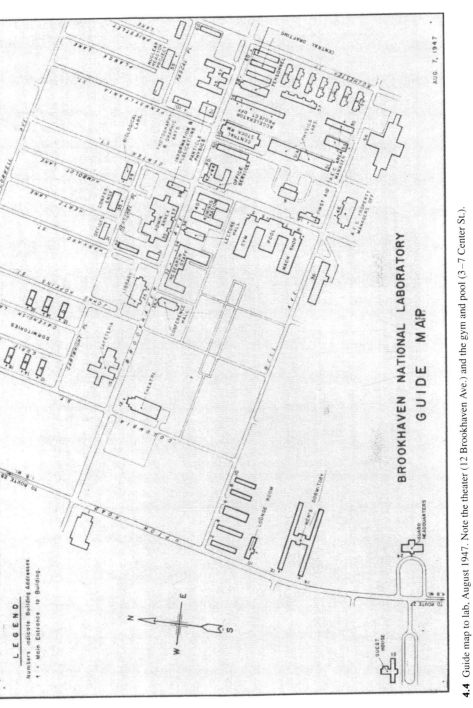

4.4 Guide map to lab, August 1947. Note the theater (12 Brookhaven Ave.) and the gym and pool (3–7 Center St.).

wake of Atoms for Peace, the lab had ten-week courses for engineering faculty to help bring them up to speed on nuclear engineering and reactor physics, while in 1949 health physics began a summer training program.[25]

The longest-running educational program at the lab was the summer student program, begun in 1952, which allowed undergraduate and graduate students the experience of direct participation in an ongoing research project. Participants over the years included many eminent U.S. researchers (such as Nobel laureate Roald Hoffman) and future Brookhaven scientists (including Nicholas Samios, who would be lab director between 1982 and 1997). Accommodations were spare: students were housed in old barrack buildings that had been crudely divided into rooms with two-by-four frames but with sheet rock, only on one side, nailed horizontally. Over the years, the sheet rock had shrunk, and a quarter-inch gap appeared wherever two pieces of sheet rock met, allowing occupants to peer down and observe a sliver of adjoining rooms. To enter the old army showers one had to enter the stall before turning them on; when one ingenious student used an umbrella, the practice quickly became universal. R. C. Anderson, the director of the summer program, was so embarrassed by the facilities that he would bring students in each barrack a basket of wine and fruit as a gift on the first day. The accommodations made natural staging grounds for practical jokes. One summer student recalled:

> Any door could be opened with a knife blade, and every once in a while you'd come back to find your furniture completely rearranged. One person came back to find all his books in his drawers and his clothes on the shelves between bookends. Another student in my dorm was famous for always complaining of not having enough light. His friends, tiring of the complaints, went in his room one day and replaced all of his light bulbs with high-powered flash bulbs they'd pillaged from physics. He came back, opened the door, flipped on the switch—which activated a double socket in which they'd screwed two bulbs—and BAM! there was a blinding explosion of light and it all went dark again. He felt his way over to his desk light, turned that on, and BAM! that went off, too. He, of course, had a lot of lights, and tried them all—BAM! BAM! BAM! until they were all out. He happened to stagger outside just as I came down the hall. When he told me what happened, I tried to persuade him that a transient electromagnetic pulse from the Cosmotron generator had probably shorted out his lights. It was awfully hard to keep a straight face.[26]

If on-site housing conditions were crude, off-site housing was difficult to find. The trustees had decided that the laboratory would not establish a "company town" or "university city." But housing available for rental in the middle of Long Island proved unexpectedly scarce and ramshackle, and the shortage threatened to hamper recruitment of the hundreds and eventually thousands of junior scientists, technicians, and other staff workers who would be needed. Most rental housing near the lab consisted of unheated summer cottages; even if available in the winter, their rents could soar five-fold during the summer months. J. G. Peter commissioned a survey of possible housing within a twenty-mile radius of the lab, an area that included the small but thriving villages of Smithtown, Port Jefferson, Riverhead, Patchogue, Bayport, and Sayville. The result was alarming: "There are literally no houses available for rent on a year-round basis." [27]

Through barrack renovations on-site and leases off-site, the laboratory barely managed to keep abreast of need. High rents and some local anti-Semitism also created problems. Gurinsky, arriving from the University of Chicago in the summer of 1947, spent several months living with friends while looking for housing in his spare time. He discovered that Bellport, Bayport, and other villages on the south shore of Long Island, west of the lab and closer to New York City, were beyond his means. Shifting his sights further east to a town called East Moriches, Gurinsky was puzzled by an odd discussion in a real estate office until he finally realized that the pleasant saleswoman "was trying to tell me that there were not many people of the Jewish faith residing in East Moriches." [28]

The Life Sciences

Physics, naturally enough, was the first department to be organized and to have research projects under way. The other departments were slower to get off the ground, especially biology and medicine.[29] The early story of the life science departments, which came perilously close to being terminated, illustrates the kinds of difficulties faced by the other nonphysics departments. One reason for the problems of the life science departments was neglect; the reactor and accelerator projects, the lab's raisons d'être, demanded more attention than anticipated. And since the life science departments would need the capacity to produce a variety of mostly short-lived radioisotopes for research and clinical use, full deployment of their programs would have to await completion of such projects. Second, the lab's founders tended to avoid involvement in the life sciences, whether out of natural caution or outright disdain. "Rabi notably, and most of the rest of the physicists," chemist Franklin Long recalled, "really were persuaded that fundamentally the job of the laboratory was the accelerators, and some associated theory, and everything else was glassblowing." [30] A third reason was expense. First-class researchers in medicine were costlier than those in other fields, and the trustees and the AEC were not sure the price was worth it, given the lab's orientation. Neglect, disdain, and expense all conspired to put the life sciences departments on the back burner, and nearly put an end to them. Next to nothing was done for the first few months after the lab opened and neither even had an acting head.

In April 1947 a clinical pathologist named F. William Sunderman, chief of the division of clinical chemistry at the William Pepper Laboratory of the University of Pennsylvania School of Medicine, became acting head of Brookhaven's medical department. Sunderman decided to keep his existing position until the department was more developed, and in the interim paid weekly visits to the lab. But seeing the large and expensive pile and accelerator projects under way, Sunderman thought that Brookhaven should have a large and important research hospital to accompany them.

He encountered resistance to the idea almost immediately. To the trustees, the medical department was an offshoot of the lab's main research. They had no intention of being in the medical care business, and were unconvinced Brookhaven could offer medical research facilities that could not be better placed elsewhere. Their suspicions were further aroused by the lack of interest displayed by researchers in the life sciences in Brookhaven—in stark contrast to the clamor of support among physicists across the nation. When the medical department was officially activated, on 1 July, Sunderman used the inauguration ceremony to issue a caustic assessment of the trustees' foot-dragging:

> I am reminded of the young husband who accompanied his wife to the delivery room before the arrival of their first baby. When the door opened, the husband became sickened by the smell of assorted anesthetic and antiseptic vapors. Turning to his wife, he said, "Dear, do you really believe we should go through with this?" . . . Like the young husband's spouse, there appears to be no way now that we can keep from going through with it. I fear that some labor pains, however, are still ahead of us.[31]

Labor pains indeed. They would be so severe that Sunderman, the intended midwife, would leave before the delivery.

Sunderman outlined a set of ambitious plans at the first meeting of the laboratory medical advisory board. His "ultimate objective," he said, was a two-hundred-bed research hospital, which would be completed by 1950 for $4.7 million, with another $2.5 million for equipment and annual operating expenses of $4 million. A week later, Sunderman left on a tour of medical organizations at other laboratories. In his absence, the trustees received numerous criticisms of his grandiose plans and voted to table the medical research program, allocating only enough funds to cover costs of medical service to employees. Furious, Sunderman and his allies charged, not without reason, that the physicists and chemists who ran the lab looked upon life scientists as second-class citizens, but Sunderman himself was present too little to prevail over them. Meanwhile, Haworth, responding to critics who accused the lab of turning its back on the medical department, faulted the accusers for being "unwilling to start from small beginnings." [32]

The medical department remained in limbo for months, with quarters that had been inadequate even in army days and with no research initiative. Its staff was neglected even by Sunderman, who used his rare visits to bend the ear of the upper administration. Sunderman put Kenneth Koerber, an acquaintance from Philadelphia, in charge of the staff, and the lab had two other full-time doctors, Robert A. Love and Wilma Sachs. The three ran a simple clinic in some old wooden barracks that had served as the old army hospital wards. Until a research program was developed, the only medical activities were those connected with emergencies, occupational illnesses, and catching up on preemployment physicals for people who had been hired months before.

Sunderman became increasingly unpopular with the trustees, and early in 1948 Morse informed him he would not be given permanent chairmanship of the department. When Sunderman turned in a letter of resignation, Morse named as acting head Captain Robert D. Conrad, a navy officer who had retired after being diagnosed with leukemia. A vigorous proponent of government funding of basic research, and a driving force in the establishment of a permanent Office of Naval Research, Conrad had volunteered to help Brookhaven out where he could, and had just stepped down as acting chairman of Brookhaven's engineering department.

Meanwhile, the biology department was also experiencing start-up difficulties. Its program had been more clearly articulated than that of the medical department, and consisted of three general fields of research: effects of radiation on living organisms, studies using tracer techniques of biological mechanisms, and general methods and techniques in biology. But the department was hobbled by acting chairman Leslie Nims's lack of administrative skills. In March 1948, the AEC blocked further planning in biology, too.

Then came a stroke of luck. When George Kistiakowsky of Harvard resigned as a trustee in November 1947, Harvard replaced him with A. Baird Hastings, professor of biological chemistry at Harvard Medical School. At his first board meeting, he discovered there was only one other nonphysical scientist on the board, a botanist from Yale who had already expressed his opposition to a medical research program at Brookhaven. Alarmed at the parlous state of Brookhaven's life sciences programs, Hastings sought his old friend and teacher Donald D. Van Slyke, at Rockefeller University, one of the most prominent medical researchers alive (fig. 4.5). Hastings persuaded Van Slyke to become part-time "consultant to the laboratory" on its life sciences program, and kept up pressure until

4.5 Donald D. Van Slyke, one of the foremost medical researchers in the U.S. and an early consultant at Brookhaven for its life sciences program, working on a Van Slyke machine.

Van Slyke agreed to become a full-time Brookhaven employee. "Hastings practically ordered me to go to Brookhaven," Van Slyke later recalled.[33]

In his four decades at the Rockefeller Institute for Medical Research (now Rockefeller University), Van Slyke had declined many tempting offers elsewhere. "I had everything I wanted to work with and complete freedom to do what I wanted," he said once. Everything, that is, but work past mandatory retirement. In March 1948, he turned sixty-five, and the prospect of being forced out severely depressed him. The depression was compounded by his wife's death after forty years of marriage. In an emotional tailspin, he was ready to hear Hastings's offers. Years later, Hastings said that getting Brookhaven and Van Slyke together was probably the most important thing he did in his life. "I not only saved the biology and medicine [departments] for Brookhaven National Laboratory by that move," Hastings said later, "but I saved Van Slyke." [34]

From the moment Van Slyke agreed to join the lab, first as a part-time consultant then full-time, both his spirits and those of Brookhaven's life sciences departments revived. In some ways Van Slyke was an anomaly among Brookhaven's founders, for he was not a young risk taker but a white-haired eminence past retirement age. In other respects he fit the Brookhaven concept, for he was skilled in forging interdisciplinary links as researcher and administrator. Van Slyke, indeed, had been a leading figure in the application of analytical chemistry to clinical practice, which revolutionized medicine in the first half of this century.

In April 1948, Van Slyke and Morse met in Washington with Shields Warren, director of the AEC's division of biology and medicine (DBM), who had put Brookhaven's biology program on hold at the recommendation of his advisory board. Warren had been happy to learn of Van Slyke's involvement with Brookhaven's program, which he wryly said was "particularly in need of wise guidance," and Van Slyke soon had Warren convinced of the need to allow the biology program to go ahead.[35]

Van Slyke immediately found himself fighting at least three battles. First, he had to convince the trustees that the medical department ought to have a clinical research program at all. Second, he fought pressure to move the medical department to New York City, which would solve the problem of isolation, making it (like the Cornell Medical Hospital, for instance) a kind of urban service facility to laboratories and hospitals there. Van Slyke argued that the chief value of a Brookhaven medical facility would be its proximity to researchers from other departments. Third, he had to struggle through the debate over whether life science research should encompass *any* research program favored by an environment whose facilities included a reactor and radioisotopes, or whether its work should include *only* research projects requiring the reactor and radioisotopes. Van Slyke argued for the former. For the moment his view prevailed, but that fundamental debate continued for decades, at Brookhaven and elsewhere.

Van Slyke also helped select permanent chairmen for the life science departments. After many disappointments in seeking a medical department head, Van Slyke contacted Lee Farr, a former student who was head of research at the Alfred I. duPont Institute in Wilmington. Intrigued by the unusual possibilities of the lab, Farr visited Brookhaven the next day, where he favorably impressed the directorate. Farr fit in with the trustees' "small beginnings" policy, and believed that a large hospital was not required for an effective clinical research program.[36] Farr was eventually recruited, and that August, Van Slyke and Farr put together a long-range plan for the department.

Meanwhile, Van Slyke also helped find a permanent head for the biology department, whose acting chairman was still the uninspiring Nims. At the end of 1949, Haworth persuaded Nims to resign, but the search for a replacement was long and arduous. Howard J. Curtis of Vanderbilt University, who had directed the biology division at Clinton Laboratories during the Manhattan Project, ultimately took over in October 1950.

That year Van Slyke was approached by the Eli Lilly Company, which sought an adviser to help expand its medical research grants program. Torn,

Van Slyke dutifully jotted down the pros and cons of various aspects of the Brookhaven and Lilly positions in a table, assigning a number to each, indicating its value to him. Although the pros of Brookhaven and Lilly came out to a numerical tie, the Lilly job scored higher on cons. Nonetheless, Van Slyke accepted the Lilly job, evidently finding life not as quantifiable as chemicals. But he was allowed to keep his lab as a "guest investigator," a title used for professors on sabbatical with no salary but all the privileges of a staff position. (Under this arrangement Van Slyke stayed on at the lab for the rest of his life, returning to full-time research at Brookhaven in 1956 as senior scientist emeritus.) He continued to have an imposing presence at the lab in those later years, improving many techniques he had developed in the 1930s by incorporating radioisotopic methods. Before leaving for Lilly, Van Slyke described Brookhaven's medical department as a "skeleton with some flesh on it." But at least it, and the biology department, were alive.[37]

Morse's Departure

By the middle of 1948, Brookhaven employed almost fifteen hundred people. Its six scientific departments—physics, chemistry, biology, medicine, engineering, and instrumentation and health physics—now existed in fact, not just on paper. Many of the old army buildings, including barracks, a mess hall, a small theater, the fence and barbed wire around the stockade, and the sentry booths, had been sold at auction, dismantled, and removed. And although none of the major new facilities had yet been completed, the designs had been finished and approved, and construction contracts were either already signed or being drawn up.

But in the meantime, fights with the AEC over clearance, classification, government control, and housing had created what Morse called a morale problem, not only for the lab staff but for himself as well.[38] Morse, who never wanted the directorship to begin with, had learned that he disliked working with large numbers of people. Even in wartime, Morse had worked in small groups whose members shared a sense of community and dedication to a common goal. In organizations the size of Brookhaven, Morse found he could not re-create that kind of experience.

> I began to find guards and clerks and shop assistants who had no idea what the laboratory was supposed to do and who therefore felt left out. . . . I found I was not fond of being the administrator of a large organization, of being always on display, of having to be careful of every statement lest it be interpreted wrongly—in short, I disliked having to deal with words rather than with facts or actions (Morse 1977, 242).

Morse's fatigue and distraction grew. He returned from meetings and was unable to remember what had taken place. He was particularly shaken when he returned from the January 1948 congress of the American Physical Society and found in his coat pocket a piece of paper with a strange message in his own handwriting. It turned out to be the name and address of a housemaid who had worked for him during a visit to Munich in 1930, but he had no idea of how or why he had written it down.

To clear his head, Morse took to working on theoretical physics and to finishing portions of a textbook he had begun a decade before with his former MIT student Hermann Feshbach. The struggle with theoretical concepts was invigorating, and by April of 1948 he had decided to return to MIT. On 18 June, Morse submitted his resignation, to be effective a month later.

Leland Haworth

Morse was succeeded by Leland Haworth, who inherited the task of leading the lab during the time when the special partnership between it and the

4.6 Leland Haworth, Brookhaven's second lab director (1948–1961).

AEC was effectively committed to a contract (fig. 4.6). Haworth was a practical, unpretentious man of Quaker background with a quiet but distinctive sense of humor. Tall and athletic, his solid physique would serve him well during long hours as lab director. Haworth felt it his duty to know the details of each project he oversaw, and did budget spreadsheets in pencil himself, often working late and expecting his assistants to do likewise. He came to trust people slowly, after working with them over long periods of time, and then never relaxed his trust, sometimes to his detriment. He had a knack in deadlocked meetings of posing the question whose answer resolved the issue. Disarmingly friendly, Haworth inspired confidence in subordinates and superiors alike. He was "such a patently honest man," one trustee said, "and his integrity was so high, that people, even if they disagreed with him, felt they had had a fair shake." [39]

Haworth was born in his grandmother's house in Flint, Michigan, in 1904 and raised in New York City, spent much of his youth in Indiana, and graduated from Indiana University in 1925. He received a Ph.D. in physics from the University of Wisconsin in 1931 and joined the University of Illinois faculty in 1938. During the war he was a division head at the MIT Radlab, and returned to the University of Illinois for a short time before Morse persuaded him to become Brookhaven's assistant director in charge of special projects; he arrived in summer 1947. Grateful to find someone who loved to supervise large scientific projects, the increasingly withdrawn Morse quickly focused on Haworth as a successor. Haworth became acting director in July 1948, and director three months later. For the next thirteen years, Haworth more than any other single individual would shape Brookhaven's character and leave a mark on its research. He would succeed where Morse had failed and find a way to commit to the lab the care and personal attention to detail that is more common in a small organization.

When Haworth officially took over as director on 14 October 1948, his first tasks were to put the two centerpieces of the Brookhaven concept, the pile and accelerator projects, back on their feet. For the building of advanced scientific instruments would pose new and unforeseen problems in the context of the special partnership now under way between the lab and the AEC.

The Pile Project

IN THE summer of 1946, when Lyle Borst agreed to head the pile project, he was just thirty-three (fig. 5.1). Visitors to his office usually found him at his desk, leaning back in his chair and contemplating a pencil held about a foot in front of his face through his small, wire-rim glasses. Borst would slowly rotate the pencil while mentally chewing over some pressing problem, an action he would continue even during conversation. Imaginative and strong-willed, he exercised excellent judgment about both scientific matters and people, a significant asset for someone charged with carrying out a technologically sophisticated, large-scale project in the peculiar climate immediately following the war. But he did not suffer fools gladly, which sometimes inflamed rather than calmed conflict. A supremely confident and independent man, Borst resented intrusion on projects entrusted to him, whether the intruders were colleagues, lab directors, or AEC officials.

Borst was born in Chicago, received a Ph.D. from the University of Chicago in 1941, and worked during the war at the Metallurgical Laboratory there. He witnessed the early stages of the Manhattan Project, as Enrico Fermi and his associates transformed the theoretical and possibly far-fetched idea of a nuclear chain reaction into reality. That work culminated in December 1942, when a sustained nuclear reaction was created inside the first nuclear pile, Chicago Pile 1 ("the CP-1").[1] A young pioneer in the new world of atomic energy, Borst became supervisor of research at the second nuclear reactor ever constructed, the X-10, built at Oak Ridge, which went critical in November 1943. The X-10 represented a substantial advance in reactor engineering over CP-1, increasing the chain reaction rate by a factor of about a million and producing about a megawatt of power.

After the war, Borst remained with the X-10 as a senior physicist. An ardent opponent of restrictions on nonmilitary atomic research, he had witnessed how security restrictions could hamper scientific activity. Borst believed that nothing could ultimately be gained by hiding information that competent scientists elsewhere could discover themselves, that true security could be achieved only by continued scientific progress and staying ahead of everyone else. He also found it distasteful not to be able to talk

5.1 Lyle B. Borst, head of the pile project.

about his work to his family and friends. "I'm thinking of changing my specialty," he told a reporter shortly after Brookhaven opened. "I've been in it since Oak Ridge was first planned, seven years ago. That's a long time to be quiet" (Lang 1947, 43). Borst was a cofounder of both the Association of Oak Ridge Scientists and the Federation of Atomic Scientists, organizations that had actively backed the McMahon bill and civilian rather than military control of atomic energy. Borst's views and activities brought him to the attention of the House Committee on Un-American Activities (HUAC), which in 1946 charged that these organizations were subversive. At the same time, his views resonated with those of most of Brookhaven's founders, including Morse and Smyth. In May 1946, Borst began to assist the reactor project and also served on the IUG's site committee during its final stages, contributing thoughts on possible pile locations. Once Upton was chosen, Borst was the obvious choice to head the reactor project and was hired away from Oak Ridge.

Borst put together a small design team of reactor physicists who worked extremely well together. His two chief lieutenants were Clarke Williams and Marvin Fox, both Columbia Ph.D.s and veterans of Columbia's Substitute Alloy Metals, or SAM, laboratories, an isotope separation component of the Manhattan Project. Other early members of the team included theorist Irving Kaplan, metallurgist David Gurinsky, and chemical engineers Robert Powell and Warren Winsche. These seven members of Borst's core group were all personal friends who enjoyed working together professionally. Their meetings were usually short and quiet, and participants prepared carefully ahead of time. Herbert Kouts, a physicist who arrived at Brookhaven in 1950, just before the reactor was completed, recalled his first reactor group meeting:

> Most of the meeting consisted of silence. The silence would be broken every now and then by Lyle Borst, who would simply turn to one of the members there and ask, "Is this aspect of what we are doing ready, and can you tell me what stage it is in?" Perhaps in two minutes, that subject was dispensed with, and Borst would sit back and think silently, and after a while would turn to somebody else and do the same thing. In a short meeting of perhaps half an hour, more was accomplished than I ever saw done in most meetings that lasted a full day or two. It was sheer competence, dwelling on the important things only, and getting only the important aspects of the answer.[2]

Brookhaven's reactor engineers started work on the design when reactor technology was in its infancy; the first reactor had been built less than four

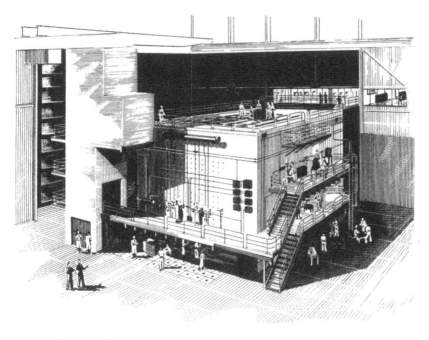

5.2 Artist's sketch of reactor.

years earlier. They were working under many constraints, some set by the AEC and some by the lab's founders, who were pressing Borst's team to have the reactor operational as quickly as possible so that Brookhaven might be on a par with Argonne and Oak Ridge. Any delays would force experimenters to go elsewhere. Borst's team therefore saw no real alternative to building an air-cooled, graphite-moderated pile fueled by natural uranium according to an existing design—that is, building an X-10-like reactor, "scaling it up as far as was cautiously appropriate."[3] An advanced reactor design could come later; the founders referred to their instrument as "Pile #1" for the first year or so, assuming that Brookhaven soon would start work on a more ambitious, technically superior "Pile #2." The scientific program of December 1946 proposed that Pile #1 could be finished in "nine to ten months" for $2.5 million, a wildly optimistic forecast of schedule and price tag.[4]

In October 1946 Borst and his associates began detailed design studies for what would be called the Brookhaven Graphite Research Reactor (BGRR). The design posed no basic problems unsolved in the X-10, which had consisted of a graphite cube twenty-four feet on a side, traversed by channels in which fuel elements were inserted. While the original aim was simply to copy the essentials of the X-10 design, Borst and company could not resist looking for improvements; nothing is further from the scientific temperament than the desire to do something the same way twice. Finding design improvements for Brookhaven's purposes was not difficult, for the X-10 had been built quickly during the Manhattan Project and with a very narrow basic goal: to create a small amount of plutonium in as short a time as possible.

Many modifications were minor, aimed at improving safety or use of the reactor as a research tool. Borst planned a cube twenty-five feet on a side. Fox, in charge of shielding, developed a new, high-density concrete that was more effective than that used at the X-10. And instead of control rods inserted into and withdrawn from one face of the graphite cube, Borst's group decided to insert and withdraw them from two of the corners, thus leaving as many of the cube's faces as possible free for experimental use (fig. 5.3).[5]

Borst's group made more significant improvements in connection with fuel handling. For the X-10, uranium slugs were welded into simple aluminum cans about an inch in diameter and four inches long. Strings of these "fuel elements" were placed in channels through which large fans blew constant streams of coolant air. A fuel element had to be airtight to keep the uranium metal from oxidizing and swelling, which might rupture the element and even cause it to become stuck inside the reactor. But as the reactor was turned on and off for periodic maintenance the alternate heat-

5.3 Marvin Fox with model of the Brookhaven Graphite Research Reactor (BGRR), showing how one set of control rods was inserted at a corner of the pile.

ing and cooling could also cause the uranium to swell and break open the can welds. A leaky fuel element was a very serious problem in an air-cooled reactor, for it could release fission products into the coolant air, though air filters would catch nearly all of these.[6] The Brookhaven scientists found a number of ways to improve this system. Gurinsky developed a novel welding technique to seal the elements. Six stubby, inch-long fins were attached to each element to enhance the cooling process by providing a greater surface area for heat transfer to the air. And the fuel elements were made eleven feet rather than four inches long. This added length drastically reduced the number of vulnerable welds though it also made the elements more fragile and difficult to handle. Each weighed about a hundred pounds, and a special "charging elevator" was built on the south face of the reactor to lift the elements up to the fuel channels. Borst added yet another improvement by attaching a one-eighth-inch-diameter piece of tube to the back of each element. The tubes were in turn connected to gauges and pressurized bottles of helium on top of the reactor. This extremely clever safety feature effectively bathed the fuel elements in a thin layer of helium. If all fuel elements were properly sealed, there would be no draw on the helium; a draw meant a leaky element that could be quickly located.

These changes in the X-10 design were relatively modest. But in the course of the design study, Borst hit on a simple yet ingenious way of making a major improvement in the way the reactor was cooled. The cooling process is of paramount importance, for the amount of cooling ultimately controls the flux, or rate of neutron production, in a research reactor (and the power in a commercial reactor). The more effective the cooling of a given reactor, the higher the fission rate at which it can be run, and the higher the flux. The low-power X-10 was cooled by air that flowed from one end of the pile to the other through the fuel channels. Borst realized that the cooling would be much more efficient if the graphite cube were cut into two equal pieces separated by a small vertical gap. Air would enter the center of the reactor through the gap and be sucked out through the fuel channels. With the coolest air flowing directly into the central and hottest part of the reactor, it would be safe to achieve a much higher fission rate and neutron flux.

For a while, Borst's team looked into the possibility of installing a power recovery system in the hot exhaust ducts of the reactor, which if successful would turn it into the world's first power-producing reactor years before that had been thought possible. But this proved more involved than anticipated, and was opposed by members of the scientific steering committee because it would interfere with the reactor's role as a scientific instrument. After an independent study found that power recovery at the BGRR would be "economically unsound," interest began to wane. Though the AEC strongly encouraged BNL to pursue the project for a while, nothing came of it (the first known production of electrical power from nuclear energy occurred at Argonne in 1951; Holl 1997, 108–9).[7]

Meanwhile, Kaplan studied how a gap would affect the operation of the reactor. One major problem was to figure out the gap's optimum size. The bigger the gap, the more air would enter and the greater the cooling. But a larger gap meant that more neutrons would leak out, debilitating the chain reaction and lowering the power. Kaplan calculated the optimum gap size at seven centimeters, but uncertainties in the calculations made him unsure whether it might not be eight. One day he remarked to an engineer that it would be nice if the gap could be moved *after* the reactor had been built so the reactor scientists could test their estimates of the neutron flux, verify the optimum gap size, and adjust it if necessary before operation began. Oh, we can do that, replied the engineer. The pile therefore was designed so that half of the cube, weighing several hundred tons, was set on a track that could be moved by a hand-operated crank.[8]

In late 1946, the MED announced that a commercial contractor would engineer and construct the reactor. Borst opposed this action, which put him at a certain distance from the project, not a position in which he felt comfortable. He also noted that a large commercial contractor would not necessarily put experienced people on the job. Borst urged that Brookhaven have more direct authority over engineering and development and perhaps even construct the reactor itself. However, Haworth told Borst not to protest too strongly, that the decision was final.[9]

Haworth's own hands were tied, for he was running up against limits imposed by the special partnership. The MED had contracted out all reactors built during the war and was not about to hear that it did not understand how to build reactors in peacetime. The agency had a firm policy requiring that architect-engineer services be performed under a prime contract with the government, and it became yet another MED practice adopted by the AEC. The policy arose from the agency's understandable desire to supervise closely a large investment. Moreover, with an eye to a future atomic power industry, the MED and then AEC wanted to encourage private companies to acquire experience in building reactors.[10]

The Manhattan District engaged Hydrocarbon Research, Inc., which had built petroleum refineries and was reputed to have a good engineering staff, to engineer and construct the reactor. Early in 1947, Hydrocarbon created

5.4 First (bottom) layer of graphite in place, showing the gap.

a wholly owned subsidiary, the Delner Corporation, to handle the reactor project. Delner produced an initial engineering design and an estimate of $5 million, which soon shot to $9.3 million. As Borst later commented, "These estimates were really not much more than guesses. . . . the contractor did not fully grasp the difficulties of the undertaking." [11]

Yet at the time Borst himself shared in the optimism. The difficulties were partly hidden from all parties by the peculiar tripartite management setup that was enacted when the AEC-AUI partnership engaged an outside contractor. AUI and lab engineers and scientific staff submitted plans to Hydrocarbon, and had the authority to spot check Delner's drawings and calculations and to initiate calculations when and where they thought best—but they rarely did, assuming that Hydrocarbon had matters in hand. Hydrocarbon officials had little reason to think that approval of the plans by AUI and lab engineers and scientific staff meant only approval of the design from a schematic point of view, rather than thorough scrutiny from the point of view of reactor engineering. And the AEC had no reason to think that the appropriate dialogue between the two was not taking place (Needell 1983, 110–11).

A few months later, in May 1947, a dawning appreciation of actual costs and difficulties led Delner and Hydrocarbon to withdraw from the project. Borst, Morse, and Shoup now renewed pleas to the AEC to give the lab full responsibility for engineering and construction, but the request was firmly denied.[12] The contract was given to the H. K. Ferguson Company, a Cleveland corporation which had built the Oak Ridge S-50 (a small isotope separation plant) on time and at cost, earning the praise of the MED. "They did a bangup job on a hurry-up proposition for us," AEC area manager and ex-MED official Van Horn told anxious Brookhaven scientists.[13] But the S-50 was utterly unlike Brookhaven's new reactor. Ferguson simply stepped into Hydrocarbon's shoes, and began acting much, too much, like Hydrocarbon. Assuming it was taking over a competently run project, Ferguson did not conduct a thorough project review and simply adopted Delner's plans, engineering decisions, and nearly all its personnel. Also preserved was the awkward triumvirate management arrangement between the AEC, the lab, and the contractor. Lab scientists supposed that Ferguson had matters in hand, Ferguson thought the scientists were getting what they wanted, and the AEC assumed that any necessary interaction was taking place.[14]

On 11 August 1947, the various parties involved held an upbeat groundbreaking event (fig. 5.5). The ceremony took place on the highest point for miles around, a 125-foot hill on the glacial moraine running down the spine of Long Island; Pile #1 would occupy the west side of the hill, leaving the east side open for the future Pile No. 2. Wearing a hard hat, Borst ceremonially scooped out a bucketful of earth at the controls of a crane, scaring

5.5 Lyle Borst breaking ground for the BGRR, August 1947.

some of his colleagues who knew that he was not authorized to actually operate the thing when his untutored actions set the machine rocking dangerously back and forth. But Kaplan, watching from the sidelines, was amazed it was happening at all. Although he had worked on calculations for large Oak Ridge facilities, the theorist had never been close to an actual reactor construction site. "My God, they're really doing it," he said to a friend.[15]

Construction proceeded slowly. With no experience in reactor construction, Ferguson had trouble hiring skilled reactor engineers and a frustrated Borst wound up training the engineers they did manage to hire. Borst was further incensed when the secrecy conditions imposed by the AEC created additional delays.

> The engineers had to fill out PSQ—personal security questionnaire—forms when hired, and then the AEC would consider granting clearance, a process which usually took one to two months. What do you do with someone during that period? You have to hire them without telling them what they were to do, but then use them conventionally until clearance comes through. So Ferguson had a "bullpen" for uncleared people to do conventional engineering—building foundations, and so forth—but there was little of it to do. . . . It was even worse when we got into construction. This was in the middle of a building boom, and workers could pick and choose their jobs. Now, this lab was fifty miles from New

York City, and trade union employees were mostly in New York City and had to come out to the lab. Once again, you had a hiatus of one to two months before bricklayers and ironworkers and so forth could be cleared for this job. . . . You had bricklayers and ironworkers who couldn't be told what they were going to do and had to work for a month or two before clearance came through. They had to be paid, but there was not much to do. . . . We had an economic drain where we were getting virtually nothing out of the people who had been hired.[16]

There were other kinds of meddling as well. At one point during its construction the reactor was visited by a congressional committee concerned about security. Borst knew that such committees do not return from visits empty-handed.

It was perfectly evident that we had to do something which would appear in one of their reports and justify their trip up here. I thought about it a little bit, and said, "Well, you know, if somebody got on the border of our reactor zone and had a bazooka filled with samarium [a material that easily absorbs neutrons], this would do a perfectly wonderful job on our reactor"—because of course the filters couldn't keep it out, and it'd distribute through the whole structure and poison the reactor completely. Well, this was just *exactly* the kind of thing that they wanted; a fancy device, a scientific—pseudoscientific—affair. So they went back and wrote up their report, and so we got a letter very shortly thereafter saying, well, we were OK except for the problem of a bazooka. And if this bazooka had samarium in it that would shut the reactor down. Bob Powell and I figured out quickly what to do. We built a barricade to protect the filters from bazooka fire. It didn't get in our way at all.[17]

As 1947 ended, spirits at the lab were high. Instrumentation department member William Higinbotham wrote a ditty to the tune of "My Darling Clementine," and it made the rounds of the lab's department Christmas parties.

> In a sandhill on Long Island
> Excavating for a pile
> Is the Ferguson contractor
> Working for our brother Lyle.
>
> Oh Brookhaven, Oh Brookhaven,
> Darling of the AEC
> With its peacetime chain reactor
> What a fine place it will be.[18]

On 26 December, a record snowfall buried Brookhaven, abruptly stopping all construction activity. In January 1948, the lab had to ask the AEC for another $5 million for the project, which the agency "very reluctantly" granted. Work progressed slowly for two months as Long Island sank into a bone-chilling winter. In March, after the final concrete pour for the foundation, Ferguson was still forecasting completion in early fall 1948, at a projected cost of $15 million.[19]

But an engineering problem surfaced, having to do with thermal stresses inside the reactor. Borst's gap innovation had allowed the reactor to be operated at a much hotter temperature than the X-10, giving rise to temperature differences that would create stresses in the materials due to thermal expansion. In March 1948, when reactor engineering staff who were testing the cooling system asked Ferguson for an engineering report to check calculations, they were horrified to discover that these thermal stresses had been ignored in the report. While thermal stresses in the X-10 were small enough to be safely neglected, in the much hotter BGRR they would be significant. To cite one example of the problem, the graphite sat on three-inch steel plates that would warm up along with the graphite, but the coefficient of expansion of the steel was about 3½ times that of the graphite. It would have been easy to accommodate these stresses early in the design and engineering process—and issues relating to the heating of the pile had been on the agenda in numerous meetings between Brookhaven and (first) Hydrocarbon and (then) Ferguson scientists dating back to December 1946. But neither Borst's group nor Hydrocarbon's subsidiary Delner (which had done the initial engineering design early in 1947) had paid attention to the temperature differentials; neither Delner nor Ferguson had noticed the need, and even Borst's usually fastidious group had not noticed the absence. Because no allowance had been made for them, the expanding steel would push against the rest of the assembly with terrific pressure. If the temperature reached a certain point, the supporting structures would buckle. Ferguson, which had inherited the problem, began to look for a solution.[20]

But a second engineering problem was soon discovered in the cooling system. To cool the reactor, ambient air entered the reactor through the gap—at the top, bottom, east, and west sides of the reactor—passed through the fuel channels, and exited the north and south faces into the set of large chambers called plenum chambers. From there, the now-hot exhaust air passed underneath the reactor into a pair of air ducts, huge cement passages that measured ten by fourteen feet, big enough to accommodate a New York City subway train. The air ducts carried the exhaust air to filters and then to a set of blowers that expelled the air up the stack. But the exhaust air in the ducts and plenum chambers would be so hot that it would eventually destroy the concrete walls. A metal insulating layer was therefore installed on the inside walls of the ducts and plenum chambers, anchored by what were known as Nelson studs, to prevent the concrete from overheating and spalling. In spring 1948, a number of these studs were found to be cracked, apparently from faulty welding, and Ferguson began a repair program that it initially thought would take "several weeks." [21] The company did not formally bring the matter to the attention of AUI, so the Brookhaven staff assumed that no special measures were necessary and that Ferguson would simply replace the bad studs.

In July 1948, studs were still cracking, and Ferguson was forced to admit the reactor had "fallen distinctly behind the original schedule." [22] But the company continued to be optimistic, blaming the delay in part on the weather and in part on misjudgments of how much work could be done concurrently rather than seriatim. In October 1948, a group of presidents and other officials of the nine AUI universities toured the BGRR construction site (fig. 5.6). A ceremony was held on 8 November 1948, when the first block of graphite was set into place before a small delegation of scientists and officials representing the AEC, the lab, and H. K. Ferguson. The reactor group staff still had confidence in Ferguson's schedule and sent a notice to the *Physical Review* about the reactor's impending start-up (Borst et al. 1948). Borst told attendees at the 1948 Thanksgiving meeting of the American Physical Society that "construction of the Brookhaven reactor is progressing well and initial operation is expected early in 1949" (Borst 1949). Borst also put together a start-up team by persuading several university reactor scientists to take sabbaticals at Brookhaven in 1948–1949. This team would build instruments, then carry out the long series of measurements and tests that would be necessary just before the reactor went critical.

But duct studs were still cracking in December 1948. Baffled, Ferguson finally raised the issue formally with the reactor staff but for a while managed to convince them it was just a particularly stubborn construction problem of the type commonly encountered in large projects. [23] In February 1949, Powell took one of the problem duct bolts to Gurinsky, who examined it with his metallographer. They saw an unusual series of tiny cracks in the weld that convinced them that the entire welding process might be at fault—meaning that all bolts, even the newly repaired ones, were likely to crack (fig. 5.7). Gurinsky and the reactor's chief project engineer Max Small contacted and then met with an emeritus professor of mechanical engineering at Columbia, who arranged for a meeting a few days later, at Brookhaven, between representatives of the Babcock and Wilcox Co. (B&W) engineering firm and members of the lab's staff. According to Gurinsky, this meeting "shook up the project and the laboratory management" (Gurinsky [n.d.]b, 29).

At that meeting, the B&W engineers expressed deep dissatisfaction with the duct bolt welding and with the basic engineering design of the lining. They found that the design for these ducts had been drawn from an engineering handbook intended for structures whose parts were all at the same

5.6 In October 1948, the BGRR's construction site was toured by (left to right): George B. Pegram (Columbia), Harold Dodd (president of Princeton), Frank Fackenthal (Columbia), Henry Smyth (Princeton), David Gurinsky (BNL), Leland Haworth (BNL), Dwight D. Eisenhower (president of Columbia), Charles Seymour (president of Yale), William W. Watson (Yale), George A. Brakely (Princeton), Edmund W. Sinnott (Yale), and I. I. Rabi (Columbia).

temperature, with no provision made for their expansion or contraction. The consequences were potentially disastrous. After enough bolts had broken, the liners would have fallen, the concrete would have peeled and piled up near the blowers, and since the area would have been radioactive, the cleanup would have been, in Borst's words, a "fiasco." [24] The gravity of the situation finally began to dawn on Brookhaven officials, for they realized that major changes were needed in the welding process and in parts of the reactor's structure. Haworth, having lost faith in Ferguson's rosy forecasts, persuaded Ferguson to engage B&W as a consultant, and in March 1949, the two companies signed a letter of intent. This action was a major turning point for the reactor project.

Finding B&W was a lucky break. The company was anxious to get into the reactor business, and assigned several first-class engineers to the task. As B&W earned the respect of the lab scientists, the scope of its consulting arrangement grew from the stud problem (the liner anchoring system

5.7 Looking into one of the exhaust air ducts, showing stud bolts used to anchor liners.

was redesigned) to the still-unsolved thermal stress issue, and eventually Haworth asked them to make a complete survey of the reactor's engineering. B&W knew about insulation and thermal expansion problems from its experience building furnaces, and was able to suggest corrections that could be implemented even though construction was virtually complete. Ferguson balked at many of the recommended repairs, but Haworth insisted that the company prepare to make all of them, regardless of expense and delay, and that the work be closely supervised by the laboratory staff. The repair work was to begin 1 March 1950, with 11 August the new target date for completion. Haworth and members of the lab business and budget offices began to work late, sometimes until one or two in the morning, to figure out how to keep building the rest of the laboratory while covering the escalating costs due to the repairs. Somehow Haworth managed to keep a sense of humor. He kept track of additional reactor costs in a column on his office blackboard, startling visitors by its heading: "stud fee."

The delay prevented Borst from tapping the skills of his carefully assembled start-up crew, who in summer 1949 ended their sabbaticals and had to return to regular university teaching positions. Borst appointed Vance Sailor, a recent Yale Ph.D., to take over start-up operations. While Borst originally wanted Sailor to be the nucleus of Borst's own research group, he arrived at just the time Borst's start-up crew was beginning to disperse, and Sailor found himself in charge of putting together a new team.[25]

In January 1950, two months before Ferguson was to begin repairs, Haworth authorized a "subcritical" loading. Following a method that had been used at other reactors, the group partially loaded the reactor, filling about a third of the eleven hundred channels with uranium. This was not enough for the reactor to go critical—have a self-sustaining chain reaction—but was enough to allow the reaction to progress to the point at which critical loading, and the behavior of the reactor in operation, could be predicted with reasonable confidence.[26]

Meanwhile, tensions were running high between the Brookhaven and Ferguson staffs. Ferguson employees resented lab officials' frequent criticisms. During a visit to Oak Ridge, Borst made barbed remarks about Ferguson and showed a cartoon of the reactor that depicted a dump truck pouring dollar bills into a hole in the ground, and word got back to company employees. Also contributing to Brookhaven's discontent was the fact that Congress had added a rider to the fiscal 1950 appropriations bill, called the Deficiency Appropriations Act, which stated that if the cost of a federally funded construction project exceeded the original estimate by 15 percent, a detailed explanation had to be made to the House Appropriations

Committee, to the Bureau of the Budget, and, in cases involving the AEC, to the Joint Committee on Atomic Energy. With the overrun on the reactor project rapidly approaching 15 percent of an earlier appropriation figure of almost $22 million, lab officials feared a time-consuming audit.

In yet another source of tension, AUI trustees asked Milbank, Tweed, their law firm, whether they could sue Ferguson for the cost overruns created by the engineering errors. In January 1950, a representative of the firm reported that "a case against The H. K. Ferguson Company probably existed," but that "it would be extremely difficult for the corporation to prevail in court." Part had to do with the uncertain nature of liability in the bizarre management scheme in the new partnership.[27] Lab officials and then the AEC tried to pursue the possibility for nearly another year before the issue was dropped.

But by the time Ferguson began its repair work in March 1950, Brookhaven had the situation in hand. The reactor project engineers were taking the initiative in checking calculations, and paying close attention to Ferguson's work. And Ferguson managed, through substantial overtime, three shifts, and, in July, a seven-day week, to stick to the schedule.

Along with the reactor, Ferguson was also building a set of supporting structures, including a hot lab for processing radioactive materials and preparing radioisotopes, a semihot lab for handling less intense sources of radiation requiring fewer protections, and two laboratory wings in the reactor building for general-purpose labs into which one could bring tiny amounts of radioactivity.[28] Plans for disposing of radioactive wastes at the laboratory were also nearing completion. These wastes would come from the reactor, the hot lab, and other facilities handling radioisotopes. Solid, liquid, and gaseous wastes would be involved and separate systems were needed for each.[29]

Solid radioactive wastes would consist of spent fuel rods, structural elements of the reactor that had become radioactive, contaminated tools, clothes, etc. Spent fuel was temporarily collected in the canal and then sent to Oak Ridge for reprocessing. All other solid radioactive waste was to be collected and stored in a waste management area, the old Camp Upton ordnance storage site, where most of it was packaged in well-shielded, concrete-lined fifty-five-gallon barrels, transported to the Brooklyn Navy Yard, and, following standard AEC procedure (until 1960), dumped at sea.

The liquid radioactive waste disposal system was the most involved and important of the three disposal systems, for while solid wastes were shipped off-site and gaseous wastes left to decay, the liquid radioactive wastes were to be concentrated, diluted, filtered, and released to the environment when reduced to small enough levels. Given later community concerns about the liquid waste disposal system, it is worth quoting a 1948 memo of its designer, reactor scientist Bernard Manowitz:

> An efficient "hot" waste disposal system is perhaps a more urgent need at Brookhaven National Laboratory than at many of the other sites because of special conditions existing here:
>
> (a) The laboratory is located in a highly populated area
> (b) The water table in this area is high and is used extensively for well water
> (c) The nearest streams all pass through populous communities before discharging to the bay. . . .
>
> Although it may not be definitely established that weak radioactive contamination is a demonstrable hazard to health, it is a fact, however, that appeals to the courts or to the public health authorities are usually based on claims of health impairment and also that public sentiment is likely to be sympathetic to health arguments. Radioactive pollution does not have to be injurious to health to be socially undesirable.[30]

Liquid radioactive wastes would arise from hot lab processes and from the reactor canal, which would be contaminated by the process of preparing damaged fuel elements for shipment off-site. The wastes would be piped from their site of origin to a waste management center in the basement of the hot lab. There, wastes of a low enough activity were routed into the laboratory's sanitary sewage system, while hotter wastes were sent to an evaporator, which would divide the waste into a concentrate and condensate by an evaporation process. The concentrate was put into drums, mixed with concrete to form a solid, shipped to the Brooklyn Navy Yard, and, along with much of the solid waste, disposed of (until 1960) at sea. The condensate or distilled liquid, which still contained trace amounts of radioactivity due to a small amount of material that had been caught up into the steam by the evaporation process, was sent to the lab sewage system, where, along with other waste, it passed through a bacterial digester and a sand filter. By the time it was discharged into a creek (the Peconic River) and reached the site boundary, the radiation level was far below the AEC limit and a fraction of background radiation.[31]

The technical design of the reactor made ^{41}Ar the only significant radioactive gaseous by-product. Argon, an inert gas that makes up about 1 percent of the Earth's atmosphere, has a relatively high ability to absorb neutrons; some argon in the air coolant would become radioactive as it passed through the reactor. But ^{41}Ar has such a short half-life (1.8 hours) that, depending on wind conditions, a large percentage of the seven thousand or so curies released each day would disperse and decay before reaching the ground. Still, with solid wastes shipped off-site and liquid wastes reduced to background radiation levels, ^{41}Ar was the agent that would potentially increase exposure of the surrounding community to radiation above background. A meteorology program was established to study the dispersal of gases through the environment once they left the stack and to monitor atmospheric radiation. Sixteen monitoring shacks were built at various places on site, rigged with a number of types of detectors that fed into a series of dials. A sixteen-millimeter movie camera was set up so that every few minutes a light would come on and a frame be taken. Once a week, the film was removed and developed, allowing any trends to be monitored. If local weather conditions threatened to cause a poorly diluted air flow from the stack so that exposure limits would not be met, the reactor would be shut down.[32]

On 11 August 1950, the target date Haworth had chosen at the beginning of the year, the Brookhaven Graphite Research Reactor was completed. The final cost was $24,980,000—just $5,000 shy of the amount that would have triggered the terms of the Deficiency Appropriations Act, and the need for lab to prepare exhaustive reports on the project and an elaborate defense of the overruns. Completion of the project was a milestone in Brookhaven's early development and the development of peacetime atomic research. The departure of vast numbers of Ferguson construction workers contributed to the feeling that the laboratory was beginning to stand on its own. The graphite reactor was the lab's first major facility, and although

5.8 The Brookhaven Graphite Research Reactor, plus stack, after completion.

its design was not advanced, nuclear research of the sort Rabi and Brookhaven's other planners envisioned finally could begin. "Everything we did we knew would be a precedent for the people who came after," said Borst. "We felt a real sense of history." [33]

With Ferguson's work done, the reactor scientists began loading the final fuel elements, an arduous process requiring almost two weeks of constant work. Members of Sailor's start-up group worked around the clock, making makeshift beds for themselves in unused offices and labs as they loaded elements and took measurements. Handling these eleven-foot elements—eleven hundred in all—was an extremely delicate matter requiring several operators, for the aluminum casing and fins were not much thicker than aluminum foil. Each element was built inside its own carrying tube. The loaded carrying tube was put inside a bridge tube that crossed the fourteen-foot-wide plenum chamber and carefully pushed from the tube into the channel. Channels in the north half of the pile were filled by pushing the element across the central gap. Final positioning of the element was made by detecting that its anchor (a small aluminum protrusion welded between two fins) had engaged the anchor slot in the graphite. A wedge-shaped orifice was then fitted into the ends of the channel to adjust the air flow allowed through each channel to be consistent with the nuclear heat production in that channel. The helium tube ends of each element, north and south, were unwound during charging so that their ends remained on the charging elevator during element charging. An electrician's snake was then passed through a conduit (a separate conduit for each channel end), down the north or south face of the reactor, into the inner end of the bridge tube, and out to the charging elevator. The snake was connected to the end of the element's helium tube and pulled out the top of the reactor, where its tube was outgassed and connected to the helium system. The charging process posed a great opportunity to make expensive mistakes, and was trusted only to the reactor operators and supervisors who had been specifically trained, tested, and qualified (Borst, Sailor, Fox, and Powell never themselves loaded fuel into the reactor).

> The work of scientists is notable for its lack of drama [Lyle Borst once wrote]. It is usually difficult to say when a piece of apparatus starts to work, and it is even more difficult to decide when an experiment is complete. The uranium chain reaction is outstanding, therefore, since the change from an inert subcritical assembly of fissionable material to a supercritical chain reactor is sudden and, to all intents and purposes, discontinuous. [34]

On the morning of 21 August, 349 channels had been loaded. The reactor engineers had calculated that the machine would go critical when about 390 channels were loaded. Given the pace of loading enough channels for criticality would be finished that evening. Brookhaven officials and several AUI trustees began to drop by to witness the event, and a party was scheduled immediately after the moment of criticality in the large, open entrance to the reactor building. The dignitaries congregated in the observation room just outside the control room, with space for about fifty people to peer inside through a glass window. Borst, with his characteristic professionalism, allowed only four reactor scientists in the control room— Powell, Fox, a shift supervisor named Jack Phillips who operated the console, and himself. He banished all the guests to the observation room, including a considerably annoyed Haworth; meanwhile, the entire operating staff (supervisor plus four operators and a number of instrument technicians) had open access to the control room. Following what would become standard procedure, Borst would inform Powell what he wanted, Powell would tell the shift supervisor to carry out a step in the procedure, the supervisor would instruct the operators, who would actually turn the switches, etc. The procedure was usually slow, tiresome, and without incident.

After members of Sailor's group carried out each loading and flux measurement, Borst and those in the control room relayed data about neutron production rate in the reactor via the building's public address system to those in the observation room. By midnight, 385 channels had been loaded, the reactor was still not critical, and Sailor's group was preparing yet another loading.

> Borst changed the call on us at the last minute [Sailor recalled]. We were extrapolating that we were about 4 channels short of critical, and were going to load 8 channels. But all these guys were waiting for the party to start, and this was already long past midnight. Lyle said, "We'll load four channels"—which would save about half an hour in putting in the fuel elements and connecting up the tubing and all that. So we loaded four channels, and it turned out that we would be right on the head. The control rods were pulled out, and we had a very strong neutron source in the center of the reactor to start the chain reaction off. Lyle announced about 15 minutes later that the reactor was critical, and all the dignitaries rushed out to the party. He then immediately scrammed [shut down] the reactor so that we could load 4 more channels and take our time.[35]

The reactor went critical at about 1:30 A.M. on 22 August 1950. While Borst, Powell, Sailor, and a dozen others stayed behind to continue the loading, the party began on the lower level of the building. Borst was a teetotaler, and at least a few of the lab staff, assuming this would be another of Borst's mild affairs, had fortified themselves before arriving. But this time Borst made an exception and had allowed Fox, who was more of a bon vivant, to concoct an innocent-tasting but powerful punch.

The principal beverage at the party [Manowitz said] was to be artillery punch, ordinarily made by pouring in various liquors, with soda water added. It was made in a fifty-gallon glass vat, and different people were assigned different liquors to add into the vat. . . . This turned out to be a very potent brew, and it *really* was a wild party. A few people passed out; how many, I don't know, because I was one.[36]

Several days of measurements followed, in the course of which Kaplan was able to calculate exactly the optimum size of the gap between the two halves of the cube. As he had suspected, it was eight centimeters. Four days after criticality, one of the operating crew, in the presence of Borst, Kaplan, and Powell, turned the hand crank that slowly moved one half of the pile. To move the several hundred tons of graphite a centimeter took over seven hours. Though the amount of movement was small, some of the mechanism's gears were wearing visibly under the great pressure. The movement was stopped, and a member of the reactor staff shop was dispatched to coat the gears with grease. Cranking recommenced, as the massive pile slowly crept away from its twin.

By the end of the year, the power of the reactor was raised to ten megawatts and the instrument came into routine operation. Ironically, it never ran at the power at which the temperature differentials would have been hazardous as the reactor was originally designed—though the problem with the stud welds would still have been serious at the lower power. Borst recalled:

It would run thirty megawatts in the summertime with all five fans, five one-thousand-horsepower blowers roaring like hell, which could be heard all over the site. In winter, the air was cool and had greater cooling capacity. But the cost of those five blowers, the five one-thousand-horsepower electrical blowers, was enormous, and we very quickly realized that we had to study an economical place to run that system. It turned out that we could run twenty-five megawatts with three fans; two of them shut down. I think we ran twenty, twenty-five megawatts all the rest of the lifetime. So never came close to what was threatening. The maximum fuel temperature that we considered safe to run was 250° C for fuel, 350° for graphite, and we ran that all the time.[37]

Several factors had contributed to the project's problems. Some, like the incorporation of improvements whose effects were inadequately understood by engineers, could have happened in any large scientific project of any period, though they were magnified by the awkward tripartite management setup. Others, including an optimism regarding large-scale research and development projects, related to the specific postwar historical context. But the principal problem, absence of satisfactory oversight and a clear-cut division of responsibility, arose from the inexperience of a fledgling organization in managing the new large scientific-technological projects that soon would be a staple for national labs. The episode was an important lesson for Haworth and other lab officials, for it led them to do some much-needed soul-searching from which they greatly profited. In the next decade and a half, Brookhaven would build two more large reactor facilities—at the AEC's insistence, with outside contractors—without the same difficulties. When, at one acrimonious meeting between Brookhaven, AEC, and Ferguson officials in the course of the BGRR project, Shoup commented that "the purpose is not to 'find goats' but to do a better job in the future," the remark was evidently sincere.[38]

In an insightful article on the building of the BGRR, historian Allan Needell remarked that the BGRR story is an illustration of a "dangerous underestimation of the technical and managerial difficulties to be overcome in large-scale research and development projects" that often took place in the wake of the Manhattan Project, even in scaling up an existing design (Needell 1983, 122). But it would be an exaggeration to characterize the story as a debacle, or even much more than an embarrassment, for the lab. Three and a half years was a remarkably short time in which to construct a reactor in peacetime conditions. Despite the newness of the technology, once problems were identified solutions were readily found. And the facility would go on to become an important center for U.S. nuclear research for over a decade.

On the day reactor construction was completed, Borst went to a hardware store in nearby Center Moriches and purchased a small, broad-blade hatchet. In large red letters he painted "BNL" on one side, and "HKF" on the other. He collected several of the Ferguson engineers, and got the group to bury the hatchet ceremoniously near the base of the stack. Ferguson made its own gesture of reconciliation by handing out little souvenirs to the Brookhaven staff who had worked on the reactor: paperweight-sized models of the pieces of graphite in the pile.[39]

Some hatchets were more difficult to bury. All concerned had been embarrassed by the delays and cost overruns, and Borst in particular had acquired enemies in the course of the project. He did not appreciate Haworth's constant involvement in tasks that Borst felt were his, nor did Haworth appreciate Borst's reluctance to let the lab director intervene. The two had sometimes exchanged harsh words. Worse, Borst had gotten on the bad side of Rabi, who held him personally responsible for the delays. Finally, Borst was still adamantly opposed to the secrecy that the AEC still insisted must surround reactors. For all these reasons, Borst felt he did not have a future at the lab—that it was indeed finally time to change his specialty, as he had told the *New Yorker*—and decided to leave.

5.9 Aerial view of site, showing reactor and stack.

Borst's group was disappointed at the news. All of them were technically attached to a department, generally the physics department, and on loan to the pile project. Many in the reactor group had assumed that they would simply be let go as soon as the reactor started. An engineering department was officially created at the beginning of 1948, partly for morale purposes, but little was done for a year. In January 1949, the pile project had been combined with the stillborn engineering department to form the reactor science and engineering department, with Borst as its head. All the members of the reactor group had been looking forward enthusiastically to continuing to work under Borst on other projects after completion of the BGRR. Kaplan remembered:

> The pile group had worked so effectively together, and we had such excellent people with us that it was clearly a shame to break the project up. It had also become clear that there was a field—nuclear engineering—waiting to be developed. We thought there was room for basic research in nuclear engineering: in reactor physics, heat transfer, nuclear chemistry, materials, shielding. These were the thoughts that led us to propose a nuclear engineering department and we were happy and pleasantly surprised when the BNL administration and the AUI trustees bought the idea. Borst was the obvious one to head up the department.[40]

They tried to talk Borst out of leaving, but in vain. "Once Lyle made up his mind," Kaplan says, "you couldn't shake him." [41]

Borst left the lab in summer 1951 for the University of Utah. Thirty-five years later, he was asked in an interview whether he had ever reflected on his role in the construction of the first reactor designed in peacetime specifically for civilian research, a historic artifact in postwar U.S. science, whose remains still sit on the top of the hill where it was built. [42]

Borst cast his eyes down and grew thoughtful. "Oh, yes. Yes. I've thought about it many times," he said, quietly, confidently, proudly. "But I'm an old man now, you see, and the world has passed me by." He laughed, a little too quickly, as if to suggest that nothing more on that subject would be forthcoming. Lyle Borst, so used to keeping secrets while fighting on principle against the need to, made it abundantly clear that he was still prepared to keep quite a bit to himself. [43]

Community Relations

COMMUNITY RELATIONS were rarely a concern for university-based, prewar scientific laboratories, nor were they a consideration for wartime labs. Things were different for Brookhaven and its sibling laboratories—big, government-funded, peacetime facilities engaged in nuclear research after World War II. Some community relations problems had nothing to do with atomic energy, such as those created by government seizures of land for Argonne (Holl 1997, 55–57). Others had to do with the fact that the national labs had tremendous symbolic value, and thus inevitably became a part of political agendas and attracted sensationalist media coverage. In the early 1950s, for instance, conservative radio commentator Paul Harvey campaigned against Argonne in "radio broadcasts, speeches, and contacts with Joint Committee staff," at one point scaling the lab's barbed wire fence in the dark to demonstrate how easy it was to penetrate the lab's perimeter and therefore how vulnerable were our atomic institutions to the communist menace (Holl 1997, 96–97). In the long run, however, concerns about radiation would become increasingly important. And while Brookhaven (like Oak Ridge but unlike Argonne) at first was relatively isolated from such concerns as it had been built in a sparsely inhabited area, the local population was growing.

Of all the changes experienced by postwar U.S. science, the increased importance of community relations is certainly one of the most dramatic. At the beginning, the AEC's so-called "Decide-Announce-Defend" approach with respect to community relations issues was still effective. Also, prior to the environmental movement, the majority of the public still adhered to what sociologists have called the "human exemptionalist paradigm," in which human societies at "*as if* they were exempt from ecological constraints" and as if technological and social progress could continue infinitely (Catton and Dunlap 1980, 25). Finally, a growing appreciation for the social role of science and technology eventually would lead to increasing questioning of its authority.

Even prior to these developments, several factors complicated relations between Brookhaven and the community. The first was that a laboratory is not a particularly "community-friendly" type of cultural institution. Museums, universities, and most other cultural institutions provide some sort

93

of regular service to their neighbors. But a laboratory is by nature a relatively cloistered environment; and Brookhaven was not only a national laboratory, beholden to the federal government, but a basic research lab, one whose work was unlikely to have immediate spin-offs that the community could easily appreciate or benefit from. The lab's scientific staff were hired, not from the surrounding Suffolk County, but from all over the country and even the world. Industrial contracts were likely to go to large companies based off Long Island, like Bethlehem Steel, H. K. Ferguson, and General Electric. Even the lab's hospital was a research hospital, with patients usually recruited nationally rather than locally. Thus the lab provided few local services to the community through which a natural rapport could develop. It had prestige, as one of the foremost scientific centers in the world. But it was hard for members of the community to maintain a feeling of pride toward an institution that, except in special circumstances, they could not visit or even drive through freely (though the lab's backwoods was crisscrossed by much-frequented horse—later dirt bike— trails).

A second complication was historical circumstance. The national labs were established less than two years after the U.S. had exploded atomic bombs over Hiroshima and Nagasaki, and the public inextricably associated atomic energy, reactors, and radiation with atomic explosions, destruction, and warfare. A third complication was the tendency of the media to sensationalize, and then fail to correct faulty information, which exacerbated fears of an already dreaded hazard. Thus, in 1946, "reports that one man from the Oak Ridge uranium works had died of radiation exposure stirred a minor newspaper sensation," though when he turned out to have died of hepatitis "the press promptly forgot about the worker" (Weart 1988, 111). This plot—the press taking seriously wild rumors about ills supposedly caused by radiation, then ignoring corrections—was enacted over and over again, at BNL and elsewhere.

Still a fourth complication was that, in the public eye, radiation is different from all other toxins, as science historian Spencer Weart has argued. Atomic energy almost immediately became a symbol: for some, reviving ancient hopes of social revolution and utopian dreams of transformation, but more often a fearsome icon of the danger of looking into forbidden subjects and tampering with nature, the inevitable punishment being catastrophe and apocalypse—"nuclear fear" (Weart 1988). A number of people, including the AEC's first chairman David Lilienthal, saw it both ways; atomic energy, for Lilienthal, had pushed humanity onto a threshold where either revolution or apocalypse was possible. The power of that symbol to attract displaced fears interfered with discussion of the real hazards of radiation at Brookhaven and elsewhere, and disturbed as well its relations with the surrounding community.

In April 1948, a memo by AUI assistant treasurer John D. Jameson, entitled "Notes Concerning Possible Public Education Activities of the U.S. Atomic Energy Commission Contractors" was sent to the office of the general manager of the AEC in Washington. Based on Jameson's conversations with some trustees, the memo is a revealing expression of the feeling of "revolution or apocalypse" in the new world of atomic energy. "The atomic energy business is one still surrounded by fear and mystery," Jameson wrote, and "the task of assisting the average man to enter and live in the atomic age successfully must be undertaken and executed deliberately. It must not be considered a by-product of the other jobs or as something which will take care of itself in time." If left unattended-to, fear and mystery might lead people into "mass hysteria":

> Under the spell of such neurosis they might take political action which would set back the frontiers of scientific development for generations. To put it in a word, there might be another "burning of the books." The fear that the average man has of the frankenstein monster of the machine or of science has never become articulate, but lies not very far below the surface. There are many evidences that Science is on trial and being allowed to continue its preoccupation with atomic and other mysteries because the people still believe and hope that great benefits will flow from scientific development. If they falter in their belief that a more or less steady flow of the good things of life, of surcease from worry, or relief from pain comes [from] the Laboratories, their fears of the "priesthood" will prevail and they might go so far as to take over the temple.
>
> A proper public education job will be founded upon recognition of such possibilities as this and upon the need to assist the citizenry to achieve peace of mind about the atom without having to go through the centuries of experience that accompanied the same achievement in the case of, for example, fire. Mr. Lilienthal has made frequent references to the analogy between fire and fission. It is a fruitful analogy. Over a period of a relatively few years it alone, properly used, might assist our generation to span an intellectual development which our forefathers took thousands of years to accomplish.[1]

The field of atomic energy is still so young that it has "not yet developed a generation of hobbyists and tinkerers," the memo continued, but it was expanding so rapidly that "atomic hams" might soon become as numerous as "radio hams." Because the activity of atomic scientists had thrust humanity into a new world, it was their duty to help guide humanity through the coming revolutionary transformations. The AEC's contractors, who would be helping to develop so much of the country's scientific knowledge, might be able to help out in assisting "the citizenry as a whole to enter and live in the atomic age successfully," not by propaganda but by producing "a small but continuous stream of unemotionalized, pure fact about the atom and atomic energy" to counteract "rumors or erroneous deductions."

Meanwhile, administrative aide Karl Hartzell began preparing a report, "Education in Atomic Energy," on past and possible lab initiatives in public education.[2]

But many trustees, including the most influential ones, thought that community relations were not germane to the lab's activity. Rabi and Zacharias in particular appear to have assumed that research into atomic energy needed no accompanying public justification or explanation. Science was inherently interesting, Rabi felt; as a youth, hadn't he and many of his friends been fascinated by the test of Einstein's theory of relativity during the eclipse of 1919?[3] He and Zacharias believed that public relations, if needed, was the government's duty to handle through educational programs (later, Zacharias became involved in restructuring U.S. science education, and Rabi eventually reversed his position). A laboratory was simply not the right kind of institution to engage in public education. Maybe there would be no "atomic hams" tinkering with basement-built cyclotrons; still, it seemed to the founders reasonable to expect that the normal educational processes would filter down enough so that the public would come to accept and appreciate atomic research.[4] Within a few years, discussion among trustees of possible public education initiatives ended.

An enormous amount of community outreach did take place during the early years (contrary to statements by certain Department of Energy officials in 1997 and 1998), including many talks by top lab scientists and administrators to local organizations, a traveling van, mobile exhibits, open houses, school demonstrations, and so forth. What the founders were rejecting was a specific kind of outreach: courses and other formal educational programs for the general public. In hindsight, however, the kind of community outreach that did take place was less effective than hoped for.

But the neighbor who threatened a lawsuit over radiation damage after seeing the plume of smoke rising from Camp Upton's old coal-burning plant was an early sign that nuclear fear might make a dialogue between scientists and at least some community members more difficult than anticipated. It was no surprise that questions about safety in general and radioactive waste disposal in particular were among the first to be asked at the laboratory's initial press conference on 28 February 1947.[5]

How serious was the danger? Two issues would be involved in answering this question: technical issues that involve scientific fact, and political/personal issues that involve community values. The first involves assessment of the actual effects of different levels of contaminants on human beings and the environment; the second the perception of risk by the community in its public aspects. Any technology, any machine, any human activity involves some degree of risk, and values are involved in assessing if the return is worth the risk. Some risks—invisible risks, such as those pertaining to radiation, or involuntary risks that expose large populations, such as those experienced by people living next to a nuclear research facility—are quite rationally deemed less acceptable than others. It is probably fair to say that most scientists believe that achieving a consensus between the public and the experts is simply a matter of mass education; but this, as George Steiner suggests, is a fantasy of modernity, shared alike by both Jeffersonian liberalism and Marxism-Leninism (Steiner 1989, 67). As Brookhaven's founders discovered, in a lesson that would be taught and retaught to them for decades, much more is involved in navigating a consensus about what constitutes a safe environment, and the obstacles are formidable. Moreover, when achieved, any consensus is fragile and easily disrupted by things like local political events and accidents at the lab.

In 1947, AUI engaged a public relations firm, Pendray & Liebert, but let it go early in 1949 on the AEC's insistence because it was afraid of angering Congress by use of public money for promotional activities. In community relations, the scientists would have to go it alone. Revealingly, they thought they could.[6]

Soon after the lab opened, Philip Morse established a speakers' bureau (which still exists), of senior scientists and administrators, including himself, willing to address local organizations and whoever else was interested. "One of the most important jobs," he said, "was to get the community around satisfied with the idea of a nuclear laboratory, with all this horrendous publicity on atomic bombs, to get them to feel that this was something to be proud of rather than afraid of." The early talks seem to have been well intentioned, somewhat successful efforts to downplay fears through reassurances and statistics. Trust us, the speakers said, we are the experts, and we have made the lab a safe place to live and work.

> Always the first question in the question period was, "Is this dangerous?" And my usual answer was, "Well, yes, there are dangers. There are dangers in any factory, but I'm willing to put my word on the line to say that the danger that we would show to anybody working in the lab or outside was very considerably less than your crossing the street outside on a Sunday afternoon."[7]

But Morse and others soon also encountered a number of wild, even bizarre fears that seemed to them to have come out of nowhere: pilots who worried about becoming sterile by flying over the laboratory, a woman who feared the radiation would make her pregnant, farmers who thought the lab had made their ducks radioactive, a man who viciously condemned scientists at the lab for tampering with God's creation (Weart 1988, 178).

These concerns, voiced in the first year, before the lab had any significant radioactive materials on hand, involved more than simple misunderstanding. One day the lab received a shipment of ordinary light-green glass, called uranium glass because it contained minute amounts of that element. Upon learning its name, a worker who had handled the glass became psychosomatically ill enough to be sent home. "He was *really* sick," recalls Franklin Long, a chemist who became a member of the board of trustees. "That psychological impact was not new. We would see it again." [8]

Many laboratory staffers discovered with dismay that friends or neighbors thought of Brookhaven as "the bomb factory," an illusion that could not be eradicated and was abetted by erroneous newspaper reports. Once a consultant meteorologist picked up a hitchhiker and idly asked what he and his friends thought of the reactor about to go up in their midst. The hitchhiker replied that he was afraid, and kept remarking, "They got stuff there that will burn right through you." Concerned how little real information about the lab had reached this person, the consultant wrote Brookhaven meteorologist Norman Beers that "[a] lot of local and not too busy legal boys with that kind of a countryside can make it difficult for the Laboratory in the courts." If a lawsuit developed, he advised planning "the defence for an appeal to a higher court rather than trust a jury picked from local residents . . . as far away from Upton as possible, preferably in Washington, D.C." [9]

At first, the reputation of science and scientists was high enough that issuing reassurances often sufficed. For the moment, stories of supposedly radioactive ducks, life-threatening uranium glass, radiation-induced pregnancy, and tampering with God's atoms were told and retold around the lab, with some amusement. The joke soon wore off.

The scientists were little prepared for handling this kind of reaction, and for handling community relations in general. Even engaging in the effort ran against the habits and beliefs of most scientists who had just returned from war-related work. Many naively behaved as though safety were purely a technical, scientific question, and that talking to the community about safety would involve no more than relaying what Jameson had called "unadulterated, pure fact." But it would have to involve, at the very least, learning to couch one's message, not as if one were speaking to a colleague, but rather in values appropriate to those to whom the message is addressed. It would also have to involve learning certain rules governing public discourse of which the scientists appeared at least to be wholly ignorant and unappreciative.

One such rule, for instance, is that immediate interests are more powerful than global perspectives. Shortly after the lab opened, one scientist told a group of worried oystermen that emissions from the lab would not affect the normal amount of radiation on Long Island, and pointed out that the amount of radiation in the Rockies—which was quite harmless—was five times as great as on Long Island. "Ain't no oysters in the Rockies," was the reply (Lang 1947, 35). Yet laboratory scientists would use that kind of argument—equally true, equally unpersuasive to many in the absence of shared background assumptions—in response to those kinds of questions for decades (comparing, say, exposure to leaks at the boundaries to numbers of chest X rays). One basic problem with such answers is that they failed to address or even recognize the real fears that had motivated the questions.

A second rule is that public discourse tends to occur in images and metaphors. Metaphors, of course, are the traditional means for translating the unfamiliar into the familiar, and a stock in trade of popular science journalism. As a new kind of institution carrying out new kinds of research, many metaphors would be used to describe the lab and its work. When *Newsday,* a Long Island newspaper, early on tried to explain Brookhaven's existence to its readers, it called the lab "Suffolk's first university." [10] Scientists tend to find the use of images and metaphors irresponsibly sloppy— but they would find images and metaphors involving values particularly disturbing, for values are exemplary things, goals toward which we aim but never reach, and images connected with values therefore tend to be extreme.

In August 1948, for instance, the *Hartford Times* got wind that lab scientists were exploring the possibility of power recovery at the BGRR. Under the headline "Atomic Power Nearly Here," the newspaper announced that Brookhaven's new reactor would use atomic energy to power a generator, for the first time in history.

> If the Upton experiment proves successful, its historical importance will equal that of Wright Brothers' invention of the airplane or any of the other great discoveries like the telegraph, the telephone, etc.
>
> Many such things are needed to offset the damage done, both physically and morally, to the world through the wartime employment of the energy. [11]

The reaction of the engineers ranged from embarrassment to fury, and one protested that the article was "inaccurate to the point of maliciousness." [12] For them, the project was highly experimental, a risk. Atomic power might well be impractical—the extreme imagery of the article not only falsely suggested that power recovery (an applied project) was a major focus of their research, but raised expectations that were certain to be dashed, which would inevitably lead to charges of incompetence at the lab. Yet the newspaper's images had done no more than connect the experiment's possible success with the values, interests, and background knowledge of its typical reader. Also, the newspaper was merely following the lead of several scientists and engineers elsewhere who had hyped nuclear power with almost the same degree of hyperbole.

This pattern—the media couching descriptions of scientific work and its value in images; scientists finding the images misleading and excessive and reacting with shock and feelings of betrayal—would be repeated over and over. Most disturbing, the same pattern would appear, not only when the accounts were attempting to describe the lab's work positively, but when they were describing potential dangers as well. In that event, the aftereffects would include an increase in community suspicion.

In October 1948, a widely publicized disaster took place at Donora, Pennsylvania, when an accumulation of smog from local smokestacks caused the deaths of twenty people, and hundreds of others suffered respiratory distress. The incident happened just as lab meteorologists were studying safety measures to be taken in monitoring the BGRR's stack emissions, evaluating ways to keep the ^{41}Ar levels low. An Associated Press story in November connected this work with the Donora episode.[13] Lab officials protested, for while there were real risks from the ^{41}Ar, to compare the situation with Donora was irresponsibly misleading. But in the loose, generic approach of the wire service story, it was comparable; both involved airborne poisons, and Donora was the most extreme instance thereof with which the reader would be familiar.

Shortly thereafter, nationally syndicated columnist Drew Pearson picked up the story. Using the skills that made him an effective journalist, Pearson described the work of the lab scientists in hyperbolic prose and with apocalyptic imagery that successfully left readers with a sense of urgency and concern. "Deadly waste gases" and "deadly atomic radiation" from the reactor, Pearson wrote in his column, *Washington Merry-Go-Round,* might create "another Donora smog tragedy just outside New York City." [14] From Brookhaven's perspective, Pearson's article consisted of irresponsible exaggerations and inflammatory rhetoric, and the article was angrily denounced by AEC officials.

But Pearson's column, like similar articles that periodically would be written about Brookhaven ever since, was often accepted at face value and widely commented on by Long Island and New York City newspapers and radio stations, giving rise to the lab's first serious public relations episode. It was a pattern frequently repeated in later years, with the analogy to Donora, which has long faded from public memory and no longer has emotional purchase, replaced by others that did: Three Mile Island, Love Canal, Chernobyl.

Part of the gap between the perceptions of lab scientists and outsiders arose because of different notions of the character of scientific inquiry.

Scientists often inquire into subjects about which they know a substantial amount, in order to apprehend things from a new angle or perspective. But to nonscientists, inquiry can suggest doubt, therefore ignorance, therefore that the danger may be greater than what the scientists know or say. Consider, for instance, the following generally sympathetic editorial, written in the wake of the controversy caused by Pearson's column:

> It is difficult to determine if the announcement made by the Brookhaven National Laboratory (atom lab) regarding ground water tests and air tests should create a sense of complete security. The report is that radiation from the lab will at no time be harmful to the people living nearby. The fact that tests are being, and have been made shows that the scientists were not certain. . . . No matter how emphatic the statements, lay minds probably will always see a question mark in front of them. It is only natural that they should.[15]

One of the most effective members of the speakers' bureau was R. Christian Anderson, the first organic chemist at Brookhaven and an easygoing, articulate young man whose ability to quote Hawthorne in his job interview impressed lab officials. Anderson did not feel it beneath his calling to engage the local community in the activities of the lab, nor did he throw up his hands in frustration at the formidable obstacles he encountered. Curious enough to be interested in the difficulties he ran into, he was also imaginative enough to learn from his mistakes:

> The horror stories began very early. Farmers east of here complained that their cows were being killed by radioactive dust, and I would go out to talk to them and point out that we don't even have a radioactive source at the lab yet. One woman came up and threw her arms around my knees and pleaded with me not to put her baby in jeopardy. There'd be a flap over radiation in the tristate area, and someone would call the lab and say, "Send a speaker!" Many times I'd have to stand up in front of groups of outraged citizens and try to deliver the message, "This is not a real problem."

Frustrated by the difficulties, Anderson sometimes adopted the backpack and hiking boots image appealing to environmentalists. At other times, he tried a hard-hat persona and wore red, white, and blue; red tie, blue and white shirt. Though uneasy with what he feared might seem to be patronizingly brazen flourishes he discovered the message did not seem to be carrying itself. Sometimes he found it effective to restate the concerns and invite the audience to mull over their origins.

> Once, at the Patchogue Rotary Club, I discussed the science of radiation protection at great length; the inverse square law, shielding, and so forth. I said, "Doesn't it make sense that the scientists who know the most about radiation protection and who work with radioactive substances daily would take the maximum protection they could, to protect themselves and their families who live in

the community?" I thought I'd made a convincing argument. But afterward a guy came up and said, "Don't give me all that crap—what kind of pill do you take?" That remark taught me a lot about what an enormous gap existed between what we knew and the public's perception of it. The irony is that the remark came from science itself, but the reference was so embedded in our culture that it wasn't alien any longer. That really taught me a lesson. It taught me that you cannot take even the simplest illustration out of a scientific background and use it in a public forum safely. I took to simply repeating that story—and it often worked to great effect. I had held up a mirror, and people in the audience recognized that that's how they think.[16]

Anderson was good—perhaps too good, for the lab often succumbed to the temptation to dump specific complex public relations issues, like the shipping of reactor fuel rods, on his platter, which Anderson was too gracious to refuse. This contributed to the problem by helping to hide from lab officials a growing, long-range gap of perspective between lab scientists and members of the community.

Even as Brookhaven matured as a national institution, community relations continued to be a problem. Local newspapers tended to give prominence to even the most incredible rumors about the lab. Consider the case of the death of Kenneth Koerber, who began work in the medical department in June 1947. In August 1948, Koerber was involved in a serious automobile accident, left the lab, and returned to Philadelphia. When Koerber died in fall 1956, the Philadelphia medical examiner, Joseph W. Spelman, called the death radiation poisoning, announcing that Koerber's bones contained "1,000 times the maximum safe concentration of radiation."

The finding was implausible on its face. The reactor was unfinished at the time of Koerber's departure, and only a few radiotracers, to which Koerber did not have access, were in use; if he had swallowed them all, he could not have acquired the reported amount of radioactivity. Still, most newspapers bought the story whole, often couching it in lurid, inflammatory rhetoric. A *New York Post* headline read, "MARTYR OF BROOKHAVEN LAB," implying that an evil institution had crucified an innocent man.[17] When specimens of Koerber's bones and liver were examined at Argonne, it was shown that he did not die of radiation poisoning and his body showed "appreciably nothing above the activities to be found in contemporary man due to fallout and natural sources."[18] A *Science* editorial observed that "[a]s atomic reactors go into action and the testing of weapons continues, we may from time to time expect to read similar scare stories, couched in equally extravagant language. We recommend an attitude of skepticism until an assessment can be made on the basis of the best information available."[19] Copies of the *Science* editorial were sent to magazines and newspapers that had run the original scare stories, but the journal was naive if it expected that periodicals intended for the general public, when handling emotion-laden health-related issues such as those involving radiation, henceforth would habitually adopt an "attitude of skepticism."

News about the lab was often phrased in the most lurid way possible. One foggy morning in fall 1960, a thirty-three-year-old physicist named John Gibson was driving to work and was approaching a small bridge over Long Island Rail Road tracks just south of the lab's entrance. A truck carrying six casks of spent reactor fuel from the BGRR was coming down off the bridge the other way. The truck skidded on the wet pavement and its driver lost control. The truck jackknifed and slammed into Gibson's car, pushing it into a ditch and killing him instantly (the spent fuel casks were intact and no radiation was released); Haworth and Powell were on the scene within minutes, and were around when Gibson's body was cut out of the twisted wreckage. The next day, the *New York Times* headlined the story, in its front-of-section "News Summary and Index," "ATOMIC SCIENTIST KILLED BY RADIOACTIVE WASTES."[20] While early on such sensationalism had little impact on the lab's activities, that would change in later years.

Public reaction to issues involving radiation is sufficiently interesting and complex as to have become the subject of study by historians and social scientists. Weart cites the "windshield pit" episode of the mid-1950s as one particularly dramatic illustration of the power of atomic energy to attract displaced, anxious fears. Beginning in 1954, shortly after an atomic bomb test in the South Seas that March, hundreds of Seattle citizens called the police to complain that a mysterious agent, probably related to fallout from the bomb, was causing tiny pits in the windshields of their cars. Soon, windshield pits were seen in other cities. "Few believed the real explanation: windshields ordinarily collect tiny pits over time but usually nobody notices. Anxiously scrutinizing their surroundings, the citizens were projecting into their perceptions a novel fear" (Weart 1988, 187). The windshield pit episode provides a good model for one type of phenomenon involving neighborhood public and media scrutiny of Brookhaven and other national laboratories involved in basic research: "Pits" (small releases of contaminants to the environment, certain operational procedures, management practices, etc.) that are always present and harmless, can, through a random series of external events, suddenly be focused on and represented as posing a serious health threat, and interpreted as representing grave and inexcusable lapses.

Community fears were easy to arouse, hard to assuage. The assumption that something really hazardous was going on, masked by its invisibility and concealed by a conspiracy of silence, was difficult to dispel. After a sonic boom over Suffolk County, the public relations office would often receive calls demanding information about the atomic explosion that just took place. Some callers refused to accept the explanation that nothing out of the ordinary happened, assuming they were the victims of a conspiracy of silence: "But I *heard* it with my own ears!"

Everyday windshield pits or health hazard? Sonic booms or nuclear explosions? Hysteria or radiation poisoning? Public safety obviously required being able to tell the difference between the two. But making headway toward consensus over safety issues was hampered because the perspectives involved were so different. From the perspective of the lab's scientists, who made it their profession to study radiation and its effects, many of whom were internationally recognized experts in the field, the laboratory was not at all dangerous and such incidents were transparent frauds. The reactors were small, expertly supervised, and carefully monitored. No machine is flawless, and occasional breakdowns and accidents occurred. The minimal amount of radioactivity released on such occasions, even by the occasional leaking fuel element, was within safe limits, scarcely above the levels naturally present in the environment, and were much tinier than the amounts that had been distributed over Long Island by atmospheric nuclear weapons testing. In retrospect, it cannot be said that early BNL scientists were unconcerned about radiation releases to the environment. "The initial safety limits adopted by BNL proved to be attainable but were so low, by at least a factor of ten, that they were embarrassing to other AEC activities," instrumentation and health physics head J. B. H. Kuper reported to the board of trustees in May 1955. "Accordingly, the BNL limits have been increased, in accordance with AEC standards but still less than the standards proposed by the National Committee on Radiation Safety (National Bureau of Standards Handbook #59)."[21]

From the perspective of the community, there were several good reasons to worry. Before the 1955 Atoms for Peace program, much information about atomic energy was classified, raising doubt as to whether the scientists were free to disclose possible dangers. Moreover, by the late 1950s, clear evidence had emerged of the AEC's prior dissembling and mendacity with respect to nuclear hazards. Former AEC chairman Lilienthal wrote in 1958, "I don't remember any instance in which a major public body has lost public confidence in its integrity to the extent that has happened to our old friend the AEC" (Weart 1988, 204). Brookhaven was an AEC lab; to what extent were officials at Brookhaven trustworthy? At the end of the 1950s, the reliability of the authorities had become suspect.

While the lab never engaged in a large public education program, it did have a number of small-scale community relations initiatives. In 1950, it began a series of annual open houses, including a Visitors' Day, an Employees' or Family Day, and, in 1954, a Students' Day. The occasions provided an opportunity to sum up and celebrate the lab's work, and were heavily promoted by Haworth, who valued employee relations. They generally took place on successive weekends in September with the same show for everybody, and held in the gym. Each department was encouraged to set up a display, and one booth featured Mr. Atom, who answered any and all questions about atoms, their constituents, and the subatomic world. The Visitor's Day in October 1958 was especially memorable for the fact that instrumentation head Willy Higinbotham, who had spent three years during the Second World War working on radar indicators, and who had patents for a number of circuits used in analog computers, had created a little game by connecting an analog computer to an oscilloscope and rigging it to display a simulated side view of a tennis court on the screen with a simulated bouncing ball. By rotating knobs and pressing buttons, two players could "hit" the ball from one side of the court to other. Crowds of people flocked around to play.[22] (Afterward, he made a version that played basketball.) Though Higinbotham, a cofounder of the Federation of Atomic Scientists, often said he most wanted to be remembered for his work in arms control, popular articles in the media rarely fail to cite his role in the invention of what was almost certainly the world's first video game—proto-"Pong"—which he never thought to patent.

Another community relations initiative was a special lecture series named after Brookhaven cofounder George Pegram, proposed in 1956 by R. C. Anderson and established at the end of 1957, shortly before Pegram's death in October 1958 at the age of eighty-one. The aim, according to the program, was to provide some prominent individual "with a forum to discuss some area of his experience and interests which deal with the broad implications of science in our times." Though a committee was formed that was nominally supposed to select the speaker, Haworth was adamant about his right to choose the first: Lee DuBridge. DuBridge gave the first of the four talks in his series on 14 September 1959: "An Introduction to Space." Brookhaven had almost nothing to do with space exploration, but the subject, for reasons about which DuBridge could not have known, proved fantastically topical. Two days before the first talk in the series, on 12 September, the USSR launched Luna 2, the first spacecraft to crash-land on the Moon, soon followed by Luna 3, the first spacecraft to circumnavigate the Moon and take pictures of the far side. The second Pegram lectureship was Rene Dubos on "The Dreams of Reason," while the third speaker, Charles Coulson, a distinguished chemist, spoke about "The Scientist and Society." Not everyone was pleased with the results. A few scientists griped that DuBridge's talks "could have been given at a higher level," while others at the lab were made uneasy by Coulson's invocation of the Divine, in his efforts to say that science, like everything else, was part of a grand design.[23] Other Pegram lecturers, like J. Robert Oppenheimer in 1963, gave talks that, while interesting to the scientists, were far above the heads of the general public. The difficulty experienced by Pegram lecturers in satisfying both scientists and the public was yet another reflection of how formidable the task of bridging the two audiences really was.

Right after the war, the national labs could afford to operate relatively independently of the communities in which they were located, inasmuch as they were sponsored by the AEC through special contractors like AUI— institutions designed to withstand popular and even, to a certain extent, political pressures. Over time, due to many factors including the greater scale and cost of modern science, heightened sensitivity to environmental issues, and antigovernment sentiment, this ability to operate in relative isolation was eroded. Overturning deeply ingrained habits, scientists would have to learn to work as if in a fishbowl. Their past practices, however much within contemporary regulations and safety standards, would be exposed to numerous lawsuits. The scientists would have to learn to act in the knowledge that some day, due to a random series of events over which they had no control, their every dollar, every chemical, and every practice might become open to scrutiny by Congress, the press, and individuals in the community. They would have to learn to express themselves openly, persuasively, and concisely when real environmental leaks, the understanding and evaluation of which required a comprehension of complex relative-risk issues, inevitably occurred. And as a community, scientists would have to learn to make concerted appeals to persuade Congress and the public of the value of their projects. While the new importance of community relations would not become fully expressed until half a century after the birth of the national laboratories, it would reach into every aspect of lab life: at Visitors' and Employees' Days in the late 1990s, Mr. Atom was replaced by a person who answered questions about the Brookhaven's impact on the local environment.

At the end of his life, shortly after the cancellation of one of the laboratory's major accelerator projects, Rabi was asked what he thought had been the lab's greatest failing. "We didn't bring the public with us," he said.[24] This has to be understood as an admission of frustration rather than failure. Nobody knew how to take the public with them, not at Brookhaven nor anywhere else. Certainly, it would take far more to cross the gap between lab and community perspectives than a few talks and visitors' days. It would have required dozens, perhaps hundreds, of lab scientists serving on PTA boards, in church groups, as volunteer firemen, and participating in other activities involving regular contact with other community members, so as to destroy stereotypes of their profession and to defuse fears of their craft. It would have required the lab management occasionally to take major safety-related improvement projects that, while not mandated by existing safety regulations and designed to address remote dangers or extremely low-level risks, would show the community that its perceptions mattered.

But in the immediate postwar era, such behavior ran against existing habits and practices. And even it would not have guaranteed success. Institutions the size and nature of Brookhaven are exposed to the stochastic nature of events: unforeseeable developments, unrelated happenings, and sudden shifts in public opinion mean that the community and government relations of such institutions cannot be treated as a "problem to be solved." The lab's political vulnerability grew, and would crest in 1997, on its fiftieth anniversary, when the Department of Energy used the occasion of a slow leak at the spent fuel pool of a reactor, posing no health threat, to cancel the agency's contract with AUI to manage the lab.

The Accelerator Project

M. STANLEY LIVINGSTON (fig. 7.1), head of the accelerator project, knew cyclotrons as well as anyone except perhaps the man who invented them, Ernest O. Lawrence (Heilbron and Seidel 1989; Courant 1997). In 1929 Livingston entered Berkeley as a graduate student in physics. There he met Lawrence, who had just arrived the year before and was attempting to work out a new method of accelerating particles by boosting their speeds in a series of small steps. He built a device consisting of a pair of hollow D-shaped electrodes or "Ds" that faced each other across a small separation, the whole assembly being placed inside a vacuum chamber. This chamber in turn was sandwiched between the cylindrical poles of an electromagnet. Lawrence then released charged particles into the gap between the Ds. Because the Ds were oppositely charged, the particles were attracted to one of the Ds and repelled by the other. As the charge at the Ds was alternated, the particles began to spiral like unwinding tetherballs inside the Ds, as they were attracted first to one side of the chamber and then the other, their paths bent into a curve by the magnetic field. Each time the particles crossed the gap, the charge on the Ds was reversed, giving the particles a small boost in energy; little taps to the unwinding tetherballs. Inside each D was a field-free region; when the particles entered they did not "see" any charge, and they coasted on a curved path until they reached the gap again. According to the equations of motion, the period of revolution, or time it took to make a complete orbit, remained constant as the particles sped up, so that when an oscillating electric field of the right frequency was applied to the Ds, the particles stayed in step, or "in resonance," and received a small boost of energy with each half-revolution. Lawrence called the device a magnetic resonance accelerator, but the name that stuck was *cyclotron,* coined by joking lab assistants.[1]

In September 1930, Lawrence put Livingston to work on the device as a Ph.D. project. In December, Livingston managed to detect resonance for what was almost certainly the first time, although Lawrence, with characteristic but premature optimism, had already told members of the National Academy of Sciences that he had seen resonance himself. Livingston rebuilt the device, whose largest single piece was a four-inch electromagnet, almost from scratch, describing the results in his Ph.D. thesis. The next summer, Lawrence and Livingston built a cyclotron with an eleven-inch magnet, with which they were able to accelerate protons past a million electron volts (MeV), putting Berkeley on the map as a world center for physics and its staff on the roster of people to watch. "Lawrence made waves in science," Livingston once recalled. "He made such big waves that we all got thrown up to the top. I coasted down the front side of that wave for the rest of my career, using what I had acquired in the way of a reputation, experience, and contacts." [2]

From then on, Lawrence and Livingston built ever bigger cyclotrons, and in the process established them as *the* tools of the emerging field of nuclear physics, the study of the characteristics and behavior of the atomic nucleus. Physicists could use cyclotrons to collide protons and other particles with nuclei to discover and study different types of nuclear reactions as well as different kinds of nuclear isotopes, or forms of the same element. "Nuclear physics just blew wide open," Livingston said. "Here was this machine, where essentially everything you touched made a publishable paper. . . . The papers just poured out of that lab." [3] According to historians John Heilbron and Robert Seidel, "The technical achievement was mainly Livingston's; the inspiration, push, and above all, the vision of future greatness, were Lawrence's. His professorship gave him the place to stand, and Livingston gave him the lever, to move the world of physics" (Heilbron and Seidel 1989, 101).

But Livingston began to feel used and unacknowledged. Lawrence's application for a patent on the cyclotron in 1932 did not mention Livingston. According to one of Lawrence's associates, Livingston once complained about not receiving adequate credit, only to be told, "If you're dissatisfied, you can drop out of the project, and I'll get someone else who can take your place" (Davis 1968, 44). Livingston grew increasingly unhappy. "I wanted to get some place where I could do it myself on my own machine," he said.[4] In 1934, he moved to Cornell, and with $800 and half a dozen graduate students of his own built the first cyclotron outside Berkeley, a 16-inch 2 MeV proton machine. Four years later he moved to MIT, where, for $60,000, he built a 42-inch 12 MeV cyclotron that came into operation in the spring of 1940. Meanwhile, the 1939 Nobel Prize for physics was awarded for the invention of the cyclotron—to Lawrence. In 1946, when Philip Morse of MIT was reluctantly roped into directing the new laboratory on Long Island and began casting about for someone to head its accelerator project, his MIT colleague Livingston was an obvious first choice. Livingston, for his part, was eager to take the necessary risk, for here, finally, he saw the chance to create a truly big wave of his own.

The accelerator project was a different enterprise from the reactor project. Reactors create neutrons with essentially one small energy spectrum,

7.1 M. Stanley Livingston, head
of the accelerator project.

and with some exceptions experimenters are interested in neutrons within
that energy range. Accelerators can boost particles to a wide variety of
energies, and experimenters had important work to do at all of these ranges.
The initial program report of December 1946 therefore spoke of several
energy regions—low (a few MeV), medium (a dozen or so MeV), high
(hundreds of MeV), and superhigh (over a billion MeV)—and wanted to
construct different machines for each.

But the accelerator project, like the reactor project, was challenged by
the special partnership forged between the AEC and the AUI. The AEC's
desire to create a technological base for the new era of atomic energy meant
farming out to industry as much advanced work as possible, and the agency
felt that the technology of the two lower-energy machines had been standardized sufficiently to permit them to be built by commercial contractors
under minimal supervision, who would gain valuable experience thereby.
But the frontier marking the area where technology transfer has taken
place, always in motion, is not clearly marked. That frontier is easily misjudged in the presence of influences such as pressure to make major new
facilities operational quickly, the incorporation of innovations whose engineering consequences were unforeseen by the contractors, post–World
War II optimism regarding rapid construction of major scientific facili-

ties—and, most important, the temptation of scientists to assume that contracting engineers think as they do, leading to failures of supervision.

Meanwhile, the fate of the accelerator project's two high-energy machines was shaped by the AEC's desire to coordinate its large projects on a national level.

Low Energy: The Electrostatic Accelerator

The Initial Program Report called for an *electrostatic* accelerator to be built for the low-energy region. These use a moving belt inside an insulated column to load electric charge generated at a base onto a terminal. When a vacuum tube is put between terminal and base, particles push-pulled by the voltage differential can be made to flow down it. Invented by Robert Van de Graaff, a research fellow at Princeton and then MIT faculty member, electrostatic accelerators (or Van de Graaffs, as they were sometimes called) were able to accelerate beams of intense, finely tunable beams of particles to energies in principle of several MeV (Herb 1959; Bromley 1974). These devices promised to be effective tools for cancer therapy, industrial radiography, and research. Van de Graaff himself had founded the High Voltage Engineering Corporation (HVEC; the company claimed the trademark "Van de Graaff") with two colleagues, John Trump (uncle of Donald, future real estate tycoon) and Denis Robinson, to build the machines commercially.

Like the pile project, the accelerator project was under pressure to get its devices operational rapidly. Speed could be gained by purchasing conventional facilities from industry where possible, to allow the lab's still-developing scientific staff to focus on more advanced machines, and the 1946 Program Report suggested purchasing a 12 MeV, proton-accelerating electrostatic accelerator from HVEC (electrostatic machines are also rated by their voltage differential in millions of volts or megavolts, MV). Buying from HVEC was not like buying from industry; more like having a trusted colleague do the job. But the HVEC, itself new, was not yet able to deliver a device quickly. The alternative was to copy a good existing machine. The most advanced design work had been done by Ray G. Herb of the University of Wisconsin. Herb was as much (or even more) of an innovator in electrostatic technology as Van de Graaff himself, and an internationally sought-after authority. A Herb-designed 4 MeV electrostatic device was under construction at Berkeley, whose staff promised to give their Brookhaven colleagues a complete set of drawings of the machine and include suggestions for improvements. Livingston assembled a group to design and build a version of the Berkeley device, at a projected cost of $300,000.

7.2 Assembling the electrostatic accelerator.

George Hoey, an electrical engineer and radlab veteran, was chosen to assemble a staff to head construction.

Meanwhile, General Electric's lab in Schenectady had been building a 3.5 MeV electrostatic accelerator on an AEC contract, and the project was suddenly canceled after about three-quarters of the design and one-third of the construction had been completed. Wanting to place the machine, the AEC put G.E. in touch with Brookhaven. G.E. said it could deliver the machine in eleven months for $308,660, or about what it would cost the lab to copy the Berkeley design but in less time and without staff diversion. This *was* a risk, but the AEC was clearly behind it. Also, the quick delivery time would make it, as lab officials noted, "the first electronuclear machine in actual operation at Brookhaven." Contract terms were reached in October 1947 that foresaw operation early in 1949. The G.E. machine, AUI secretary Shoup wrote Morse, was a "happy windfall" (fig. 7.2).[5]

It was instead an ill windfall. Though an electrostatic accelerator is simple in principle, engineering state-of-the-art models is difficult and develops through a combination of long experience, extensive trial and error, and much luck; through "black magic," as engineers say. The machines create huge voltage differentials, and slight imperfections can cause a "flashover," as sparks leap from one part of the machine to another and damage it. To curb flashovers, the devices depend heavily on insulators and are housed inside large tanks in high-pressure atmospheres. Ordering a machine from G.E. seemed safe enough, not only because it had built other types of accelerators, but because the company was famous for its insulators, used worldwide for high-voltage transmission lines and transformers. The company had invented and patented an insulating material called Herkolite, which it was using in the support structures, made by layers of paper fiber soaked in shellac. The shellac was partly derived from tree resin; tiny insects were sometimes embedded in it. But the real flies in the ointment would lie elsewhere, in almost everything but the insulation.

At first, none of the scientists in the accelerator project had any suspicions about the instrument under construction. As in the pile project, the company thought it was producing what the scientists wanted, the scientists simply assumed that the engineering project could be left safely in the company's hands, and the AEC supposed that the necessary cooperation was taking place. Without anticipating any real problems, Livingston formed a group to install and operate the machine, and to head it selected Everett Hafner, a twenty-seven-year-old doctoral candidate from Rochester, who arrived in February 1948 (fig. 7.3).

Hafner had a daredevil streak and quickly acquired a reputation for scandalous behavior. Once, before turning in his radiation badge to health physics for the weekly check, he hung it on a neutron source until it turned black to see if they noticed. They did, of course; an irate health physicist reminded him that someone with a badge that black was supposed to go in for a spinal tap. Another time Hafner was caught in flagrante delicto with a woman on the Brookhaven grounds by a security officer, and threatened by an outraged Haworth with loss of security clearance. Hafner was given the services of David Alburger, a young Yale Ph.D. assigned to the group, and a technician, Robert Lindgren, and began readying the place for the machine's delivery.

But ominous signs began to appear. Berkeley's newly completed electrostatic machine was already flashing over at a mere 2.4 MeV; high-voltage engineering was evidently more difficult than anyone thought. Still more disturbing news from Berkeley followed a visit from Herb. Herb had brought their machine up to 3.9 MeV and then offered to show G.E. engineers how he had done it, but they refused. Herb, and everyone else who knew electrostatic machines and the black magic they required, was astonished. But the clearest sign of looming disaster appeared when Hafner visited G.E.'s Schenectady plant early in 1949.[6]

You need to build the things in a clean environment, and they were building it in a big warehouse with the end doors wide open—and Schenectady is not a clean

7.3 Everett Hafner working on tank for the electrostatic accelerator.

place, or it wasn't then. If you wiped your finger across the electrodes, you could feel the grease on them. Conditions were filthy, and that's *death* for an electrostatic machine.[7]

Though the device was not ready for delivery, Hafner insisted it be brought to the cleaner lab environment as soon as possible. When it finally appeared on-site, in July 1949, Hafner's group was shocked. Nearly everything except the Herkolite insulating columns had been almost mindlessly assembled. The thirteen-ton tank, which had to be withdrawn from the base plate for maintenance, was operated by a hand crank. The windows, which had to be opened periodically, were installed with hundreds of Allen head screws that took a long time to remove. That was only the beginning. The vacuum engineering, so critical to an electrostatic device, was inexcusably poor. Leaks sprang up all over, frustrating the engineers and delaying efforts to get beam.

In October, the accelerating tube itself began to leak.[8] Brookhaven engineers tried coating the tube with extra cement, to no avail. This leak was especially frustrating because it did not show up when the tube was tested in the open air; it appeared only after it had been installed in the tank, all the studs bolted, and the tank partly pressurized. This made the leak next to impossible to pinpoint along the eight-foot acceleration column, which consisted of hubcap-like metal electrodes separated by inch-thick white ceramic insulating rings. Someone idly suggested that the only way to detect it would be to find a person foolhardy enough to be sealed and pressurized along with the column inside the tank. Hafner promptly volunteered.

The other group members thought he was out of his mind. Risks included subjecting him to pressurization or depressurization too quickly, making him suffer what scuba divers call the bends. If something went wrong, it would take half an hour to depressurize and unseal the tank before rescuers could get to him. If a valve failed, and the pressure rose to maximum all at once, he would be killed.

The morning of 11 October 1949, an ambulance waited outside of the west end of building 901, which was to house the device, and the first new building completed at the lab. Inside, a dozen firemen and medical and safety personnel watched as the now-motorized drive slowly pushed the thirteen-ton tank against the face plate, sealing Hafner within. Clad in a white safety suit, he stood inside the tank armed with a flashlight, a hose to supply him with oxygen, and another hose that sprayed helium. The idea was this: when the vacuum pumps connected to the accelerating tube started operating, indicating a leak, Hafner would begin spraying the helium slowly over the accelerating column. When analyzers connected to the vacuum pump registered the presence of the helium, Hafner had hit on the leaky spot. In case a freak electrical failure made it impossible to retract the tank by motor, the firemen attached a huge hook to the wall and connected chains and pulleys between it and the tank so they could jerk it open by hand if necessary.

"Began pressure leak hunt at 0900," reads Hafner's log book.[9] For a moment, even Hafner felt fear. "I was pretty calm until I heard them put on the bolts—48 two-inch bolts, each tightened by hand. I kept thinking that if something went wrong, they would all have to be *undone* by hand." [10] Bolting the face plate took half an hour. Then pressurized nitrogen from a tank atop building 901 slowly filled the tank. Hafner felt as a scuba diver does while descending on a dive. He popped his ears.

When the pressure reached eight pounds per square inch (half an atmosphere, or about what a scuba diver receives at a depth of sixteen feet), the vacuum pumps automatically switched on, indicating a leak. Hafner slowly began spraying the helium over the acceleration tube. In a few minutes, the analyzers revealed that Hafner was at the leak. The tank was slowly depressurized. Hafner felt like a scuba diver rising to the surface, and popped his ears again. Each of the forty-eight bolts was opened by hand, the motorized drive pulled the tank from the face plate and Hafner stepped out into the open air. He had been inside a little over an hour.

Just for fun, one firemen suggested, let's see how easy it is to open the tank using just the pulleys. They couldn't budge it.

The acceleration tube was removed, and the tiny leak found in an all but inaccessible place. An entire section had to be taken back to Schenectady for renovation.

But no sooner was this defect repaired than others appeared: incorrectly calibrated instruments, inadequate installation, bad seals, misaligned parts, poor materials, and so on. It was, the Brookhaven engineers wrote, a machine full of "cut corners." [11] And in a repeat of what was happening at about the same time in the pile project, G.E. continued to adopt a confident posture that a final solution was right around the corner even as a steady stream of faulty parts and shoddy workmanship made it ever clearer that G.E. had been over its head in building the machine. Meanwhile the ultra-high-energy accelerator group had purchased for its own use an electrostatic accelerator from HVEC, a 4 MeV machine with a twelve-foot acceleration tube that was HVEC's first big order. While Hafner's group struggled with its device, the ultra-high-energy group purchased, delivered, and tested its own, which worked well. For a while, the members of the electrostatic accelerator group toyed with the idea of abandoning theirs and starting over by purchasing a twin of that machine from HVEC, but HVEC was already well booked with orders.[12] Adding insult to injury, the chemistry department purchased *its* own electrostatic accelerator, which ran at 2

MeV (about what seemed to be the peak capability of the G.E. machine) for a mere $97,000.[13]

Haworth reported the matter to AUI trustees, who raised the matter with Milbank, Tweed. The firm said that G.E. clearly had breached the contract and that the lab had two options: rescind the contract and force G.E. to refund money already paid out or accept the machine and press a damage claim. Replacing the machine would take over a year and a half, and in its present form it did have some uses as a research tool. The trustees decided to try to persuade G.E. to drop the price of the machine by about $100,000, equal to the amount of money required to compensate the lab for costs already incurred plus the amount it would require to bring it to specifications. Fearing damage to its image, G.E. at first refused to concede that its engineers couldn't make the machine meet specifications, but in October 1950 the company finally capitulated. In exchange, the laboratory agreed to accept the machine as is: operating between 2 and 2.5 MeV. "The agreement," Haworth wrote, appears to be "the best available solution for an admittedly unsatisfactory situation."[14]

One day shortly thereafter, early in 1951, two engineers from the G.E. plant in Schenectady unexpectedly arrived at the lab. Their equipment: screwdrivers. Their purpose: remove the General Electric nameplate with the company logo from the device, a job that took them all of five minutes. Mission accomplished, they returned to Schenectady.

Meanwhile, the electrostatic accelerator group began borrowing the services of Clarence Turner. Turner had worked on most of the high-energy electrostatic accelerators to date. His Ph.D. work (1943) at the University of Wisconsin had been spent helping Ray Herb build his first successful 4 MeV device. Turner spent two years in a Van de Graaff group at Los Alamos before joining Luis Alvarez's accelerator group at Berkeley as supervisor of the 4 MeV machine there. In August 1949, Turner came to Brookhaven to supervise the final construction and installation of its HVEC-built electrostatic injector for the ultra-high accelerator. Upon learning of the troubles with the G.E. device, he began dropping by building 901 to help out. His experience and deep knowledge of the principles of electrostatic accelerators allowed him to succeed where Hafner's heroics had not.

After a long, hard look at the machine, Turner spotted two fundamental problems. One involved an effect known as *sputtering*, the tendency of the electrodes to splash atoms off their surfaces. This slowly coated the porcelain insulators separating the electrodes, eventually producing a conducting path that drained voltage from the terminal. To fix this problem, Turner developed a new kind of electrode lined with stainless steel, a nonsputtering material, at the tips.

The second problem involved *electron loading*. As an electrostatic accelerator is charged, stray electrons begin to flow up the accelerating tube in the reverse direction from protons. These electrons collide with parts of the machine, releasing more electrons, and eventually an avalanche of electrons flows from base to terminal in a steady current that nullifies much of the voltage, draining or "loading down" the charge being pumped onto the terminal. While other electrostatic accelerator builders knew of this effect, Turner studied it carefully, and found that electron loading rose steeply with the accelerator's energy, by a staggering fifth power of the terminal voltage. Moreover, during their trip up the accelerating tube, the electrons were accelerated and would slam into the terminal with an energy of however many MeV the machine was running at, producing powerful X rays. These X rays passed through the insulating gas inside the tank, ionizing it and creating a conducting path that encouraged flashovers. To reduce this effect, Turner installed a tapered beryllium disc at the terminal end of the accelerating tube with a small hole for the proton beam; beryllium, an element with a low atomic number, gradually slows down electrons that pass through it so they produce fewer X rays. Turner's work inspired engineers at HVEC and elsewhere to devise other ways to cut back electron loading, among them an inclined field tube to sweep away the electrons before they reached the terminal (Turner 1951).

Under Turner's direction, the electrostatic accelerator group eventually would rip out and rebuild everything except the Herkolite support structure of the machine. The group would finally get the device running at 3.5 MeV, but only by rebuilding it entirely. From the 1950s until 1970, when a new electrostatic accelerator was completed, the "3.5 MeV Van de Graaff," as the machine was usually called, was the main instrument of Brookhaven's low-energy nuclear physics program, and it would continue to be used until 1994. Until the day the 3.5 MeV machine was removed from the lab, in 1996, one could still spot the screw-holes where the G.E. nameplate once was fastened—a mark bearing silent testimony to the early troubles of the AEC-lab partnership.

The Sixty-Inch Cyclotron

The initial program report called for a sixty-inch cyclotron to cover the medium-energy region of a dozen or so MeV, and do a variety of tasks that Livingston called the horsework of nuclear chemistry.[15] At first, Livingston and the accelerator project staff planned to build the cyclotron themselves. As Morse wrote in March 1947:

> The cyclotron is essentially a research tool in a developmental stage. It involves first of all research and development items for which the specifications are

worked out as the design progresses. Only secondarily does it consist of commercial items for which the specifications can be set down in advance. In the provision of the research and development items, we feel it is absolutely essential for Dr. Livingston and his staff to have direct contact with and supervision of the suppliers.[16]

But again the AEC intervened. It wanted technology transfer of as many atomic energy–related techniques to as many industrial concerns as possible, and insisted that a commercial outfit build the sixty-inch. The idea was not far-fetched. By that time, cyclotrons of that size had been in use for seven years (Berkeley had one working in 1939), and the technology seemed fairly standardized. The lab protested, but won only the right to select the company. It ultimately accepted a bid from the Collins Radio Company, of Grand Rapids, Michigan.[17]

Collins Radio, like G.E., seemed a safe bet. Its scientific director, W. W. Salisbury, had worked on Berkeley's sixty-inch cyclotron, and the company agreed to use a similar design for Brookhaven's machine. A contract was signed in 1948, which put the cost of the approximately 20 MeV machine at $580,000 plus a $30,000 fee. Completion was expected in July 1949 and operation in September 1949. The lab proudly announced that its sixty-inch cyclotron would be the *"first commercially* constructed research instrument of its type ever built," and that it would have the highest combined power and current of any cyclotron ever made (fig. 7.4).[18]

But by summer 1948, significant cost overruns had appeared, prompting Van Horn to warn the lab it would have to pay the costs itself. Additional cost overruns the next year provoked another stern warning by Van Horn.[19] Many arose from the informal way Brookhaven and Collins Radio had been doing business, confirming AEC fears about accountability. Brookhaven's staff would remark that such-and-such would be a "good idea" or "nice thing to do," and Collins Radio's staff would act on the suggestion, charging the contract.[20] But the most critical budgetary problems arose from a number of design improvements that Brookhaven and Collins Radio had agreed to formally, but that involved unanticipated problems. The project ran afoul of the Deficiency Appropriations Act (which had loomed so menacingly over construction of the BGRR), requiring the AEC to submit a detailed explanation to various agencies if the cost of one of its construction projects exceeded the original estimated cost by 15 percent. By fall 1949, the total estimate for the sixty-inch cyclotron project had climbed to $980,000, and Van Horn had to ask Haworth for the required explanation. It was the only Brookhaven project for which this was necessary.[21]

Installation of the cyclotron was completed in spring 1950, but leaks in the vacuum system delayed beam hunting until summer. Collins Radio

engineers began to look for a beam, without success. In May, with a satisfactory vacuum yet to be achieved and a beam not yet detected, Dunbar
consulted Milbank, Tweed about the possibility of terminating the contract
and taking over the machine. Collins Radio insisted on trying to fix it, and
sent Salisbury to Brookhaven for a few weeks, where he succeeded in
eliminating major vacuum leaks. Troubles persisted, the expense kept
climbing, and the AEC balked at providing supplements. In March 1951,
with the cost of the machine nearing a million dollars, the lab decided it
had "come to the end of the road on funds and supplements," and stopped
payments to Collins Radio.[22] Company president Arthur Collins visited the
lab a month later, and volunteered to send Salisbury himself to fix the machine at no charge. Still, defects plagued nearly all major parts, and the
company's ability to improve the machine was clearly exhausted. Moreover, Collins Radio's research department was rapidly disintegrating; Salisbury shortly left, and about half a dozen others had also left or were planning to leave by January of 1952. Impatient to be done with Collins Radio,
Haworth declared in April that the machine was "operating substantially
in accordance with the specifications laid down in the contract," and formally ended its contract with the company.[23] The machine had been de-

7.4 Erection of sixty-inch cyclotron, built by Collins Radio Company (late 1948).

layed about three years, had a 50 percent increase in cost, and was working poorly. Like the Van de Graaff, it was shut down for major renovations. The files of the cyclotron group include this poem, written sometime in the fall of 1952 by an anonymous wag:

> I think that I shall never see
> A cyclotron perverse as thee.
> A cyclotron whose tubes are lit
> From L.I.L.'s expensive tit;
> Who with this vital juice delights
> In feeding naught but parasites.
> A cyclotron whose magnet's field
> No perfect symmetry will yield,
> But with its chamber lids will nod
> When asked: "Are your harmonics odd?"
>
> A bastard, built but not designed,
> Conceived in Salisbury's mind;
> A mental contraceptive would,
> I think, have done a lot of good.[24]

The 240-Inch Synchrocyclotron

In 1946, the 500 MeV energy region had suddenly become extremely important. At such energies, researchers would be able to produce mesons, particles originally discovered in cosmic rays. One type of meson, called pi-mesons or pions, were thought to carry the strong force, or "glue" holding the nucleus together, and an accelerator powerful enough (about 500 MeV, Brookhaven's planners thought) to produce copious numbers of pions might enable researchers to address the all-important question of nuclear structure. The authors of the Initial Program Report therefore called for building an accelerator whose energy was a comfortable margin above that: 600 to 1000 MeV.

Ordinary cyclotrons could not reach that energy. As particles approach the speed of light, their mass increases so that they take longer to make one revolution, and consequently get out of synch with respect to the regular oscillations of the accelerating system. In 1945, a way around this limitation was announced in the U.S. by Berkeley physicist Ed McMillan and independently in the Soviet Union by Vladimir I. Veksler. Their idea involved decreasing the frequency of the accelerating voltage as the energy of the particles rose, so that the particles would still be in phase when they reached the gap. An accelerator based on this idea, called a *synchrocyclotron*, accelerates particles in bunches and relies on what is called phase stability; particles leading the bunch receive less push, and particles

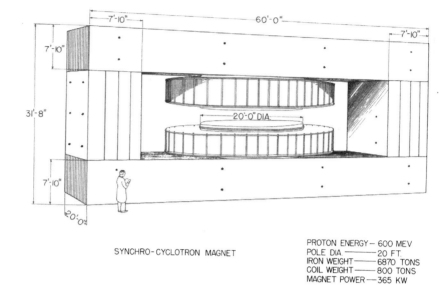

SYNCHRO-CYCLOTRON MAGNET

PROTON ENERGY— 600 MEV
POLE DIA. ————— 20 FT.
IRON WEIGHT ——— 6870 TONS
COIL WEIGHT ——— 800 TONS
MAGNET POWER —365 KW

7.5 Diagram (by Plotkin) of proposed 240-inch synchrocyclotron, never built.

lagging behind more push, so the group stays in phase together (Veksler 1944; McMillan 1945).

Brookhaven's synchrocyclotron was to be Livingston's baby (fig. 7.5). In 1947, the largest cyclotrons were the 130-inch, 250 MeV machine at Rochester and the 184-inch, 380 MeV cyclotron at the Berkeley Radiation Laboratory (UCRL as it was called then, now the Lawrence Berkeley National Laboratory or LBNL). Livingston wanted to outdo even Berkeley by scaling up its design. In April 1946, during the first discussions about the still siteless lab, Livingston proposed a 240-inch, 500+ MeV synchrocyclotron; a lower-energy machine, he said, might not be able to produce pions, while a much higher energy machine would be too costly. To help with the design Livingston hired engineer Martin Plotkin. Since the lab had not yet opened and Livingston was still at MIT, Plotkin moved to Boston, where he worked on the design and built and tested a one-fifth scale model. When Plotkin wrote up and mimeographed some of his early work on 11 January 1947, it was a milestone of sorts for the lab, for it bore the title "Brookhaven Technical Report #1."[25]

Livingston's pet project soon encountered two difficulties. One was the announcement that Columbia was to build its own synchrocyclotron, a $1.2 million 400 MeV device with a 145″ magnet, at Nevis, an estate twenty miles north of Columbia on the Hudson River, which was originally built by Alexander Hamilton for his son and bequeathed to the university by the du Pont family. Brookhaven's proposed synchrocyclotron would not be as much of a jump over existing machines as Livingston had planned. A more significant difficulty was Rabi's interest in a more advanced type of accelerator.[26] For as McMillan and Veksler had shown, particles could be accelerated, not only in a spiral, but also in a racetrack-shaped orbit of constant radius provided the magnetic field, and the frequency of the accelerating voltage, was increased in strength to keep the particles in the same orbit. This made possible a kind of accelerator called a *synchrotron*. Instead of needing two huge magnet poles to cover the entire diameter of vacuum chamber, like slices of bread a sandwich, a synchrotron could use numerous smaller magnets around the orbit, strung like beads in a necklace, which would keep the particles in an orbit of fixed radius. This made it possible to build more powerful accelerators at lower cost: while for a cyclotron the rule of thumb was that the cost was roughly proportional to the cube of the radius, for a synchrotron the cost was proportional to the radius.

On a trip to Berkeley at the end of 1946, Rabi learned of Berkeley's enthusiasm for proton synchrotrons and that the scientists there were drawing up plans for a 10 BeV machine, and he was given a sketch of the design by Berkeley accelerator builder William Brobeck. Rabi became convinced that Brookhaven should forget the synchrocyclotron, which was about to become obsolete, and throw all its resources into the proton synchrotron on which they had already begun working. In science, Rabi liked to say, "You always go for broke." [27]

The day he returned, Rabi dropped in on Morse, who telephoned Livingston and asked him to come by. "I can remember that evening very clearly," Livingston said years later. Rabi passed around Brobeck's sketch. The only realistic action, Rabi said, was to immediately begin work on a 10 BeV synchrotron. "We would have to live with this competition with Berkeley—that was part of what I learned that evening with Rabi," Livingston recalled. But still he was reluctant to take this course.

> It was much higher energy than I had myself ever had any experience with. And I didn't have an Ed McMillan at my elbow to stimulate me. . . . [But Rabi] gave the impression to me that that was my job—to build a multi-BeV accelerator for Brookhaven. . . . That was how the Cosmotron got into Brookhaven, by Rabi's personal introduction of the need for a multi-BeV machine.[28]

But Livingston had a deep personal commitment to cyclotrons and his stubbornness had been hardened by his bitter experience of being all but ignored when it came to prizes, patents, and even credit for cyclotrons. Admitting that synchrocyclotrons were already a dying breed and could never reach 10 BeV, Livingston pointed out that the lab could build one sooner than it could a synchrotron. While willing to help draw up designs for the proton synchrotron, he wanted to avoid putting all of Brookhaven's high-energy eggs in the synchrotron basket.

The outcome was a yearlong battle of wills between Rabi and Livingston. The first skirmish occurred at a planning committee meeting in October 1946, when Livingston advanced a detailed proposal to build his 240-inch synchrocyclotron at a cost of five million dollars over three years. Many of the fundamental problems had already been worked out at Berkeley's 184-inch, Livingston said, and a synchrocyclotron would be more reliable and create a more intense beam than a synchrotron. Turning this observation against Livingston, Rabi argued the scheme was not bold enough. A synchrocyclotron would sap laboratory resources; the lab needed to forget the modest step and go all-out for a 10 BeV machine to seize leadership in physics. The others present were equally divided, and the committee decided not to decide.[29]

Livingston and Plotkin pressed ahead with the synchrocyclotron project in the spring of 1947, but under Rabi's relentless pressure the project began to lose priority. At a scientific advisory committee meeting in April,

> it was agreed that the design of a 10 billion volt proton-synchrotron should be the prime effort of the accelerator project with further design on the 600 million volt cyclotron being carried on only in such a manner as not to delay the prime effort.[30]

At Berkeley, Lawrence, Brobeck, and company pressed ahead with their ambitious plans. At Brookhaven, in August 1947, twelve physicists gathered to try to make a final decision on which accelerators to build at the lab. They included the principals of Brookhaven's accelerator project (Livingston, Kenneth Green, Hartland Snyder) as well as Haworth, Rabi, and some prominent accelerator physicists from the outside. Livingston gave a historical background, and Green took notes. The group agreed that Brookhaven was to build large proton accelerators—big and soon! All agreed a 10 BeV machine was "the important accelerator for Brookhaven."[31]

But opinions differed regarding what, if any, intermediate steps to take. Livingston described his 240-inch project, which he defended as practical and quick.[32] While a proton synchrotron looked appealing in principle, no large version had successfully been built. Synchrocyclotrons were still the most reliable and powerful method of acceleration.

Then Rabi held forth. Even the minutes make clear the sheer force of his personality. "Be bold," he said. Repudiate "the safe path and little steps" and strike for 10 BeV. "Be a little wild." Still, no conclusions were reached. "We agree to disagree," read Green's handwritten notes. But a showdown obviously loomed.[33]

Morse and Haworth decided to seek outside advice, and on 20 October had a lengthy conversation with AEC commissioner Robert Bacher and James Fisk, the AEC's first director of research. The Brookhaven officials discovered the AEC was unenthusiastic about the 240-inch, for it was only a factor of two above existing machines, and they wanted a larger step—a factor of five or more. Taking this advice to heart, Morse and Haworth then decided in favor of the synchrotron project alone.[34]

Livingston was furious, and banged out an angry memo of protest, claiming the decision compromised the integrity of the lab:

> The original concepts, before the Atomic Energy Commission was activated, were based on the assumption that the research program would be largely determined by the planning of the Brookhaven research staff. . . . It now becomes evident that the program must be based primarily upon the policies of the Atomic Energy Commission and that the Laboratory staff and the Universities' advisory groups have relatively little significance. As such, it is not the "free" laboratory for fundamental research which had been visualized, but is now directly controlled by the national interests of the Atomic Energy Commission. . . . The decision . . . will postpone for several years longer the maturing of an instrument at Brookhaven which is capable of producing mesons. To just that extent I believe that it will postpone the effective development as a research laboratory in this field.[35]

For the moment, Livingston gracefully surmounted his anger, and began work on the design for the proton synchrotron. But he seemed to have lost his enthusiasm for the accelerator project. He began to take long absences, and six months later informed the lab that, in September 1948, he would be returning to MIT.

Until the 240-inch synchrocyclotron, accelerator development had proceeded in short, confident steps. The scale, the cost, and the interests of the AEC made that no longer feasible. Henceforth, accelerator builders would have to learn to gamble, to take larger steps knowing that the risks involved in a misstep were correspondingly greater. All the while, scientists take risks, planning projects involving problems whose solutions they do not yet have, assuming that they will be able to solve them in time; this is precisely the kind of engineering that an industry cannot afford to do. The demise of the 240-inch made it clear that scientific leadership would have to involve making still bigger gambles, and numerous illustrations of this turned up in the Cosmotron project.

The Cosmotron

Brookhaven accelerator physicists quickly found themselves involved in another conflict, over how ambitiously to design the synchrotron.[36] With the AEC's two accelerator laboratories each planning a 10 BeV machine, Fisk arranged a meeting between Berkeley and Brookhaven personnel to coordinate their programs.[37] The meeting took place in March 1948 at Berkeley: Morse, Haworth, and Livingston came from Brookhaven, while Lawrence, McMillan, and Brobeck represented Berkeley. Fisk told the assembly that the AEC had a limited pot of money to be allocated between the two labs for building accelerators—enough for one 10 BeV machine, perhaps, but definitely not two. If the money were split between the two machines, the agency could fund one smaller 2 to 3 BeV accelerator, suitable for producing pions, and one 6 to 7 BeV machine, which would be suitable for producing antinucleons. Who wanted the smaller?

Neither did. Livingston recalled, "[O]ne of us had to give or neither would get the Atomic Energy Commission to approve our work. . . . It was clear we were at an impasse." [38] Each party retreated to a side room to talk over what was happening. In Livingston's recollection, it was Haworth who began to lead the group into the decision to choose the smaller machine:

> It had to do with the fact that Brookhaven was a new laboratory. We were using a new style of designing. We were competing with one of the most experienced laboratories in the country in building accelerators. And we should recognize that fact—that if we took away the big one from Berkeley, and left them with essentially nothing but a much smaller one to build, we would make enemies of them without any question at all for the rest of the lifetime of science. . . . I tended to agree, and eventually so did Phil. We wanted, however, some concessions: Just as soon as we got the first 3 GeV [as BeVs were later known] machine running, we wanted to have the AEC's commitment to back us in the design and start of the next size larger than the 6 GeV that Berkeley was going to get. And we got that kind of assurance. . . . This left a very satisfying feeling of agreement and mutual cooperation that I think was one of the very important things that we did at that meeting at Berkeley, and gave us the chance to expect them to cooperate.[39]

Most of the rest of Brookhaven's team eventually came to agree with the wisdom of the decision. Rabi did not; he had even tried to convince the AEC not to give Berkeley anything. Decades later, he was still angry: "I could have spit blood." [40]

When Livingston returned with the news, he also reported that Fisk had asked for an agreement on names for the two machines. For the Brookhaven machine, the first to surpass a billion volts, "Bevatron" was the AEC's name of choice. The name "wasn't perfect," Livingston said, but it "may become accepted unless a better name is suggested." [41] At one staff meeting, about ten names were written on the blackboard, including such complex Greek bastardizations as *hypertachytron* (*hyper* for over or above, and *tachy* for speed). One by one they were discussed and erased—except for *Cosmitron,* which Plotkin had proposed because the machine would be doing the work of cosmic rays. For euphony—and after the form of words like *cyclotron* (and perhaps with the word *cosmos* in mind)—the *i* was changed to *o.*[42]

The name took. A week later, Livingston called Fisk to ask him to drop the ugly word *Bevatron* and start using *Cosmotron* instead (Berkeley's 6 BeV machine would be named the Bevatron). The AEC formally approved the Cosmotron in April; the first construction funds arrived in June.[43]

Thus began a certain kind of rivalry between Berkeley and Brookhaven in the building of their respective accelerators. "A certain kind," because one must distinguish between what might be called *political* and *performance* rivalry. In political rivalry, the aim is a particular prize, and gaining it requires vanquishing one's opponents. The more disabled the opponent the better, and the strength of one's opponent is a measure of one's own weakness. Performance rivalry—the kind that, say, the members of a baseball team might have in vying for a most valuable player award, or that members of a drama group might have in competing for unspecified future roles—is much different. In performance rivalry, the aim is to achieve a superior level of performance; but, in seeming paradox, one increases one's own performance by maximally sharing information and cooperating. One derives pleasure, and benefits oneself, from seeing the overall level of performance rise, and the strength of others is one's own strength. Performance rivalry requires a specific kind of creative community whose members are engaged in enacting performances, and science is of that kind. Frequently, scientific rivalries are portrayed by outsiders as ruthless quests for ego glorification, prestige, or prizes, where contestants aim to vanquish each other. And it is true that the victors in scientific rivalries often win more of the spoils, that there is frequently a strong element of political rivalry in science, and that the nuclear sciences in particular have something of an "Olympic quality" about them (Holl 1997, 176). But what transpired between Brookhaven and Berkeley would be unintelligible if pictured solely as a political rivalry; all information was shared, frequent visits took place, and each group mailed copies of memos and the minutes of its weekly staff meetings to the other, so that each lab ensured that the other could derive maximum benefit from its own work and experience.

Performance rivalry has long been an integral part of science. But because scientists themselves were no longer entirely able to decide which machines to build and where to place them, the stakes were rising to new and higher levels, which subjected the friendly performance rivalries of the past to new kinds of tensions.

Brookhaven's accelerator physicists, indeed, regarded Berkeley, home to the best and most experienced accelerator builders in the world, with a feeling of "awe, and actually with something of a feeling of doom," John Blewett said. "We felt up against an outfit like that, we really didn't have much of a chance." The Brookhaven group even toyed with dropping out; the main reason it stayed in was that it had done some first-class design work and assembled a competent group.[44]

The two institutions had dramatically contrasting styles. Berkeley had a separate engineering division that was large, experienced, and renowned for its conservative style of engineering. "They built their machines like battleships," Rabi once remarked, "really solid things that would last through the ages." [45] Brookhaven's accelerator physicists perforce adopted a different approach.

> What they had that we didn't have [Blewett recalled] was a tremendous power-house of experience and engineering, administration and backup in the form of machine shops and all sorts of standard facilities; we were starting from scratch with Camp Upton as the army delivered it to us. So we decided that we would emphasize the theoretical approach much more than had been done at Berkeley. . . . It appeared [to outsiders] that they were designing by the seat of their pants. Actually, their designs were based first on experience, and second on model studies. We decided to go more deeply into theory than they had.[46]

Brookhaven's was a "university style," which Livingston once defined as "where you pick the cheap, simple, quick way of getting there, in order to get there fast and get the job done." [47] Brookhaven's scientists compensated for their small engineering staff by relying, more than Berkeley, on theory and on their sense for the "good enough."

> At Brookhaven [Plotkin recalled], there was little hierarchy within the Cosmotron group to speak of. If you had a problem, you'd mention it at the weekly staff meeting, and somebody would say, "I have some time—I'll work on it." [48]

The management of the Cosmotron project went through a period of turmoil after Livingston left, with Haworth stepping in twice as interim head until George Collins, of the University of Rochester, became accelerator project head. What stability the project had was due to a pair of accelerator physicists, John Blewett and G. Kenneth Green. In some respects, these two were opposites. Blewett was primarily interested in the general theory and concept of accelerators, Green in technical design and hardware. Blewett typically would start out with a grand picture and fill in details as they became necessary; Green felt he had to understand every little component first. Blewett liked to reflect while fishing; Green was an electronic brico-leur who built an electronic organ at home in his basement (as far as anyone knew Green never finished it, but worked on it endlessly as a way of keeping up with electronic circuitry and how various newly developed electronic components could be utilized). But Blewett and Green had something special in common that has become rare among those of their profession: they were equally competent at theory and practice. Moreover, they served a desperately needed role in synthesizing the projects' various elements. The Cosmotron involved a number of different subassemblies, each as large as or larger than the Van de Graaff or the sixty-inch cyclotron project. These subassemblies were worked on concurrently by overlapping and ever expanding groups of scientists, engineers, and technicians. Blewett and Green were members of several of these groups, and served to coordinate the work of these groups with each other.

One group consisted of theorists who laid out the overall design and parameters of the machine. The Cosmotron's three principal theorists, besides Blewett, were Ernest Courant, Hartland Snyder, and Nelson Blachman, and they faced several major design problems. One was to determine how large to make the aperture, or opening between the magnet poles. The aperture had to be as small as possible (magnet costs shot up with gap size) but big enough to accommodate the various oscillations and perturbations of protons in the beam. The Cosmotron's theorists ultimately chose a $36'' \times 9''$ aperture; once the vacuum chamber was installed, a net area of $25'' \times 6''$ remained for the beam.[49]

Another, and related, theoretical concern was the shape of the magnetic field. In synchrotrons the magnetic field has to be slightly tapered to produce restoring forces for particles that stray from the central orbit. The field has what is known as a *gradient*, quantitatively characterized by the value of a number n, which is a measure of the degree of taper. Livingston and Lawrence in their work on the cyclotron first found that n has to be positive to produce vertical restoring forces; later Kerst and Serber showed that n has to be less than 1 to prevent centrifugal force from making the particle orbit blow up radially. So to make vertical focusing strong one would like large n, for strong radial focusing one needs large $1 - n$, and to strike the necessary compromise between Scylla and Charybdis one needs the right value for n between 0 and 1. The value chosen for the Cosmotron was set at a little more than 0.6.

Once these theoretical decisions were made, the engineering design of the major components could begin. In terms of cost and weight, the magnets were by far the largest component of the Cosmotron, and nearly all the principal figures in the accelerator project worked on them at some point. The magnet engineering provided a classic example of Brookhaven's university style, for numerous simplifications were made to make them good enough, at some cost to experimental programs, for the sake of economy and speed. John Blewett had the idea to design each magnet in the shape of a C instead of the more standard H figure. The Cs would all point away from the center of the circle, and particles would orbit in the gap between the poles. Such C-shaped magnets were easier and faster to make, involving little in the way of welding and bolting, and also made the vacuum chamber readily accessible from the outside, vastly simplifying injection and general maintenance, inspection, and repair. The drawback was that only negative particles would be easily extracted from an internal target, for the field would bend positive particles inward.

The Cosmotron consisted of 288 such C-shaped magnet blocks, each weighing 6 tons, in an 8′ × 8′ octagonal shape, with a 9″ gap cut 3′ deep in one face. The magnets were arranged in a ring divided into four curved quadrants of 30′ radius separated by 10′ "straight sections." Testing of magnet models began early in 1948, and full-scale magnet blocks were ordered from the Bethlehem Steel Company. Because the magnets began

7.6 Surveying base plates for the magnet blocks, using a radius bar attached to one of four pre-located posts. The person at right is William Moore, who was in charge of magnet testing.

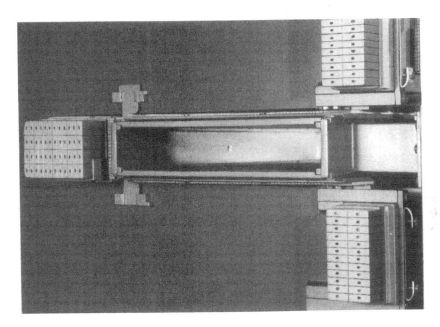

7.7 Model of Cosmotron magnet gap. The holes in the coils allowed cooling water to pass through the coils. The steel pieces making up the roof of the vacuum chamber sat on but did not touch the edges of the chamber, to reduce eddy currents.

to arrive before the Cosmotron building was complete, a "test shack," a windy and rickety 100′ square structure with a concrete floor, was thrown up behind the building, where scientists measured the field character of each magnet. Early in 1950, workers began moving the 288 magnet blocks from the test shack into the Cosmotron building, surveying them, and welding them in place (fig. 7.6).

While the magnets were being installed, the Cosmotron team had a scare having to do with the magnet gap and their Berkeley rivals (fig. 7.7). The Bevatron's initial design, true to Berkeley's conservative style, called for a huge 4′ × 14′ aperture. But also in keeping with Berkeley's style, its engineers built a quarter-scale test model, which was equipped with shutters that could be used to close down the aperture size to see if the beam was lost. In 1949, the model was built and tests began. At an aperture bigger than Cosmotron's, the beam was lost. At Brookhaven, which in a gamble had based its aperture size on theory rather than models, this news was met with "fear and trembling," especially after rumblings from Washington about cutting off funding, but the team had no choice but to plunge ahead anyway, trusting to their calculations.[50]

In the second half of 1950, the coils were installed and tested. Electric current would flow through these to create the magnetic field to control the particle beams, and they were made from single strips of copper 50' long and with a 3/8" hole through which flowed coolant water. The heavy copper pieces were shaped in a gentle curve, insulated and installed in a resin mold so they would stay rigid under the tremendous magnetic forces to which they would be exposed. The coil design, too, embodied some university-style compromises in engineering. At the end of each quadrant, the coils cut across the last magnet so that they could be connected to the electrical power generator. The copper terminals of the coil joined the copper leads to the power source at a right-angle turn where the pieces were silver-soldered together (fig. 7.8). This was risky, for it took place in the field of the magnet, which would pulse every few seconds, day in and day out,

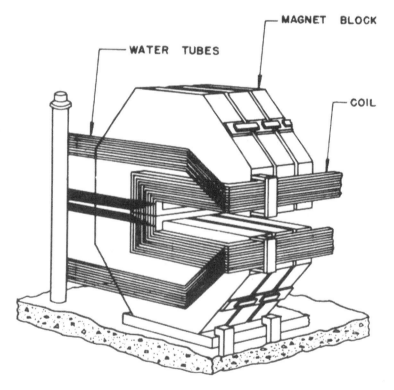

7.8 Diagram of Cosmotron magnet. Note that the coils cross the face of the magnet at the end of the block and make right angles there, which subjected them to strain when the magnet pulsed.

hammering at the right-angle joint in the coils. If the coils moved at all, metal fatigue could set in, resulting in leaks and failures of the coils. Worried, senior mechanical engineer David Jacobus tried to reduce the coil's motion as much as possible, but could not eliminate it entirely. It seemed good enough, and it was one design feature among hundreds that had to be tended to. Nobody seems to have realized just how much of a threat this design element posed, how risky this judgment call was.

Another gamble was made in the radiofrequency system which would give the "kicks" to the protons in an rf cavity located at one of the straight sections. As the protons run around the ring more and more energetically, the frequency of the accelerating kicks has to rise (for the time between crossings of the cavity gets shorter), converging on a constant time, a limit imposed by the speed of light. The rf system, too, embodied design trade-offs. If the particles are accelerated over a long time, the machine could be built with a lower peak power; if they were accelerated over a short time with high power, losses due to scattering would be less and the average power consumption would drop. Balancing these factors, the designers settled on a time of one second to accelerate the particles.

But it was difficult to construct a broad-band accelerator cavity able to cover the required frequency range in the short length of an accelerating station. In the Cosmotron, it would have to accept a frequency change by a factor of about thirteen. At high frequencies, where the radio frequency magnetic field in the core reverses very rapidly, so-called hysteresis and eddy currents induced in the core of the cavity are so large that they can heat it up, draining power. To prevent this, the designers had to build the cavity core from a material with high magnetic resistivity; some material with the wonderful magnetic properties of iron, but which also acted like an insulator. In 1948, the availability of magnetic materials for radio frequency work was limited. Eddy currents precluded use of metallic iron, except in the form of powdered iron held together in a binder, a method that would not work in the Cosmotron.

Then John Blewett, during a visit to Bell Labs, heard of a new material called ferrite that had been developed in the Philips Laboratories in Eindhoven, the Netherlands, before World War II. It was a ceramic material with high resistivity that did not suffer severe losses up into the required MHz range. Blewett took a hasty trip to Holland, and decided that ferrite was just what the Cosmotron needed.

In May 1948, Livingston, Blewett, Plotkin, and Haworth met with Philips U.S. representatives, who imparted distressing news. When the Brookhaven delegation said they needed samples on the order of square feet, the Philips officials threw up their hands, saying that the biggest pieces they could manufacture were $1/2'' \times 1'' \times 2''$—the size of a matchbox. At the scientists' insistence, they tried again, and eventually succeeded in making somewhat larger pieces, though at a cost far above what the lab could afford.

Haworth asked Lawrence if Berkeley would use ferrite in its Bevatron, sharing production costs. Lawrence refused. Meanwhile, Blewett and Plotkin visited the G.E. plant in Schenectady, and General Ceramics and Steatite Corporation in New Jersey, all of which were experimenting with ferrite. The next time Philips officials came to Brookhaven, Blewett had a sample of U.S. produced ferrite sitting conspicuously on his desk. Philips dropped its prices. The lab ordered ferrites from both Philips and General Ceramics, pieces began to arrive in June 1949, and about half a ton of them were installed in a frame around the vacuum chamber at the accelerating station and surrounded by a coil. When the rf system was finally finished in mid-1952, it was the first important application of ferrite in the U.S. The effort to make them in sufficiently large sizes at low cost opened the door for their use in many other kinds of devices.

The ferrite by no means solved all difficulties with the rf system. Indeed, the Cosmotron group was startled to discover one morning that Blachman, working extremely late the previous night, had written in thick, bold letters on the conference room blackboard the words WE'RE DOOMED! The machine, an accompanying scribble explained, would not work because of noise in the rf system. Several anxious hours were spent trying to figure out what particular demon Blachman had run into this time. Blachman arrived late that day to explain that he had asked someone working on the accelerating system how much noise that system would generate, and had been given a figure. Running a few calculations, Blachman discovered the noise would destroy the beam and make the machine impossible to operate. Fortunately, the figure proved incorrect.

By mid-1950, the Cosmotron's most serious engineering problem, and the greatest source of unexpected construction troubles and delays, was the vacuum chamber. Here, too, a gamble had been taken. At first the engineers had no idea how to build it, and construction began on the Cosmotron without even a design for the vacuum chamber. The problems were formidable. The chamber walls would have to be as thin as possible and built of a material that did not develop eddy currents (which would distort the magnetic field); at the same time, the walls had to be strong enough to stand up to atmospheric pressure. It would have to cover a large area, but without "outgassing"—emitting vapor into the vacuum—a problem that increased with surface area. The Cosmotron's senior mechanical engineer David Jacobus tried glass, ceramic, and (presumably the least desirable because of the eddy current problem) steel. In April 1948, Jacobus approached Corning Glass Works, but their price was too high. For a while, Cosmotron engineers studied both ceramic and metal. General Ceramics made them a model vacuum chamber, which was put in a steel vacuum tank, and the atmospheric pressure reduced. Suddenly the engineers heard a colossal sound, "as though all the china had fallen off the top shelf," recalls Blewett. The chamber had collapsed, ending the idea of using ceramic.[51]

Irving Polk, a twenty-seven-year-old mechanical engineer who had started his engineering career in the MIT Radlab engineering group, was in charge of investigating the possibility of using steel. Early in 1950, the vacuum chamber was the one major subassembly seriously behind schedule, and with glass and ceramic suddenly ruled out as materials, his problem was not to find something "good enough," but to find anything that would work at all. For steel to work he had to find a way to reduce eddy currents, but still make the three-foot-wide chamber strong enough that it would not sag under the atmospheric pressure. Jacobus and Polk looked for a while at high-resistance metals like Inconel, and at the possibility of cutting slots in steel and supporting the chamber with rods, but nothing worked.

In summer 1950, Blewett and Jacobus tried building the chamber of ribs of stainless steel, with each rib separated from its neighbors by a gap that the eddy currents would not cross. The ribs would then be covered by some yet-to-be-determined nonmetallic sealant (a low-vapor-pressure rubber was ultimately found), and the eddy currents induced in each individual rib would not be large enough to create field disturbances in the chamber (fig. 7.9). Jacobus persuaded the rest of the department to go ahead with the idea, but partly because he had made a whopping mathematical error and had underestimated the sag in the vacuum chamber roof by a factor of thirty. But by the time this was discovered, a way around the problem appeared. By the end of 1950 he and Polk finished their design and construction was started. To create the vacuum in the chamber, twelve twenty-foot oil diffusion pumps were installed at intervals in a trench in the concrete surrounding the magnet ring. Like so much else on the Cosmotron, the pumps were recycled; they had come to Brookhaven via Oak Ridge, where they had been used in the uranium separation project.[52]

IN JUNE 1950, the component construction of the Cosmotron was about to go into full swing (fig. 7.10). Meanwhile, a number of accelerators had just been built or were under way, in the U.S. and Canada, able to reach a few hundred MeV, and Columbia had just dedicated its 400 MeV cyclotron. "The springtime of Big Physics," *Physics Today* wrote, "has arrived."

John Blewett's thoughts turned momentarily to the history of accelerators. He idly graphed the energy of accelerators against the years over a

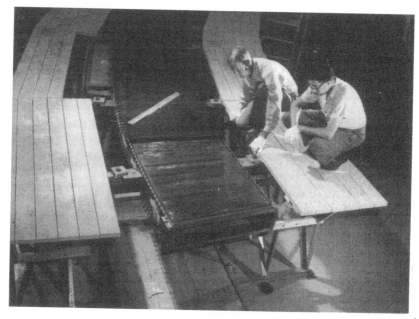

7.9 Engineers David Jacobus and Irving Polk assembling the vacuum chamber. The steel roof ribs, slightly separated from each other, are covered with a rubber sealant.

twenty-year period from 1930 to 1950. The results were astounding. Logarithmically, with tenfold energy increases, the result was a straight line. Blewett then extended the line to the Cosmotron's energy of 2.5 BeV, and it intersected the line at 1952. He circulated the graph with a memo (fig. 7.11):

> The attached graph is a self explanatory plot of reports in the literature of successful attempts to accelerate charged particles to high energies. Two conclusions present themselves from this plot.
>
> 1. The curve is approximately exponential. The peak energy has increased over two decades by a factor of ten every six years.
> 2. Whenever the popular accelerator of the moment has shown signs of reaching an upper limit a new invention has arrived just in time to save the exponential characteristic.
>
> It should be noted that the dates plotted are the dates of publication of results which may have been achieved as much as a year earlier.
>
> The writer will not be responsible for any predictions made on the basis of this graph.[53]

7.10 Final magnet block being assembled, prior to insertion of vacuum chamber, July 1950. The vacuum pumps were installed in the trench, covered by planks in the foreground.

Blewett wondered if the Cosmotron would indeed be finished in 1952. He also wondered what the next new invention would be. Then he passed the memo and plot on to his colleagues.

AT THE beginning of 1951, construction of the Cosmotron's components was almost complete (except for the vacuum chamber), and the Cosmotron team was anxious to begin testing components. Blewett and Green concocted a temporary vacuum chamber of glass segments joined by rubber socks so a beam could be introduced. In the summer of 1951, with this rudimentary system, a low-current beam was injected into the machine and tests performed.

Meanwhile, the chamber was manufactured in quadrants. As each of the fifteen-thousand-pound curved sections was completed, late in 1951, it was picked up by the crane and swung over next to the magnets. In October,

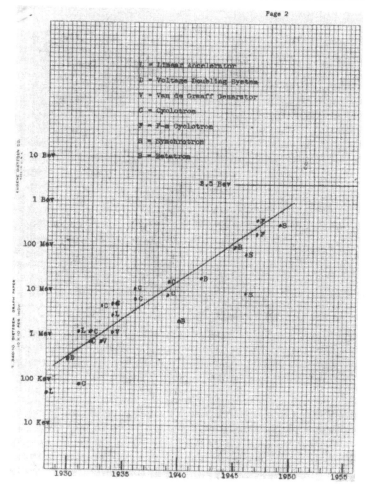

7.11 Blewett's plot of accelerator history.

Jacobus built a dozen screw pushers operated by twisting a handle, spaced them along the chamber every few feet, and ordered members of the Cosmotron team to operate them. At a call of "Stroke!" each person gave the screws half a turn, causing the huge slice of vacuum chamber to slide into the gap a fraction of an inch (fig. 7.12). Some people complained only half-jokingly that others were not pulling their weight, putting an unfair burden on their neighbors. Thus did the Cosmotron's vacuum chamber—its last major component to be finished—creep into place, using this most ancient of methods for moving heavy objects, and accompanied by this most ancient of complaints.

7.12 Setting the vacuum chamber into place, using screw pushers (October 1951).

The power supply, consisting of a 1,750-horsepower alternator, ignitrons, and a forty-five-ton flywheel to act as an energy storage device, was also hooked up. At the beginning of the acceleration cycle, the flywheel would drive an alternating-current generator—slowing down in the process—and the ignitrons would rectify the AC power produced, transforming it into a pulse of DC power flow through the coils. At the end of the one-second acceleration period, the ignitrons would run as inverters, and be switched over to convert the DC energy stored in the magnetic field back into AC energy to drive the generator, which would now be used as a motor to help speed the flywheel back to its original level. Thus the energy stored in the magnetic field would be put back into the flywheel, saving an enormous power drain, and the actual power used would only be the heat loss due to current in the conductors. The process would be accompanied by a distinctive THUMP! every few seconds as the load was suddenly put on the flywheel, a noise that could be felt and heard all over the Cosmotron building and which became a familiar background sound for every Cosmotron user and department member as an indication the machine was working.

At the end of 1951, Hildred Blewett proposed an outline for a BNL technical report on "Cosmotron Design and Construction," with the chapters to be farmed out to different groups of individuals. The idea grew, and

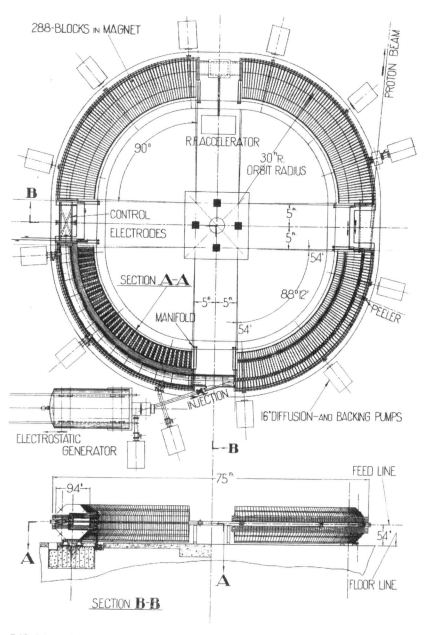

7.13 Schematic layout of Cosmotron.

eventually appeared in a Cosmotron issue of *The Review of Scientific Instruments* (volume 24, number 9), edited by Hildred Blewett and with an introduction by Haworth, containing twenty-eight articles on the Cosmotron and its various subsystems.[54]

In March 1952, the injector was connected to the vacuum chamber and the experimenters began to look for beam. Recalls Cosmotron staff member Ed Dexter:

> After we first injected the beam, it should have made a number of traversals but didn't; it made something like a third the number of revolutions and then just disappeared. It was very discouraging. Days were spent with people looking and looking trying to find out what was wrong. It was the first time protons had been isolated for that long, and someone jokingly said that maybe protons didn't really live that long. Other people thought maybe Berkeley had been right and that the aperture was too small. It turned out to be a faulty sensor in the accelerating system, and as soon as we fixed it the problem stopped.[55]

"BeV Day" was on 20 May 1952. That morning, the Cosmotron's physicists produced beams of 200 and then 400 MeV. The crew broke for lunch. Afterward, they came back and turned it on again, and straightaway the machine went past the 1 BeV mark, the first accelerator to do so in history. "Well, I'll be damned," the laconic John Blewett remarked. "A billion volts" (fig. 7.14).[56]

The occasion gave rise to much rejoicing. For a few days, the Cosmotron hosted a number of journalists who took pictures and wrote awed stories; "At Last, Man-Made Cosmic Rays!" was one headline. The Cosmotron was also visited by astounded members of the Berkeley crew, who returned to reduce the Bevatron's aperture to 1' × 4', comparable to the Cosmotron's. A month later, the Cosmotron reached 2.3 BeV, the predicted limit until coil windings on the magnet pole face could be turned on.

Shortly thereafter, BNL's accelerator builders—along with Livingston, often a summer visitor at Brookhaven—learned that they would soon be visited by a delegation of accelerator scientists from a new laboratory in Europe, then named the European Council for Nuclear Research and now the European Laboratory for Particle Physics (known then and now as CERN, after its original name).[57] CERN, which did not yet even have a site, was jointly operated by a consortium of European nations in a way explicitly modeled on Brookhaven and its operation by a consortium of U.S. universities; CERN was a "super-Brookhaven," said one European scientist. In June 1952 delegates to CERN endorsed the idea of building a proton synchrotron; as German physicist Werner Heisenberg put it, CERN's proton synchrotron "could be constructed along the lines of the Brookhaven Cosmotron," but scaled up to 10 to 20 BeV (Hermann et al.

7.14 Hildred and John Blewett, 20 May 1952, BeV Day. Protons accelerated over one billion electron volts for the first time in history.

7.15 Accelerator department head George Collins with CERN visitors at the Cosmotron console, August 1952. Left to right: Collins, Odd Dahl, Rolf R. Wideroë, Frank Goward.

1987, 89). In July, a team of three CERN scientists led by Norwegian Odd Dahl began to draw up preliminary plans for the European accelerator. Dahl scheduled a trip to Brookhaven along with his deputy, British physicist Frank Goward (who had built the first working electron synchrotron), and Norwegian physicist Rolf R. Wideroë (designer, in 1928, of the first particle accelerator; it had been Wideroë's paper that had inspired Lawrence's idea of the cyclotron). The team sent ahead a copy of its preliminary souped-up Cosmotron design (fig. 7.15).

Faced with an impending visit by eminent guests, Livingston felt that the host scientists should do more than show and tell. He organized a study group to brainstorm for improvements on the Cosmotron's basic design for the Europeans to incorporate into their machine. For instance, one major drawback of the Cosmotron was a curious consequence of the fact that the gaps in its 288 C-shaped magnets all pointed outward, a university-style engineering decision that permitted easy access to the vacuum chamber in the gap, but had some undesirable side effects. The most serious of these was that a limit was set to the energy achievable in the Cosmotron by magnetic saturation of iron that, in the Cosmotron's magnets, resulted in an increase in the n-value (see above) to the point where the beam was no longer stable in the radial plane, although strongly focused in the vertical direction.

A modest improvement was made possible by addition of "poleface windings"—current-bearing copper bars imbedded in flat sheets on the faces of the magnet poles. Thus the energy was pushed to 3 BeV, but then iron saturation took hold and the n-value coasted to values greater than 1.

Livingston made a novel proposal to defeat this phenomenon—locate some of the magnetic sectors with the Cs facing inward. When these sectors saturated their effective n-value would go to high negative figures. But with the right number of reversed magnets the *average* n-value could be kept between 0 and 1. Reversing some of the magnets would also address another problem: When a bunch of protons struck a target inside the machine, it created a shower of new or "secondary" particles; but because the magnets bent the paths of *all* positively charged particles inward toward the magnets, physicists had a much easier time observing negatively charged particles, which the magnets bent outward, away from the machine. Reversing some of the magnets might make positive secondaries also accessible.

Livingston asked Courant to investigate whether the local high values of n could safely be averaged out. Courant, an expert on the theory of orbit stability in accelerators, was initially skeptical, arguing that turning some of the magnets inward would disrupt their focusing, or ability to keep the particles in orbit. It would involve repeatedly altering the shape, or gradient, of the magnetic field, or its change in strength from the center outward.[58]

But when Courant began to calculate the effects exactly, he was astonished to find that switching the magnets—alternating the gradient—focused the particles even better than before. Magnets that focus particles strongly in one direction (say vertically, with a large gradient index n) also defocus them in the other direction (horizontally, requiring a large positive value of $1 - n$ for focusing). Until that time, accelerator builders had compromised and used magnets that focused weakly in both directions. What Courant discovered was that the net effect of arranging a series of magnets, each of which strongly focused in one direction and strongly defocused in the other so that the magnets alternated direction, was strongly focusing in *both* directions.

Hartland Snyder then recognized and developed an extended analogy between the way magnets focus beams of particles and the way lenses focus light beams. The analogy proved extremely powerful, and allowed the wholesale transfer of an entire network of equations and a systematic way of approaching and solving problems from a well-articulated field into the new context.

In a few exciting days, the Brookhaven accelerator physicists realized that their attempt to alternate the direction of the magnets had led them to a fundamentally new method for focusing particles more strongly that would make possible n values not only greater than one but perhaps even into the hundreds. The discovery would also allow them to build much narrower beams of particles, which was important because the cost of accelerators depended heavily on the cost of the magnets, which in turn was a function of aperture size. The Cosmotron's aperture was $8'' \times 12''$, the Bevatron's aperture $12'' \times 48''$. Calculations now suggested that it was possible in principle to reach over 30 BeV with an aperture of one or two inches, while saving millions of dollars in cost. The possibilities seemed almost limitless, and in the enthusiasm of the moment Livingston even began talking of 100 BeV machines with two-inch apertures that could be draped over the natural terrain rather than having to be precisely engineered on a level surface.

The CERN scientists arrived on 4 August for a week at Brookhaven. They expected the usual tour and set of informal meetings, during which they might receive a few helpful suggestions to take home. Instead, they were startled to discover Brookhaven's accelerator physicists abuzz with excitement. On learning the news, they scrapped their idea for a 10 BeV Cosmotron scale-up and began to work on plans for a 25 to 30 BeV strongly focusing proton synchrotron (soon to be called the PS). (Courant's notebooks of this time are now at the Smithsonian Institution.)

Soon after the European visitors departed, Courant, Livingston, and Snyder mailed a paper about their discovery to the *Physical Review*. They mentioned an anticipation of the principle they had discovered in a 1938 paper by Columbia physicist L. H. Thomas in what was known as a "Butterfly cyclotron." The authors mentioned certain applications of the strong focusing principle that arose "by analogy to lens optics." One of the most significant of these was to a kind of four-pole magnet, called a quadrupole, that proved indispensable in subsequent particle accelerators. A few days after Courant, Livingston, and Snyder sent in their article, Blewett mailed in one by himself, entitled "Radial Focusing in the Linear Accelerator" (fig. 7.16). In it, Blewett described how quadrupoles could remove a major stumbling block obstructing the development of linear accelerators. The two papers appeared back-to-back in the 1 December issue of the *Physical Review* (Courant, Livingston, and Snyder 1952; Blewett 1952).[59]

But the Brookhaven scientists were flabbergasted when Berkeley scientists, and the AEC, asked them to keep details of the discovery a secret and not to distribute preprints of their articles about it. They strongly urged Brookhaven to withhold information at least until a meeting, in September, of an AEC committee called the "Senior Responsible Reviewers," which would pass on the need to classify the work. It turned out that Lawrence and the Berkeley staff were involved in a project, ultimately abandoned, to

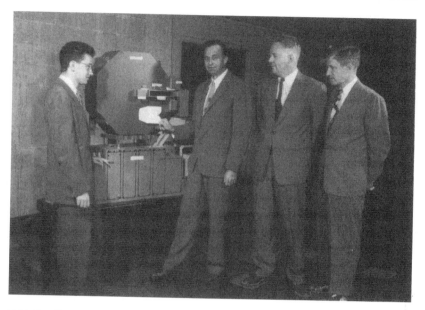

7.16 Four Brookhaven scientists who worked on strong focusing. From left to right: Ernest Courant, M. Stanley Livingston, Hartland Snyder, John Blewett. Livingston holds a model strongly focusing magnet next to a model Cosmotron magnet for scale comparison.

build an accelerator to create weapons-grade nuclear material for the military, which was so top secret that participants had kept it even from Brookhaven colleagues. The project involved using a cyclotron with the Butterfly type of alternating field configuration, and since there was a close relation between that concept and the new one, the Berkeley team and the AEC thought it, too, ought to be classified. Though as far as Brookhaven's accelerator physicists were concerned "the cat was out of the bag," as Courant put it, Haworth had to argue with the AEC and Berkeley to keep the work unclassified, and ultimately prevailed.[60]

In September 1952, Haworth told the board of trustees that, according to the gentlemen's agreement between the AEC, Brookhaven, and Berkeley, the next U.S. accelerator after the Bevatron, which could be built on this new principle, would be at Brookhaven. He added that he intended to submit to the AEC, early in 1953, a proposal for a $25 million, 75 to 125 BeV machine. The discovery, Haworth told the committee, even raised the possibility that accelerators would be small and cheap enough that accelerator development could be returned to universities and not be the exclusive prerogative of national labs.[61]

But Haworth knew he needed to act quickly if Brookhaven were to keep its hand in the accelerator game. Strong focusing was such a radical discovery that other labs were jumping on it. CERN's PS was a strong-focusing machine, while accelerator physicists at Berkeley were building strong focusing lenses, and accelerator physicists at Cornell (Robert Wilson), MIT (Livingston), and Princeton were working on proposals.[62]

Therefore, only a few weeks after discovery of the strong focusing concept, before the Cosmotron was even operating, Haworth began pressuring lab accelerator physicists to design a strong focusing machine. This also helped keep them active at just the time their work on the Cosmotron was winding down, heading off tension between them and the experimenters anxious to move in and take over. In September, he established a working group of Blewett, Courant, Green, and Jacobus, with himself as head, to meet once a week to plan a "Multi-Billion Volt Synchrotron." In October 1952, he figured a way to reassign thirty-five hundred dollars earmarked for the Cosmotron to a new "Accelerator Development" account to support the project.

In November, the group received a telegram from Alvarez at Berkeley reminding them that they were not alone:

> . . . HAVE JUST APPLIED A 4 SECTION, ELECTROSTATIC STRONG FOCUSING SYSTEM TO A 4 MEV PROTON BEAM . . . ORIGINAL BEAM IS PARALLEL AND HAS DIAMETER OF ALMOST 2 CM. WHEN CALCULATED $\pm$ 22 K VOLTS IS APPLIED TO SYSTEM, BEAM DIAMETER DROPS TO LESS THAN 1/2 MM.[63]

As soon as he had read Blewett's paper, Alvarez had built some electrostatic quadrupole magnets and put them in his proton linear accelerator, the first application of strong focusing in an accelerator, less than four months after its discovery.[64]

The Cosmotron Dedication

Meanwhile, the Cosmotron itself had been overhauled, and by the end of the year was running at its design energy of 3 BeV. It had cost a total of $8.7 million.

Completion of the Cosmotron created a small quandary for the AEC and AUI. The staffs of both agencies felt that, given its size and significance, the Cosmotron's dedication should be marked with a ceremony. But was it really appropriate to inaugurate a scientific instrument with speeches, politicians, balloon waving, food, and other fanfare?

Of course, thought AUI officials. The Cosmotron's dedication ought to be treated as analogous to a ship christening or keel laying—a public celebration of the completion of a device paid for by taxpayers and devoted to social good. Moreover, it would provide an excuse to expose Brookhaven to the approving gaze of newspaper reporters and important statesmen, including, with luck, the president of the United States. The admiration of politicians and journalists could help publicize Brookhaven and promote continued funding for basic research.

Absolutely not, ruled AEC chairman Gordon E. Dean. The Cosmotron's dedication should be a purely "scientific and academic" affair. Explaining Dean's logic, a fellow AEC commissioner wrote AUI trustees that the Cosmotron "could not be well understood by a sufficiently large percentage of the population to permit it to be used as a basis for a presidential pronouncement. It differs in this respect from the keel of a ship." [65]

Reluctantly, AUI officials acquiesced. They restricted the guest list to scientists, a few politicians, including members of the Atomic Energy Commission, and a small number of journalists. These sent simple printed invitations:

The United States Atomic Energy Commission
and
Associated Universities, Inc.
invite you to visit Brookhaven National Laboratory
on the occasion of the dedication
of the Cosmotron
December 15, 1952

Niels Bohr and Werner Heisenberg—prominent physicists who had made their reputations in the years before World War II and the biggest scientific celebrities on the guest list—sent regrets. But many noted American scientists and politicians came, including Enrico Fermi, the Nobel laureate in physics from the University of Chicago, and Lewis Strauss, soon to replace Dean as AEC chairman. The only presidential timber at the gathering was Harold Stassen, president of the University of Pennsylvania (one of the AUI nine) and ex-governor of Minnesota, who four years before had campaigned unsuccessfully for the Republican nomination for president. The real celebrants, however, were the dozens of young scientists who had worked on the Cosmotron or were preparing experiments to be performed on it. For four years they had worked long days, occasionally sleeping in cots beside lab benches. They knew that, with the Cosmotron, Brookhaven would jump into the ranks of the world's forefront scientific institutions. For them this prospect was worth the labor and low salaries. A dozen or so of the ambitious researchers at the dedication ceremony would become Nobel laureates, others would become heads of laboratories, and still others key scientific or administrative players in the U.S. science program.

Their enthusiasm was undampened by the cold, drab dawn of 15 December 1952. The temperature stood just above freezing and a chilly wind blew through the pitch pine and scrub oak surrounding the lab. Around 10 A.M., chartered buses bearing Washington dignitaries pulled to a halt in front of an old wooden army theater that had been pressed into service for the introductory speeches by lab director Haworth, AUI president Lloyd Berkner, and AEC area manager Emery L. Van Horn. Lunch followed, and then aides escorted the dignitaries over to the Cosmotron for the actual ceremony.

The event was held behind the machine in the test shack, the thrown-together corrugated steel building whose portals were still littered with construction debris. Stepping gingerly through plaster dust and lath, the 200-odd guests took their seats in rows of temporary chairs placed on the concrete floor. One waggish Cosmotron department member had painted "THIS SIDE UP" on top of the 2,200-ton machine. Not everyone was in on the levity. A journalist sidled up to theorist Robert Serber and asked, "When does it move?"

The ceremony, which began shortly after 2:00, included a speech by AEC commissioner Henry D. Smyth. The Cosmotron, Smyth declared, was built "for the enlightenment and benefit of mankind." The phrase was happily caught and repeated by journalists, and a few months later was engraved on a steel plate mounted over the accelerator—though the first time around, Smyth's name was misspelled.[66]

The Cosmotron was set up to produce a burst of neutrons, triggering a detector that would give off some noise to prove that particles had indeed been created. But the still-new machine had not been working well, and the Cosmotron's proud parents had rigged the machine so that it could *seem* as if it were functioning properly even if, like some obstreperous infant brought out for the first time in public, it decided to misbehave. This precaution proved unnecessary, and the newborn performed impeccably in front of its parents and admirers. Afterward they walked back from the Cosmotron building, past row upon row of wooden barracks, to the research staff lounge, a converted army headquarters, for cocktails at 6:30, and then were marched next door for dinner in what just a few years before had been the army camp gymnasium and was now the lab gymnasium, complete with basketball hoops and wooden floor. Courant found himself seated at the same table as Edward Reynolds, AUI's first president. After introductions all around, Reynolds peered at the slight, bespectacled, and very young-looking Courant and avowed as to what remarkable work his father had done with Livingston and Snyder in the discovery of strong focusing.[67]

Mariette Kuper supervised the dinner arrangements, and her efforts succeeded, where others had failed, in wrecking Dean's intention to keep the dedication "scientific and academic." She had set up several dozen tables, covered them with paper tablecloths, and in the center of each placed pitchers of her signature martinis. The pitchers had been chilled beforehand in the freezer, and on the tables they glistened temptingly with frost and tiny rivulets of dew. They looked for all the world like pitchers of water, which is how they went down. As the evening wore on Kuper's martinis slowly took effect, disarming the reserve of the young risk takers who were savoring their achievement. Behavior loosened, voices grew loud, and things got boisterous.

When the final speaker, Detlev W. Bronk, president of Johns Hopkins University, took the podium, his talk, "Changing Patterns in the Furtherance of Science," was delivered to no one in particular. Luis Alvarez, a forty-one-year-old Berkeley physicist sixteen years away from his Nobel Prize, set his tablecloth on fire. A representative from Wellesley, finding her mixed drink too strong for her taste, attempted to dilute it by pouring in water from the pitcher on the table, inadvertently lacing it with martinis. She shouted encouragement to Bronk before passing out on the table.

Bronk, for his part, was lucky that his audience was not paying attention. He had mixed up the text of his speech with one he was scheduled to give in Canada a few days later, puzzling those still compos mentis with references to "your king." [68]

But nobody thought the revelry excessive. The young risk takers had indeed performed well. Brookhaven's team, which had initially thought the odds against success so great that they had considered abandoning their efforts, had succeeded in building the most powerful tool of particle physics in the world. They had gambled in several ways—in the magnet design, the rf system, and vacuum chamber materials—and won. They had outguessed the eminent and experienced Berkeley engineers on the key question of aperture size, and finished the Cosmotron two years ahead of Berkeley's Bevatron (though Brookhaven scientists benefited from a two-year delay in the Bevatron due to diversion of personnel and resources to the top secret project). And the Cosmotron was the only major completed project in the course of which Milbank, Tweed had not been consulted.

But the event was a milestone, too, for the lab. In six years it had undergone a huge transformation. It now had two major instruments and attracted a renowned staff and set of visitors. With the Cosmotron's completion, all of the instruments planned by the founders were in operation (except for the canceled 240-inch), and a project to build another forefront accelerator was underway.

The Brookhaven experiment, in short, was working. Its staff had shown that it was indeed possible to have government fund a facility in which good, university-type basic research could be carried out. In just a few years, starting with nothing, Brookhaven's scientists had managed to do work that was the equal of the best of the other labs in the country, as evidenced not only by the existence of the Cosmotron, but of the development of a principle soon to become the basis of all large accelerators.

But the character of this experimental community was about to change. Previously, it was unified by the centripetal focus of completing the major projects. Now it would experience a centrifugal force as individuals with different training and interests began to take advantage of the new freedoms and capabilities opened up by the new machines to pursue their own agendas. Such a transition is similar, in general outline, to what happens in, say, a political community following a successful struggle to create a new institution, but a research community poses its own special demands. And the specific kinds of forces acting on the community differed depending on the character of the instrument that served as their focus. The following two chapters will look at the communities working around the Brookhaven Graphite Research Reactor (chapter 8) and the Cosmotron (chapter 9).

Reactor Research
in the 1950s

FROM THE outside, the Brookhaven Graphite Research Reactor (BGRR) building dominated the Brookhaven landscape. The light brown, boxy brick structure, with its red-and-white striped smokestack, occupied the highest spot on the site, and its image was used in an early lab logo (fig. 8.1). This was appropriate enough, because the need for a research reactor had impelled the creation of the lab in the first place. From the inside, the cube-shaped reactor towered over thirty feet tall in the middle of a hangarlike room, amazing first-time visitors. "When I first saw it," recalled Vance Sailor, head of one research group, "I had just come from a university cyclotron project, and I remember how stunned I was when I first walked into the building. The size was astonishing—a facility that was to be used for research just couldn't be so big!" [1]

The faces of the reactor resembled great pegboards, with holes lined up in regularly spaced rows and columns—for experiments on the east and west faces (experimenters without security clearance could use the west face), and fuel elements on the north and south faces (fig. 8.2). A large overhead crane ferried equipment and shielding around the room, while an elevator rode up and down the south face to install fuel rods. Atop and beneath the reactor were large openings for experiments that required especially wide beams; including these facilities, the reactor had about sixty-seven experimental stations. At one face, one set of pneumatic tubes blew samples in and out of the reactor, while another set whisked the irradiated samples to counting rooms and lavishly equipped laboratories in an adjacent wing of the building. When the reactor was in operation, the room hummed and buzzed as motors drove vacuum pumps, experimental equipment, and the flywheels for the control rod drives, while workers shouted out conversation. The reactor itself emitted a pronounced, steady hiss as air coolant was sucked into cracks between the top shielding blocks and into beam holes, and sped through the fuel channels at up to four hundred mph. "All the noise reverberated through the building, pulsing off the walls and roof," recalls Sailor. "It was full of life and activity, with a continuous atmosphere of excitement." [2]

A research reactor exists for one purpose only: to create vast quantities of neutrons. Yet just this makes it one of the most versatile of scientific instruments. Nuclear physicists, starting to develop quantitative models of the nucleus, needed neutrons to obtain data on nuclear behavior. Solid-state physicists wanted neutrons for discovering the locations and motions of atoms in materials. Chemists could use neutrons to examine many properties of matter of interest to them. Life scientists used neutrons to study the effects of radiation on organic tissues and to create radioisotopes for research and treatment. Reactor engineers needed information on various fuel types, moderators, coolants, and control systems, as did naval and aircraft engineers designing nuclear-powered submarines, ships, aircraft, and, later, spacecraft.

The atmosphere at the BGRR was freer and looser than at the other nonweapons reactors at Argonne and Oak Ridge, where researchers had to have clearance and security regulations often severely hampered university relations (Holl 1997, 81). Haworth managed to secure agreement from the AEC to open a face for uncleared experimenters provided that they had no access to, nor view of, the rest of the reactor. A thirty-foot-high steel partition, inevitably called the iron curtain, was built to isolate the west face, which became declassified in May 1951. The atmosphere at the BGRR was too loose for the AEC's taste; scientists, even department chairmen, often invited colleagues from Soviet-bloc countries, ignorant of or deliberately flouting the AEC's strict rules on such visits. Haworth periodically found himself in the awkward position of having to hastily inform the AEC that a visit was impending or had already taken place, and sometimes had to issue stern memos to department chairmen reminding them of the rules.[3]

Completion of the BGRR brought about significant transformations in the community working at the reactor. The most obvious and immediate was a division of labor that sprang up between operators and users. The former belonged to the reactor department, headed by Marvin Fox. The latter, scattered throughout several departments, consisted of clusters of experimenters who worked in close proximity on widely varied research programs using different strategies and types of instruments.

Brief recapitulations of the careers of several such groups will illustrate the numerous and varied aspects of the BGRR's user community. Donald Hughes, head of the neutron physics group and dean of the BGRR's experimenters, was an exemplary group leader whose skills included the ability to manipulate the AEC's interests when necessary to advance his own ends. The Goldhabers were a remarkable husband-and-wife team whose work consisted of systematizing nuclear states in research that could be carried out only at a reactor. Vance Sailor's crystal spectrometer group illustrated the rewards, and dangers, of taking a huge risk by jumping to a new technology. George Dienes arrived at Brookhaven to conduct research that was essential to the BGRR's operation, but wound up developing an important basic research program. Herbert Kouts did the reverse; his basic research interests led him to a program of enormous applied importance. Finally, the early story of boron neutron capture therapy, an attempted cancer treatment in which patients were lowered into a pit in the BGRR's shield, is a particularly dramatic case of how hopes for quick and dramatic applications of atomic energy in the life sciences often turned into disappointments.

Donald Hughes: Neutron Physics and Cross-Sections

Any portrait of life at the BGRR during the 1950s must begin with Donald Hughes, dean of Brookhaven's neutron physicists, the single most influential experimenter at the reactor and leader of its largest group (fig. 8.3). He literally wrote *the* book on the subject, *Pile Neutron Research,* a textbook used worldwide in the 1950s (Hughes 1953b).

Hughes was born in 1915 in Chicago of typical American hybrid ancestry, his mother of Swedish and father of Scotch-Irish origin. He was an

8.1 Early lab logo (late 1950s) with image of reactor and stack.

8.2 West face of BGRR, which could be used by uncleared experimenters.

8.3 Donald Hughes at neutron reflector on east face of pile.

intense, active youth interested in music and art, and while a physics major at the University of Chicago was on the wrestling team (Palevsky 1962, 3). He stayed on at Chicago, receiving his Ph.D. (for cosmic ray cloud chamber work) in 1940, joined the faculty there and during World War II was an assistant director of the Manhattan Project's Chicago activity. But administrative work made him miserable, and he persuaded Enrico Fermi to let him do research in the evenings at CP-1, the world's first reactor, an innocuous pile of graphite blocks stacked in a squash court. At the war's end, Hughes was director of the nuclear physics division at what was to become Argonne National Laboratory.

Hughes was unhappy, in 1948, to hear of the AEC's decision to make Argonne a center for reactor development. He sensed (correctly) that Argonne's program in basic nuclear physics would suffer, and clashed with lab director Wally Zinn about his role there. Not one to accept marginalization, Hughes looked around for a new worksite. Hughes's decision, early in 1949, to relocate to Brookhaven was "big news," according to one scientist, heralding Brookhaven's arrival as a major basic research center (Brockhouse 1986, 37).

Hughes brought along Harry Palevsky, who would be a loyal collaborator for the rest of Hughes's life, and a few others to form a neutron physics group at the soon to be completed BGRR. As a group leader, Hughes was politically shrewd, ferociously hard-working, and demanding. He thought out his ideas carefully, and in executing them had little tolerance for variation and input from others. Those who valued their independence did not seek to work with Hughes. But those willing to sacrifice self-determination were richly rewarded by work on a wide range of fundamental and engrossing projects on the forefront of physics, and were well supported and protected by Hughes in turn. Palevsky, for instance, had not finished his Ph.D. work due to a falling-out with his advisor, but this made no difference to his career as long as Hughes was alive. Palevsky once recalled:

> His casual manner was as far as could be from the truth. Hughes drove himself unmercifully. . . . [O]ne of his greatest talents was picking the right problems to pursue. . . . Hughes never hesitated to tackle any branch of physics if he felt some fundamental knowledge would result. . . . He was very sure that his own research was among the most important in physics, and so he was primarily interested in his own problems. (Palevsky 1962, 7)

Hughes and his group participated in a number of fundamental nuclear physics topics that had begun to be explored at Argonne and Oak Ridge during or right after the war. One was neutron optics; the techniques of reflecting, refracting, and deflecting neutrons in an analogous manner to photons. He would write *the* book on that subject, too: *Neutron Optics*

(Hughes 1954a). Another specialty was neutron cross-sections, or the likelihood that an interaction will take place between a neutron and a nucleus at a particular energy. Indeed, the neutron group's largest single project involved measurement and compilation of neutron cross-sections.

A cross-section, symbolized by the Greek letter sigma (σ) and measured in units called barns, is how "big" the nucleus appears to the neutron: the larger the cross-section, the greater the effective size of the nucleus and the likelier it is that the nucleus will absorb a neutron of that energy. The process is analogous to what happens when an atom absorbs a photon. Just as absorbing a photon causes an atom's electrons to rise to a higher, "excited" state, absorbing a neutron raises a nucleus to an excited state. Just as atoms have only a limited number of possible excited states at low energies, so do nuclei. By changing the energy of incoming neutrons, and noting at what energies neutrons are absorbed, experimenters can map the energies at which nuclei have excited states—resonances, they are called.

To separate neutrons of different energies as they emerge from a reactor, experimenters use devices called choppers, which were first developed at Oak Ridge. A chopper consists of a drum of neutron-absorbing material, traversed by slots, that spins at high velocities, cutting the beam into packets. The target material is placed at some distance from the chopper, and as the neutron packet travels that distance it spreads out as a pack of runners do, with the faster ones in front and the slower ones behind. The time it takes each neutron to cover the distance to the target thus yields a method, called time of flight, for determining the energy of a neutron and thus the cross-section of the target nucleus at that energy. In the 1950s, much experimental nuclear research consisted of determining characteristics like the widths and spacings of resonances (how broad or sharply focused they were on a particular energy, and how far apart they were in energy), while theoretical nuclear research consisted of trying to discern clues to nuclear structure from this data.[4] "Mapping the new world [of atomic energy] at the time was conceived largely in terms of measuring 'cross sections'" (Holl 1997, 76).

Because nuclear physicists, reactor physicists, and weapons builders all needed information on cross-sections, such work involved, for a time, a relatively happy marriage of basic and applied interests in AEC-sponsored work at reactor labs like Brookhaven. Lack of adequate cross-section data had delayed an experimental breeder reactor in Idaho, and the Korean War sharply intensified the AEC's desire for work relating to military programs. Viewing the matter as urgent, AEC officials met at Brookhaven in July 1951 and promised to bolster cross-section programs at all three national labs (ANL, BNL, and ORNL) but to centralize work at Brookhaven, "the only one of the national laboratories whose main function is research."[5] (But Argonne, to which the AEC sent more applied tasks, bore a heavier burden in terms of demands on its overall priorities; Holl 1997, 95–96.) Haworth exploited the AEC's escalating needs to boost support for instruments used to measure cross-sections. He got the lab a new eighteen-inch cyclotron, which, while acquired for producing high-energy neutrons in cross-section research, actually saw more use as a nuclear physics instrument. The lab also received increased support for choppers and neutron detectors. Hughes, who had been head of the AEC reactor development division's neutron cross-sections committee when it was established in 1948, became director of an informal Brookhaven-based program called the Sigma Center, a subgroup of the neutron physics group, which was assigned to collect and collate cross-section data.[6]

But the marriage of basic and applied interests in cross-section research had its rocky moments. C. R. Russell, of the AEC's division of reactor development, was a prophet of doom, prone to announce dramatically at meetings that his division was "in a state of war," that it needed results on a crash basis, and that all basic research interests had to wait. Hughes patiently pressed the interests of basic researchers, arguing that, for the sake of morale and efficiency, people should not be removed from work for which they went to the labs.[7]

In August 1951, Hughes (aided by Irving Kaplan, Sailor, and others), circulated a preliminary table of the "best values" of a number of neutron cross-sections drawn from several sources. The table, which was fifteen pages long and listed available cross-sections for just over three hundred isotopes, had the effect of spotlighting a number of disagreements and suspicious measurements. There followed an AEC compilation entitled *Neutron Cross Sections,* or AECU-2040, and a Brookhaven publication, BNL 170 and 170A, and later BNL 325, the famous Brookhaven "Barn Book," on the cover of which was a drawing of a farmer's barn.[8] Sailor recalled:

> The art of compiling turned out to be a sensitive process that took a few years to fine-tune so as not to cause adverse reactions. Sometimes Don would compile before the experimenter was able to publish, and sometimes this would cause resentment by the experimenter—particularly if his manuscript got rejected by a journal on the basis that the data had already been published in BNL 325. Don did not have the time for fine-tuning, i.e., to seek permission to include pre-publication data in his "Barn Book." Later we got better at avoiding hostile encounters.[9]

Many cross-sections were classified, including all isotopes of uranium and plutonium, and any isotopes involved in reactor or bomb design. The Sigma Center issued two handbooks, one classified (edited by Kaplan) and one unclassified (edited by Hughes). Since the classified group could not share its work in open scientific meetings, the AEC organized a parallel series of classified meetings.

One of the neutron group's first projects was to build a fast chopper for its own cross-section studies (a slower one was for other purposes), which came into operation in 1953 (fig. 8.4).[10] One of its first important cross-section measurements was of ^{235}U and ^{238}U, of obvious interest to the military. But theorists also found the ^{238}U cross-section fascinating. It had been measured with crude instruments during the Manhattan Project, and thought to have two large resonances at 6.7 and 50 eV. But the neutron group found that the 6.7 eV resonance was much sharper than had been thought, and that the 50 eV resonance was actually composed of many smaller resonances. The result could not be described with the Breit-Wigner formula, a famous description, dating from 1936, that described the shape of a cross-section in the neighborhood of a resonance, or how the cross-section rises steeply as the energy of the incoming neutron approaches the resonance energy, peaks, and then equally sharply falls off above the resonance energy. An active interplay took place as many theorists, including Breit and Wigner themselves, visited Brookhaven to look over the shoulders of the experimenters.

Charles Porter was one theorist who closely followed the work of the neutron group, and whose work was studied by members of the neutron group in turn. For his doctoral thesis at MIT, Porter had contributed to a model in which the nucleus was treated as neither clear like a refracting glass ball nor opaque and all-absorbing, but as *slightly* opaque, or "cloudy." The model was called the Cloudy Crystal Ball model after a famous humor department in *The New Yorker*. In a major step forward in nuclear physics, Porter worked out the model in detail with MIT physicists Hermann Feshbach and Victor Weisskopf to explain a large set of medium-energy neutron cross-sections measured at Wisconsin. When the model appeared in 1953, Hughes's group further explored its application. That year, Porter received his Ph.D., came to Brookhaven, and became one of the lab's most important nuclear theorists, and one who worked closely with experimenters. Another of his contributions, with Robert Thomas of Los Alamos, was a well-known formula called the Porter-Thomas distribution. The Porter-Thomas distribution describes a nuclear process in which excited nuclei decay via certain reaction channels, by providing the probabilities of the decays into the width amplitudes that characterize the channels—perhaps an early example of what today would be called chaos theory.[11]

President Eisenhower's Atoms for Peace initiative, which began with a speech in 1953, by 1955 allowed Brookhaven's staff to publish work that

8.4 Fast neutron chopper at the BGRR, 1953.

had previously been restricted to secret reports; it also allowed BGRR research groups to accept foreign guests (though not guests from behind the iron curtain). The BGRR's own iron curtain "fell" (to save money it was not dismantled, but its doors were opened and have been left propped open ever since) and the security booth removed. A "United Nations Conference on the Peaceful Uses of Atomic Energy" was organized in Geneva as part of Atoms for Peace, at which nuclear scientists from around the world discussed their work publicly for the first time. Brookhaven literally left its stamp on the event when the lab's graphic artist George Cox won a competition to design the conference logo which was used, among other places, on U.S. postage stamps, a copy of which hangs in the stairwell of the old reactor building (fig. 8.5). (A decade later, Cox would attract a different kind of attention when he was, briefly, artist Yoko Ono's father-in-law, before she left Cox's son Anthony for Beatle John Lennon. In the resulting struggle for custody over her child with Anthony, Yoko Ono for a time had detectives shadow George Cox.) Before the conference, the Sigma Center asked the Soviets to send them their cross-section data, hoping to compare measurements. The Soviets surprised the Brookhaven scientists on two counts: first, by complying, and second, by revealing that they had copied their Western counterparts to the extent that they also measured

8.5 Postage stamp bearing George Cox's logo. The logo won an award at an international Atoms for Peace conference, Geneva, 1955.

their cross-sections in units they were calling *barns,* though spelling it in Cyrillic letters and pronouncing it with the distinctive, rolling heavy Russian "r." The Soviet information was incorporated into the "Barn Book," which became an international standard and went through many editions.[12]

The fast chopper, with which Hughes's group made most of its cross-section measurements, was only one of several large instruments built by the neutron physics group. Another was a slow chopper, for use with neutrons from 0.001 to 0.2 electron volts for research of a different sort. It was the chief BGRR instrument for studying inelastic scattering, the collective motions of atoms in solids.[13] Members of the neutron physics group used inelastic scattering techniques at the slow chopper in a number of different types of experiments.[14] This work, too, attracted the attention of theorists. In 1953, Belgian theorist Léon Van Hove, who was spending time at the Princeton Institute for Advanced Study, spent a summer at Brookhaven. Working on ways to describe inelastic scattering, he paid close attention to the work of the neutron physics group and proceeded to compose a classic treatment of neutron scattering in two papers, one on magnetic scattering, the other on the relation of neutron scattering to spatial and temporal correlations.[15] Van Hove's work was significant, for it showed how theorists using statistical mechanics could indeed explain a complex object treated by experimenters. It was also a classic example of the interaction between theory and experiment for which Brookhaven was well suited (Van Hove 1954). Oak Ridge, however, remained the principal place where such scattering work was conducted, and where the first definitive study of critical magnetic scattering, by Shull and collaborators, would confirm Van Hove's basic treatment.

Hughes's neutron physics group hotly competed with groups at other reactor labs, and Hughes was tireless in promoting the abilities of his own group and ingenious at stealing limelight for it. In 1952, he had Walter Kato, one of his early graduate students, prepare a drawing that showed how five choppers differed in their resolution of the cross-section of an isotope of silver. The drawing consisted of a series of curves ranging from the (then) ill-defined squiggle of Columbia's cyclotron to the finely detailed and obviously superior landscape of his own fast chopper (fig. 8.6). Someone not in the field would have been unable to guess that Columbia's resolution, on the rise, was soon to overtake that of the Brookhaven fast chopper. The drawing was calculated to show off Hughes's group at its very best, and he flashed it about in talks to the APS and published it in several places, including *Scientific American* (Hughes 1953a).

> It was pretty dramatic [Kato said], and other groups were annoyed. We were all in competition, and what Hughes did was to say in effect, "We're doing the best work, therefore you ought to support us the most." It was a very political drawing. But it was the sort of thing Hughes knew how to do well.[16]

Hughes was an asset to the neutron physics community. Not all of his methods were original, and not all of his work was definitive. But because scientific experimentation is a skilled, creative activity, it does not thrive in a vacuum and the existence of a community of performers fosters the activity of each. For those who derive inspiration from the presence of competition, it provides that. But it also creates the framework for a performance rivalry, with all the benefits that brings: The existence of several groups of practitioners at different institutions means that numerous ideas and techniques are invented and developed and expanded simultaneously. It means that a topic can be explored from many sides and in many ways. It provides an incentive to work with care, given that any work is exposed to examination and question by a community of competent critics.

But despite, or perhaps because of, his dominating personality, not to mention his political cunning, Hughes was a particular gift to Brookhaven. Some of his measurements represented definitive work from which Brookhaven's reputation benefited. He had the authority and prestige to withstand pressures from the AEC. Most important, at a time when the lab was young and much of its staff inexperienced, Hughes kept abreast of all the interesting topics that could be addressed with neutrons and put members of his group on them, ensuring that, in pile neutron research at least, Brookhaven was playing in whichever games it could.

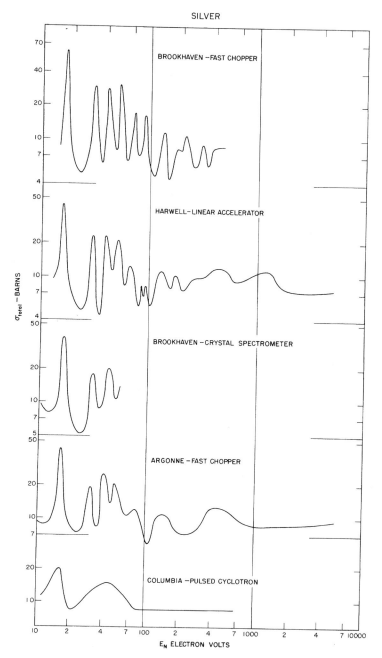

SILVER

BROOKHAVEN – FAST CHOPPER

HARWELL – LINEAR ACCELERATOR

BROOKHAVEN – CRYSTAL SPECTROMETER

ARGONNE – FAST CHOPPER

COLUMBIA – PULSED CYCLOTRON

σ_{total} – BARNS

E_N ELECTRON VOLTS

8.6 Walter Kato's diagram of a neutron cross-section of silver, used by Donald Hughes to promote Brookhaven fast chopper over competing instruments.

The Goldhabers: Nuclear Physics and Isomers

Working at several holes on the west face of the reactor, on the opposite side from Hughes's group, was the remarkable husband-and-wife team of Maurice Goldhaber and Gertrude ("Trude") Scharff-Goldhaber, leaders of the nuclear-structure group. While these two were far from the only foreign-born scientists at Brookhaven, their route to it had been more than usually roundabout.

Maurice was born in 1911 in Lemberg, a city destined to change political hands several times; it belonged successively to Austria, Poland, and the Soviet Union, and now, known as Lvov, lies in Ukraine. Maurice underwent a series of political dislocations himself. His father, in the travel business, took the family to Egypt in 1914, but two years later, the British authorities in Egypt declared Austrians enemy aliens. They interned Maurice's father and deported the other family members, who were taken in with his mother's parents to a hamlet now in the Czech republic. When his father was released at the end of World War I, the family moved to Germany. By 1930, when Maurice graduated from the *Real-Gymnasium* (equivalent to high school) he had decided on a career in science. At that time, *Real-Gymnasium* graduates could attend any university they pleased. Goldhaber chose the University of Berlin, where Max Planck, Albert Einstein, Max von Laue, Walther Nernst, Erwin Schrödinger, Otto Hahn, and Lise Meitner were often found in the front row at the weekly physics colloquium.

In January 1933, Hitler and the Nazi party came to power. Goldhaber, who was Jewish, soon decided to leave Germany. An ambitious twenty-two-year-old, Goldhaber wrote letters to the three most famous physicists outside Germany—Niels Bohr in Copenhagen, Wolfgang Pauli in Zürich, and Ernest Rutherford in Cambridge—asking to be accepted as a student. Rutherford was first to reply, and Goldhaber was on the boat by May.

After the Nazi destruction of German universities, the Cavendish Laboratory became, by default, the world's most important physics institution. Regulars at its colloquia included Rutherford, James Chadwick, P. A. M. Dirac, John Cockroft, Ernest Walton, Francis Aston, Peter Kapitza, Subrahmanyan Chandrasekhar, and Homi Bhabha, and sometimes J. J. Thomson and Arthur Eddington. Despite the eminent company, the lab was an informal, collegial place, a perfect atmosphere for a young physicist overflowing with ideas. "Ideas could be quickly translated into action," Goldhaber once recalled, "and progress was often measured in days or weeks." [17]

Goldhaber worked at first with Chadwick, who won the 1935 Nobel Prize for discovery of the neutron. In one effort, they discovered several hitherto-unknown types of nuclear reactions, including one with an extremely large cross-section (3,990 barns) in which ^{10}B, through neutron capture, forms ^{11}B, which then breaks apart into $^4_2$He (an alpha particle) and $^7_3$Li (a lithium nucleus); a generation of neutron detectors and an attempted cancer treatment would soon be based on this reaction. In 1936, Goldhaber submitted his first eight published writings on this and other work to receive his Ph.D. To earn money, he tutored; one of his students was the young Norman Ramsey.

While at the Cavendish, Goldhaber became reacquainted with a fellow refugee named Gertrude Scharff. Born in Mannheim, Germany, in 1911, Scharff had studied at several universities, including Freiburg, Zürich, Berlin, and Munich. In 1931, she and Goldhaber both attended a seminar run by Nobel laureate Max von Laue. The two dated several times and on one occasion went to one of prewar Berlin's notorious risqué cabarets. The following semester, continuing her tour of German universities, she left Berlin for Munich, where she began work on a Ph.D. under Walther Gerlach. Her reserved demeanor masked her stamina, which carried her through risky times. When Hitler marched into Munich in early 1933, she, too, understood the danger to Jews, but wanted to complete her Ph.D. before leaving. Gambling that concealing her Jewish identity would buy her enough time, she revealed her "dark secret" to Gerlach, who at first professed disbelief because she lacked a crooked nose, then promised to protect her until she finished her thesis. But fellow students discovered she was Jewish and thereafter refused to speak with her; the circle of excommunication spread, and her predicament grew more perilous. Early in 1935, as she neared the end of her thesis work, Gerlach found her a postdoctoral position at Imperial College in London, and she left the country as soon as she could pack her bags.[18] Scharff's parents also gambled, but lost; they were apparently rounded up and killed shortly after World War II began. In London, Scharff found Goldhaber, still a year away from completing his Ph.D., in nearby Cambridge, and the two became close friends.

In 1937, Ernest Rutherford, the head of the Cavendish, passed away. His replacement was not a nuclear physicist, and Maurice sensed change in the winds. Yearning to see the U.S., he visited there in spring 1938. At the April meeting of the American Physical Society, in Washington, D.C., he was recruited by Wheeler Loomis, head of the physics department at the University of Illinois at Urbana. In the midst of the Depression, Loomis was in the lucky position of having the money to build up his physics department, and besides Goldhaber his hires that academic year included the young Leland Haworth from Wisconsin and theorist Robert Serber from Berkeley.

Immigrating to the U.S. was not easy when a flood of refugees were fleeing Germany. Immigration officials were not forthcoming until they asked him if he had taught; teachers received preferential treatment. "I've tutored," Goldhaber said, remembering Ramsey, and got the visa. In May 1939, Goldhaber returned to London, married Gertrude, and the two returned to Urbana. Due to an antinepotism policy they could not both hold positions at the university, and Gertrude could not get an official job. But Maurice assigned some of his students to her, and when he was out of town she would give his lectures.[19]

After the war, the Goldhabers stayed on at Urbana, where they carried out a remarkable desktop experiment demonstrating the identity of beta rays and atomic electrons (Goldhaber and Scharff-Goldhaber 1948). Maurice worked as a consultant one day a week at Argonne. There he formed a group that included Ed der Mateosian from Argonne and his students Robert Hill (who came along from the Cavendish), Andrew Sunyar, John Mihelich, and Mike McKeown (all from Illinois). Goldhaber set them to work at Argonne's CP-3 reactor studying isomers, excited states of nuclei with measurable lifetimes. At that time, little was understood of the systematics of nuclear excited states. Goldhaber had become interested in isomers while at the Cavendish, where he had helped to show the role of nuclear spin, and set his group to exploring the subject at the CP-3 (built during the war as part of the Manhattan Project), whose plentiful source of neutrons meant a huge leap in the ability to create isomers, especially short-lived ones. Goldhaber and his young collaborators spent the first years after the war collecting data on isomers and studying their distribution.

> We studied seven-second activities [Der Mateosian recalls], isomers that lost half their strength every seven seconds. When we did them that short, we'd first clear the halls and one of us would stand by the reactor, while another, wearing tennis shoes, stood outside the laboratory door. The guy by the reactor would grab the container right after the sample was irradiated and throw it down the corridor to the other guy in tennis shoes, who would rush it to the detector, and we'd start counting.[20]

In 1947 Norman Ramsey, then at Brookhaven, realized that the Goldhabers' work was perfectly suited to the lab and offered positions to both of them. But when they visited in February 1948, both the reactor (whose flux would be an improvement over that of the CP-3) and the accelerator existed only in the form of two holes, a square one for the reactor, a round one for the accelerator. "I was a physicist, not a machine builder," Goldhaber said. "I remember thinking, 'What can you do with a square hole and a round hole?'"[21] Still, they returned to the lab in summer 1948, and discovered an isomer in tellurium in a sample shipped to them from Oak Ridge, showing that they could be productive even though the new facilities were not yet available. They returned the next summer and remained through the fall, Maurice having the semester off from Urbana. When they left Brookhaven early in 1950, the offer of a permanent job was still open.

> Then came one of those little incidents [Maurice recalled], which happen in life as well as physics, which finally push you clean over a threshold. Trude and I had talked about the Brookhaven offer during the drive back to Illinois, and still hadn't decided whether to take it. When we reached the Urbana campus, we parked in front of the physics department, as we often did, to put our books back in the office. It was a weekend afternoon, and we were inside all of five minutes. When we came out, there was a parking ticket on the car window. There had never been any restriction against parking there before. That decided us: "We're leaving!" [22]

When they arrived, Trude became the first female Ph.D. on Brookhaven's scientific staff. Goldhaber also brought along several members of his Argonne group, including Der Mateosian, Sunyar, Mihelich, and Hill. The group was given one lab in the barracks, two more in the lab wing next to the BGRR, and several beam holes on the west face of the reactor. The group moved in the day after completion of a fancy pneumatic tube system, which was intended to blow "rabbits"—containers with samples of irradiated materials—quickly to labs next door, eliminating the need for sneaker-clad students. These rabbits, about six inches long, were fashioned after department-store change rabbits, into which a salesperson would put a customer's money and bill to send pneumatically through a system of pipes to a main office, with the change returning in the same mode. Unfortunately, the pneumatic tubes that shot the rabbits into the lab invariably also blew a small amount of radioactive dust in as well. Because the group was doing very low-level counting, even that tiny amount of background was intolerable. Goldhaber abandoned the pneumatic system and went back to sneaker-clad assistants.

Like Hughes's group, the Goldhaber group ran several programs concurrently. They found new isomers and several hitherto unknown types of transitions (Goldhaber and Sunyar 1951). They developed new terminologies and classifications for isotopes (Goldhaber and Hill 1952). They also studied photodisintegrations, another interest of Goldhaber's since the Cavendish days. Photodisintegrations and isomers were, in effect, two aspects of the same nuclear physics subfield, electromagnetic transitions; the one involves photon absorption, the other photon emission. Goldhaber also contributed to an influential overview article on electromagnetic transitions (Goldhaber and Weneser 1955).

Gertrude Scharff-Goldhaber, meanwhile, set out mapping regularities in isomers based on the group's experiments. At the time, most nuclear physicists tended to study nuclei with an odd number of protons or neutrons, whose outermost particles behaved like orbital electrons when the nuclei are excited. Scharff-Goldhaber was especially interested in even-even nuclei (nuclei with even numbers of both protons and neutrons). Such nuclei, of which the alpha particle or helium nucleus is the simplest example, are much more tightly bound, so that when the nucleus is excited, the result is a collective motion.

Her most important work arose from her studies of low-energy gamma radiation in even-even nuclei. She found the first low-lying excited states in even-even nuclei; until then, it was generally assumed that even-even nuclei had only very high-energy excited states. She also noticed that the first excited states of the even-even nuclei she investigated were $2+$ states (angular momentum and positive parity), a rule with few exceptions (Scharff-Goldhaber 1953). She examined the first excited states of even-even nuclei throughout the periodic table, and made a model to exhibit the regularities of the excited states, a large, three-dimensional wall chart. She took a chart of the isotopes, and placed on top of each isotope a block of wood whose height represented the energy of its first $2+$ excited state, with each centimeter of altitude representing about a hundred kilovolts. The result dramatically illustrated the observed regularities. Some blocks were as high as 2½ feet, others a fraction of an inch. The tallest blocks tended to cluster around nuclei which were unusually stable, and had so-called magic numbers of protons or neutrons—2, 8, 20, 28, 50, 82, and 126—at which the outermost shell was closed, and in particular around doubly magic numbers such as at ^{16}O (8 and 8), ^{40}Ca (20 and 20), and ^{48}Ca (20 and 28).

The patterns found in the vast amount of information collected by the Goldhaber group, and groups at other labs, helped Aage Bohr and Ben Mottelson create a sophisticated nuclear model. The Bohr-Mottelson model, as it was called, substantially clarified the field of nuclear physics and provided at last a theoretical description of collective states and a set of parameters for defining and predicting them—a vast and, at the time, surprising consolidation of an immense amount of information. Much work in nuclear physics was then spent testing the Bohr-Mottelson model. At one of the first large nuclear physics conferences, which took place in Chalk River, Ontario, in 1960, Victor Weisskopf gave a summary talk on the model, and kept returning to the same theme, that the model is "surprisingly good" or works "surprisingly well" (Weisskopf 1960).

By the mid-1950s, Maurice Goldhaber began to feel he understood isomers and his interest shifted toward particle physics. Gertrude Scharff-Goldhaber continued to pursue her studies of the excited states of even-even nuclei into the 1990s, when declining health kept her at home virtually all the time. Her model hung on the wall of her now unused office (see fig. 9.3).

8.7 Gertrude Scharff-Goldhaber.

"See the patterns?" Maurice Goldhaber asked a recent visitor. Late afternoon sunlight streamed in the window, illuminating a set of file cabinets in which Scharff-Goldhaber neatly kept the notes she took at every conference she attended. Row upon row of books covered the walls. The isomer chart, hanging over a bookshelf, resembled a three-dimensional model of a futuristic city. "It brought a little regularity into a lot of vague data," Goldhaber continued. He paused a moment. "It was all quite surprising at the time, but it is now all folklore." Patiently waiting for the visitor to depart, he closed and locked the door as the lengthening shadows continued to stream into the quiet room (fig. 8.7).

Vance Sailor: Cross-Section Measurements with the Single Axis Spectrometer

Another person working on the same face of the reactor as the Goldhabers was Vance Sailor. Sailor had arrived at Brookhaven in 1949 as a member of Lyle Borst's research group in the reactor department. Because Borst's struggles to complete the BGRR absorbed nearly all of his time and energy, Sailor became de facto head of the group until Borst left the laboratory in 1951, when Sailor became official head. This group, like Hughes's, worked on cross-sections, but used not a chopper but a crystal spectrometer to select neutrons.

Choppers, one recalls, opened and closed periodically like camera shutters, letting through short bursts of neutrons. The bursts had to be widely spaced in order to prevent "frame overlap," or the faster neutrons in one burst overtaking the slowest in the next. As a result, the shutter has to be closed most of the time; it has a poor "duty cycle," researchers say, and they can use only an inefficiently small portion of the neutrons in their beam. Crystal spectrometers, by contrast, were much more efficient. A crystal spectrometer contains a crystal, mounted on the axis of a circular table, that reflects neutrons of specific energies (wavelengths) in specific directions, letting most of the rest pass on through. Swiveling about the same axis is an arm on which a sample is located, as well as either a detector or another axis.[23] Its disadvantages are two. First, it can examine just one specific energy at a time. Second, the energy levels are restricted by the structure of crystal itself—the size of its atoms, and the spacing between them. Beyond about a dozen electron volts, it is difficult to get a good reflection.

When Sailor arrived, Borst had been working on a high-precision, single-axis spectrometer. The central axis was mounted sturdily so the seven-foot cantilevered arm could support the sample, detector, and up to five hundred pounds of shielding without bending. The crystal was on a table attached to a gear train able to set the angle of the crystal and counter arm to within three seconds of arc (Borst and Sailor 1953). When Borst departed, he left the device in good hands, for Sailor showed himself to be obsessive about detail and about improvements. Sailor soon became frustrated, for instance, by the fact that the spectrometer was labor-intensive, and frequently had to be reset by hand.

> Initially, it was all done manually, and we took data with stopwatches and pencils. We'd change the angle by hand, start the counter and the stopwatch at the same time, stop the counter when the stopwatch said ten minutes had passed, and write down the data. I began to get frustrated. It seemed a hopeless situation: the reactor was running twenty-four hours a day, in which time we'd do about fifty cross-sections. Someone had to sit at the reactor to shift the spectrometer every ten minutes as well as work up the data and make specimens, the workload was piling up, and there were only four of us.[24]

Sailor devised automatic devices to change the samples, stop and restart the counters, and print out the count; some of the first printout devices employed were gadgets adapted from highway department devices that counted the number of cars going past a particular point on the road. Somewhere around the middle of 1952, Sailor began to wonder whether he could interest the lab administration in an on-site computer.

Brookhaven was much slower than other national labs at realizing the importance of computers, and ineffective at developing its own (for computers at Argonne, by contrast, see Holl 1997, 122–25). In 1950, and again in 1952, the AEC offered to install a UNIVAC at the lab, and the subject was brought up at a meeting of senior scientists. Only reactor theorists Chernick and Kaplan expressed serious interest. Haworth figured the lab would only use 20 percent of its time, and had no interest in running a service essentially for outside users. The UNIVAC went to NYU. When Brookhaven scientists wanted to use it, they had to take a long train ride into New York City, wait to run the data, and then take the train back. When Sailor approached Haworth to argue the case for a more accessible computer, Haworth gave Sailor permission to approach IBM on the condition that no commitments be made.

> I took the train to Penn Station, got a telephone directory, looked up IBM, and took a taxi up to their office on Madison Avenue to see what kind of equipment they had. The store had a lot of typewriters, accounting equipment, stuff like that—nothing like what I was asking for. The salesmen kept passing me on to higher-ups. After about two hours, I finally reached the top. The guy was aghast. He said, "You mean the *atomic energy laboratory* out on Long Island doesn't have any computers"? He wanted to send out the next day—but, boy, Haworth would've killed me if I'd made a commitment like that! I went back and cleared it with him first, then he set up a meeting between IBM people and scientists at the lab.[25]

Sailor again met with Haworth, who took the traditional step, when dealing with someone proposing a radical new idea that might involve a lot of work, of appointing him chairman of a committee. Haworth was reluctant to buy an expensive machine unless others were interested, and the committee began seeking potential users. Palevsky was one—he, too, was reading counts by eye and recording data by hand, and he knew about computers because his brother Max had founded a company that built computers—but the only other sizable group they could attract was payroll, which needed a computer for its accounting. Payroll's support was important but did not come cheap; they demanded the computer be located in their building—building 134—and that they have first priority between 9 A.M. and 5 P.M.

Committee members then approached the Remington-Rand Research Lab, in Norwalk, Connecticut, solicited an offer, and began to play that company and IBM off against each other.[26] The committee ultimately took Remington-Rand's offer of a plug-board-programmed computer with punch card input called a 409-2R, able to store ten words in its relay-operated memory. When deputy director Gerald Tape wrote Van Horn at the end of June 1953 to order the computer, the user priorities he cited, in order, were: payroll, bookkeeping, accounts payable, fixed assets, inventory records, and scientific departments.[27] An agreement to rent the machine was signed in July. It was installed by the end of the year, and with it came an engineer named Stuart Rideout, who delivered a series of lectures to the staff on how to program it. (Rideout would remain at Brookhaven until his retirement in 1992.)

The machine itself was about 3 feet thick and 5½ feet high, and came in two sections, one about 6 feet long, and the other, with the punch card input reader, about 5 feet long. Inside were hundreds of vacuum tubes that dissipated enough heat to warm the room significantly. The scientific groups that used the machine set up equipment that created punch cards out of the data. During the day, these cards would mount up. After working hours, a member of the group would gather the cards, walk from the reactor down the hill and across the field to the payroll building. Each group was given a large electronic wiring board, about the size of a big kitchen cutting block, that a member of the group would have to wire and solder by hand to program the machine to process its punch cards and translate the data.

Though Sailor's and Palevsky's groups were the first scientific groups to use the machine, they were soon joined by other departments, and the era of insatiable computer demand descended on Brookhaven in force.[28]

While Sailor's spectrometer group was generally interested in lower-energy cross-sections than Hughes's group, the two teams' efforts groups overlapped enough to create competition. As there was plenty of work to go around, the rivalry was genial, at least initially. Sailor, Hughes, and Palevsky carpooled to the lab together; twice a day, they shared their latest discoveries and ideas about instruments, techniques, and cross-sections. Later, some friction developed as Hughes's powerful group began nibbling away at Sailor's support.

Sailor's team studied rare earth cross-sections, but he soon became most absorbed by the careful description of the cross-section of uranium—a subject of enormous importance to the then-burgeoning field of atomic energy production.[29] In 1954, Sailor discovered that several resonances in the ^{235}U cross-section did not fit the classic Breit-Wigner formula. Instead, a multilevel formula was necessary to explain the shapes of the resonances. Charles Porter also spent much time following Sailor's work, which was incorporated into the Porter-Thomas distribution, which described the shapes of such cross-section curves. As data on ^{235}U were still classified secret, they were not published until declassified at the time of the first Atoms for Peace conference.

Another problem, Sailor realized, was that the original Breit-Wigner formula assumed the particles in the target nucleus were stationary. But even in the 1930s, Hans Bethe had realized that the effect of the motions of the particles in the target nuclei would blur the contours of the Breit-Wigner shape, a phenomenon known as *Doppler broadening* (Bethe 1937, 140). By running cross-section checks of samples at different temperatures, Sailor was able to observe Doppler broadening effects directly, establishing that there were two kinds of contributions to the deviation of an observed resonance from the idealized Breit-Wigner shape: one an experimental artifact, arising from imperfect instrument resolution, the other a physical phenomenon of Doppler broadening. These limitations on the use of the Breit-Wigner formula were of vital practical importance to reactor designers who were relying on the formula for ever-more precise applications (Sailor 1953; Wood, Landon, and Sailor 1955).

These discoveries did not end Sailor's difficulties with the ^{235}U cross-section. When excited nuclear states have the same spin and are close together, they interfere with each other, changing the form of the resonance so that the Breit-Wigner formula no longer fits. In a famous paper, Bohr and Wheeler had argued that this problem would occur in the case of uranium (Bohr and Wheeler 1939). In order to understand the ^{235}U cross-section, Sailor realized, he would have to identify which spins were associated with which resonances. To identify the spins of resonances, he would have to find ways of polarizing (lining up the spins of) both the neutrons and the target nuclei, then build a detector with a polarization analysis capability.

An early method of polarizing neutrons that used a magnetic mirror, developed by Hughes and collaborators, worked only with low-energy neutrons. In 1951, Clifford Shull at Oak Ridge had discovered that certain crystals of cobalt could be used to produce beams of highly polarized neutrons, and Sailor was able to obtain some of these crystals from Shull. Polarizing the target nuclei was much harder. The magnetic moment of a nucleus is too small to allow it to be directly polarized by the usual method, applying a strong magnetic field. Instead, physicists must employ the *hyperfine interaction,* that is, the interaction between the magnetic moments of the nucleus and of the atomic electrons, which are far larger and more easily manipulable. When atoms are cooled to a temperature near absolute zero, a magnetic field can align the magnetic moment of the atomic electrons, whose magnetic field in turn exerts a force on the magnetic field of the nucleus, helping to align it in turn. But this technique, known as the Rose-Gorter method after two scientists who developed it independently, both requires a refrigerator capable of cooling materials close to absolute zero, and works only with magnetic materials with a large hyperfine interaction and small magnetic *anisotropy,* or resistance to having their nuclear spins swung around. Spins can tend to point in certain favored directions, and depending on the type of crystal it may be harder or easier to swing them around; magnetic materials are said to be soft or hard depending on whether they are easy or difficult to get around. An anisotropic material is hard in this respect.

To pull this off, Sailor's group had to learn cryogenic (low-temperature) techniques, a major shift in orientation. Indeed, of all the BGRR group leaders, Sailor took the greatest career risk. Whereas Hughes, Goldhaber, and others used or developed fairly well understood techniques and equipment, and focused their considerable talents on scientific questions, Sailor alone committed himself and his group to developing a novel technique that had to be learned mostly by trial and error. It would be a rocky road.

In the late spring of 1957, reactor department head Marvin Fox left the lab, and Haworth took the opportunity for a reorganization, dissolving the reactor department and reconstituting it as the reactor division, headed by Robert Powell. Some of the research groups remaining in the department, including Sailor's, were transferred to physics, the remainder to reactor engineering. The organizational reshuffling coincided with the Sailor group's own shift in research orientation, and he took the opportunity to rename it the nuclear cryogenics group.

Sailor built a cryogenic refrigerator and a new spectrometer for his studies (fig. 8.8). By 1961 the group had explored the spin dependence of a number of resonances of holmium, a magnetic material known to have a large hyperfine interaction (Postma et al. 1962). The success of this work encouraged the group to attempt more ambitious and difficult resonances in other elements—but uranium, a formidable technical challenge, remained the ultimate objective.[30] The group spent another few years struggling with complicated methods for polarizing uranium nuclei, and eventually succeeded. In the waning days of the BGRR, the group at last managed to accomplish its original goal—measuring the spin dependence of the cross-section of ^{235}U (Schirmer et al. 1968). The quest had taken a dozen years. Group member Lawrence Passell recalls:

In science, you do what you can. What you can do is not always important; *Phys. Rev.* is full of such papers. But the objective's to find something you can do which will make a contribution—which will expand knowledge—though you are always limited by the techniques that you have. Now, some experiments may not look important, but they do have a certain value as demonstration experiments. Why did Vance do holmium? Based on what he knew about its hyperfine field, and its magnetic properties, he knew he could probably polarize the nuclear spins. When he was successful, it showed him he was on the right track, that he could make the method work for other nuclei. Why did Vance do uranium? That

8.8 Vance Sailor (standing) at his cryogenic refrigerator at the BGRR, 1961.

was different—that was a contribution. It not only helped to straighten out the confusing cross-section picture, but also revealed some of the limitations of the existing theoretical models of the fission process. That paper was Sailor's pièce de résistance—it was his grand finale at the BGRR.[31]

Solid-State Physics and Critical Experiments

Brookhaven's solid-state physics effort, one of whose early champions was George Dienes, is an example of a basic research program that grew out of a practical one. Dienes was born in 1918 in Budapest and finished his *Gymnasium* in 1936. An outspoken youth, he was openly opposed to the growing Nazi influence in Hungary, which often landed him in trouble. The rector of the local university, a family friend, told him that although he could matriculate (Dienes was not Jewish), he was likely to be severely beaten. Leave the country, the rector advised. Fortunately, the Dienes family had relatives in the U.S. who helped him enter what was then Carnegie Tech in Pittsburgh, from which Dienes received a B.A. in 1940 and a Ph.D. in 1947. Subsequently, he got a job at North American Aviation, in its atomic energy research department, studying radiation damage, that is, the effects of various types of radiation, especially neutrons, on materials. But Dienes learned that he did not care for the constant scramble for contracts in commercial work, and secured an interview at Brookhaven.

Brookhaven's reactor engineers were looking for a specialist in radiation damage in graphite, a subject of urgent practical application to the BGRR. The reactor was almost entirely composed of stacks of graphite blocks, through which an intense flux of neutrons streamed whenever the reactor was in operation. As Eugene Wigner had first noted during the war, fast neutrons have a tendency to knock carbon atoms out of their ordered position in the graphite lattice, and push them into interstitial locations. The result, known as the *Wigner effect* or *Wigner disease,* is a distortion in the lattice, rather like introducing pebbles between the pages of a book. The distortion drastically reduces the graphite's thermal conductivity and causes it to expand and store energy. Such expansion was potentially dangerous. Not only could it put the fuel channels out of alignment, permanently disabling the reactor, but the energy storage and drop in thermal conductivity could potentially make graphite catch fire. Hanford engineers had developed a procedure for relaxing the graphite, called annealing, which involved running the reactor for a short, controlled period at high temperatures. At high temperatures atomic mobility increases, and the dislodged atoms tend to return to their original positions, reversing the Wigner effect. But much research remained to be done on the problem both from a fundamental and practical standpoint. Moreover, Hanford's experience with graphite was highly classified and Brookhaven scientists knew little about it.

Shortly after his arrival in September 1951, Dienes began to study radiation damage in graphite (Dienes 1952; Kosiba, Dienes, and Gurinsky 1956), and helped draw up plans for an annealing experiment at the BGRR. Because the largest Wigner effect occurred in the center of the reactor next to the gap (where the air flow cooled the graphite most and natural annealing would be minimal), the reactor operators had to find a way to warm up those better-cooled areas more than others. Ultimately, they did this by reversing the air flow direction; they let the air flow from the surfaces into the gap and then out.[32]

The urgency of finding a workable annealing procedure was underscored when a sudden, unexpected release of stored energy took place inside a British reactor at Harwell late in 1952. British authorities kept the incident a tightly guarded secret. But Chadwick, now a prominent science adviser to the British government, thought it important that Haworth know and broke strict secrecy by relaying the news to Haworth; to protect the source, Brookhaven's reactor engineers referred to it guardedly even in classified documents.[33] The incident was not revealed officially to U.S. AEC personnel until January 1953. The BGRR's annealing experiment took place later that year. Not knowing what to expect, the lab prepared elaborate safety precautions, including a plan for evacuating the region. Annealing began one morning and continued into the evening, with the temperature of the reactor monitored by thermocouples that had been installed in the graphite when the reactor was built. Dienes and the other principal members of the reactor department—and Haworth himself—stayed well past midnight until it was clear that the operation had, indeed, reversed the Wigner effect. Afterward, the reactor was annealed regularly.

In October 1957, disaster struck a British production reactor at Windscale, when the graphite caught fire during an annealing operation, spreading radioactive ^{131}I over the surrounding countryside. (Iodine is generally the most significant poison involved in reactor accidents because of the large amount that tends to be released and its volatility—it vaporizes not far above room temperature and is metabolized by the human body.) Several Brookhaven scientists were called over as consultants to help figure out what happened, including Powell in one group immediately after the accident, and Dienes and Gurinsky, who by then had become recognized experts in annealing, several weeks later. The accident scared the AEC, and the agency debated seriously and at length whether it should stop annealing the BGRR and even shut it down. While at first it appeared as though the accident had been caused by an annealing procedure similar to that used at the BGRR, contributions from Brookhaven scientists helped to show that the annealing procedure was not at fault, and in the end, the cause of the Windscale accident was attributed to a broken fuel cap (Schweitzer 1993).

But Dienes's greatest long-range service to Brookhaven was not in annealing but in putting the solid-state program on its feet. Andy McReynolds, an imaginative scientist who was the first head of the solid-state group, was personally and scientifically a loner, and had failed to establish the broad-based program that Haworth had envisioned for the lab.[34] Seizing Dienes's arrival as an opportunity, Haworth and physics department chairman Sam Goudsmit instructed the collegial Dienes to reinvent the solid-state physics program. A growing number of theorists and experimenters were interested in neutron diffraction, a technique (analogous to X-ray diffraction) by which the patterns created by neutrons as they pass through a specimen are used to examine its structure. The technique had been pioneered at Oak Ridge by Clifford Shull and Ernest Wollan, who adapted X-ray diffraction equipment to neutrons. The instrument used with neutron diffraction was a double-axis spectrometer. As in Sailor's single-axis spectrometer, a mono-chromating crystal was mounted on an axis, around which swung an arm with a sample; unlike Sailor's device, the sample was mounted on a second axis, around which swung a detector, permitting experimenters to determine the angular profile of the scattered neutrons. Lester Corliss and Julius Hastings of Brookhaven's chemistry department set up a double-axis spectrometer at the BGRR, incorporating for the first time Soller slits, or multiple parallel slits in the collimators (see fig. 8.9). Raymond Pepinsky, an X-ray crystallographer from the University of Pennsylvania, was the first visitor to establish a permanent neutron crystallography program there. Pepinsky's graduate students included future neutron diffractors and BNL staff members Chalmers Frazier and Gen Shirane. Among Brookhaven's numerous other visitors was Shull himself, who worked at the BGRR for a year and a half.

The work of Corliss and Hastings can serve to illustrate the successes of this program. Corliss was a chemical physicist interested in the magnetic properties of heavy elements, Hastings a physical chemist skilled in diffraction methods. The two were brought together by chemistry department chairman Richard Dodson, who was seeking to boost the programmatic presence of his department at the reactor. Neutron diffraction seemed a good bet. Unlike X rays and electrons, neutrons could be used to determine the magnetic structure on an atomic level—something of special interest to both chemists and physicists, who theretofore had only been able to infer the magnetic structure of solids by measuring bulk properties.

The first significant work by Corliss and Hastings on their double-axis spectrometer involved the magnetic structures of nickel and zinc ferrites. Ferrites, magnetic materials that do not conduct electricity, were both

industrially important and scientifically interesting. The industrial importance lay in the many applications of ferrites in high-frequency devices.
The scientific interest involved a theory about the structure of ferrites
recently proposed by French physicist Louis-Eugène-Félix Néel (Néel
1948). Whereas ordinary magnetic materials are ferromagnetic—that is,
the atoms in their lattices all have their magnetic moments aligned in the
same direction—Néel had been examining antiferromagnetic materials, in
which the moments point in opposite directions, in effect canceling each
other out. Néel also proposed a third kind of magnetism called ferrimagnetism to account for the then-baffling behavior of the ferrites whose formula is $XO \cdot Fe_2O_3$, where X is a 3-d cation. The ferrites crystallize with
the spinel structure, which consists of a face-centered cubic sublattice of

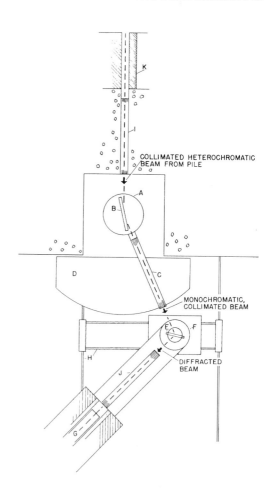

8.9 Basic design of double-axis
spectrometer used by Corliss and
Hastings. Note the tiny parallel
slits in the beam collimators.

oxygen ions, while the magnetic ions are found in both tetrahedral and octahedral interstices within this sublattice. Néel postulated that the tetrahedral array of ions was antiferromagnetically arranged with respect to the octahedral one. Since there are twice as many occupied octahedral sites as tetrahedral ones, the result is a net ferromagnetic moment.

The first real demonstration of Néel's ideas about antiferromagnetism came from Shull and Smart's groundbreaking study of MnO using neutron diffraction. Corliss and Hastings turned their attention to Néel's theory of ferrimagnetism by studying nickel and zinc ferrites as a good test case for Néel's theory, leading to their first important publication, in January 1953, entitled "Neutron Diffraction Studies of Zinc Ferrite and Nickel Ferrite" (Corliss and Hastings 1953). In it, Corliss and Hastings described neutron diffraction experiments in which they obtained results in accord with Néel's theory. Thanks to such confirmations, at Brookhaven and elsewhere, Néel received the Nobel Prize in physics in 1970.

Corliss and Hastings were familiar figures at the reactor, sharing equipment, techniques, and information with many of their colleagues in the physics department. Though fulfilling the interdisciplinary ambitions of the Brookhaven concept, the two eventually ran afoul of the AEC, for which their work looked too much like physics to warrant money earmarked for chemistry. Their work illustrated how reactor research blurred boundaries between these two disciplines by allowing exploration of the contributions of atomic structure to solid-state properties.

Throughout the 1950s, Brookhaven's radiation damage research continued, but began to be overshadowed by that of the neutron diffractors, who were involved in a new field of unknown capabilities and promise. "We knew the basic questions [in radiation damage research]," Dienes said, "and were never particularly interested in practical problems involving other reactors." [35] In 1957, the solid-state group acquired its own building, a converted barracks, and the radiation effects researchers took the first floor, the neutron scatterers the second. As the years went on, especially when a new, higher-flux reactor came on the horizon, the second floor seemed more and more the exciting place to be.

Herbert Kouts: Critical Experiments

Herbert Kouts arrived at Brookhaven in 1950, became head of the reactor department's experimental division, and bore primary responsibility for the experimental program in nuclear engineering for nearly two decades. But his most significant work would grow out of his own free interests even against the AEC's wishes. Kouts had a strong background in theoretical physics, but soon became interested in reactor design. In the 1950s, he had almost a free hand to change his research program. "I started up many programs," he recalled, "just because I decided to start them up, and usually the money just followed. I thought a thing would be good to do, the lab judged that I knew best, and that was good enough" [36] While the AEC had tapped Argonne as the central facility for reactor research, Kouts took a hole atop the BGRR that had been abandoned by the medical department, turned it into a facility where he could expose large areas to reactor neutrons, and began to explore the behavior of various arrangements of uranium rods, moderated by ordinary water.

By 1954, Kouts's studies had progressed enough for him to ask for additional lab support in the form of a critical facility—a tiny reactor small enough that heat removal was not a problem. Over the next few years, he built two additional critical facilities, which he used in thousands of experiments testing different features of reactor design. Eventually, water-moderated reactors proved most efficient, and Kouts's work grew into a substantial program whose results were eventually incorporated into the Westinghouse-developed light water reactor power program that the U.S. adopted.

> My group and I built at Brookhaven three [critical facilities] that we did not tell [formally] the Atomic Energy Commission about until much later. We just went ahead and built them because they were a useful part of our experimental program. We didn't run them at very high power—they were not meant to be run at very high power—we did a very thorough safety analysis of what we were doing, a very thorough engineering job, put them together, operated them, got a lot of useful experimental work out of them. We didn't ask for a lot of money—cost a lot less to do things that way than it does when you have to go through a lot of formalism—but you got a lot more results. It's a different world today. I'm struck by the fact that people coming into my field these days will have no perception whatsoever of how that used to be done. They think that the world really consists of making a proposal, writing it, rewriting it, presenting it and defending it, arguing, going back, maybe going through a process that takes six months to a year, and then if it's approved, going into a relationship in which you make monthly reports, quarterly reports, and the process is reviewed thoroughly from the standpoint of, "Was it done right?" and "Was the money spent right?" That's the world to these people and they think it never was any different. Well, it sure was.[37]

Lee Farr: BNCT

Medical department chairman Lee Farr led the largest and most daring use of atomic energy for radiotherapy in the 1950s, the Boron Neutron Capture Therapy (BNCT) program. BNCT is based on an ingenious technique (described in principle first in 1936) in which a nonradioactive isotope of an element (here, ^{10}B) is made to accumulate inside tumors, where it is bombarded by slow neutrons. These neutrons bring about a nuclear reaction in the ^{10}B that drastically enhances the radiation damage to nearby cells—ideally, only the tumor tissues.

Three parts would go into eventually making the technique work. The first is to find a target element with a large cross-section for thermal (slow) neutrons. The second is to attach the target element to a compound that, when administered to a patient, would be taken up preferentially by the tumor rather than healthy tissue in the zone to be irradiated. The third part is to irradiate the tumor with a large dose of thermal neutrons. When these low-energy neutrons react with the nuclei of the target element, they trigger a reaction releasing millions of electron volts of ionizing radiation energy. If all three parts of the technique can be made to work, the radiation damage is confined to a short distance of about a cell diameter, meaning that only tumor cells are destroyed and healthy tissue is spared (Slatkin 1991).[38]

The ^{10}B reaction had been discovered at the Cavendish by Maurice Goldhaber and James Chadwick in 1934. Shortly after moving to the University of Illinois in 1938, Goldhaber remarked half-jokingly to a colleague that the reaction could cure cancer—if boron could be put inside tumors and irradiated. The colleague, P. Gerald Kruger, took the remark seriously. Together with another researcher, Kruger irradiated boric acid–bathed mice tumor cells with neutrons in vitro and showed that these had reduced viability when transplanted to other mice. Kruger wrote up his experiments in an article that is the first to describe medical research involving boron neutron capture. In 1940, another group performed the first in vivo BNCT irradiation (Zahl, Cooper, and Dunning 1940). The method looked promising enough in principle to be one of the first programs that Farr and Van Slyke put on the medical department's research agenda. The BNCT program was in many respects tailor-made for BNL, for it was reactor based and multidisciplinary, involving contributions from physics, chemistry, biology, and medicine.

Early in 1950, Farr began to study compounds with an affinity for malignant tumors to which target elements might be attached, and focused on uranium as a target element, considering boron too toxic. A young doctor named Winton Steinfield was assigned the task of finding a way to modify bismark brown, a compound for which malignant tumors reportedly had a remarkable affinity, for neutron capture therapy. After months of struggle, Steinfield excitedly told Farr that he had discovered a way, but that before writing up the discovery he wanted to leave for Baltimore to pick up a boat he and his wife had just purchased and sail it back to Long Island. Although Steinfield and his wife made radio contact with the Coast Guard en route off New Jersey, the boat vanished in a storm and the couple was never heard from again. Though Van Slyke and Farr scoured Steinfield's notes, they never managed to decipher the secret. What the youthful and ambitious Steinfield discovered and whether it would have been workable is a mystery to this day, making it possible that a major research initiative was substantially altered by a pleasure boat shipwreck.[39]

Meanwhile, William Sweet of MIT, who was independently studying the possibility of neutron capture therapy, fastened onto boron as a target element and was convinced that less toxic ways of administering it could be found. Sweet, a neighbor of Brookhaven trustee Baird Hastings, was naturally led to consider the BGRR as a source of thermal neutrons, and began to collaborate on the project, convincing Farr of boron's value.

Though the BGRR was nearing completion, time remained to modify the shielding on top of the reactor (it seemed too difficult to use a neutron port on one of the reactor faces) to create a BNCT irradiation facility. The facility consisted of a pit formed by removal of several shielding blocks, in which a patient could be placed next to a small, rectangular 5×10 cm neutron port that looked through the shield directly into the reactor core. Meanwhile, William Hale of Brookhaven's division of bacteriology and immunology developed a transplantable brain tumor in mice that made experimental feasibility tests possible. Studies of boron neutron capture effects on mice, dogs, and pigs followed. Farr later coauthored papers on these experiments, which were the first demonstration that neutron capture therapy was capable of eliminating a tumor successfully without recurrence in mice that otherwise would have been killed by the tumor in a matter of weeks (Farr and Konikowski 1967). Toward the end of 1950, the director of the AEC's division of biology and medicine, Shields Warren, gave the go-ahead for clinical trials of BNCT on patients with advanced malignant brain tumors (gliomas).

A major incentive for the BNCT program was the abysmal prognosis for the glioblastoma multiforme (the most malignant form of glioma) patients accepted for treatment. Nothing else could be done for them, and death was sure and swift. In the 1950s, when virtually all the BNCT trials took place, the average survival after diagnosis for cerebral malignant glioma patients at the Massachusetts General Hospital (where many of the BNCT patients originated) was only several months. Today, the median survival after diagnosis is about a year. Thus the Brookhaven experiments seemed a way of studying a conceptually attractive, somewhat safe, and possibly effective method of treating patients for whom death was imminent (Slatkin et al. 1986).

On 15 February 1951, half a year after the commissioning of the graphite reactor, the first BNCT patient was treated (fig. 8.10). It was the first BNCT attempt on human beings, and the start of a trial involving ten patients over two years. In each case, neurosurgery had been attempted but had failed to arrest tumor growth. These early treatments were rather dramatic events. First, the reactor was shut off completely and radiation shielding blocks temporarily removed from the neutron port in the floor of the facility. The patient was taken by ambulance from the Brookhaven hospital to the reactor building, and then by stretcher to the top of the reactor, where a half dozen men would lower the barely conscious patient into the coffin-like pit. Farr and his assistants would administer ^{10}B-enriched sodium tetraborate (borax) intravenously for a minute or so, and fix the patient's head into position over the port. He would then signal for restart of the reactor. As the control rods were removed, Farr and the assistants climbed out of the pit and sprinted for the balcony. It took eight to ten minutes for the reactor to come up to full power. At that point during the first treatment, an ominous, stentorian voice boomed through the loudspeakers, "The reactor power is now critical!" Upon first hearing the word "critical," Sweet, who until then had little experience with reactors, assumed that something had gone wrong, but was assured that everything was OK. During the time it took to achieve criticality, according to Sweet's studies, the tumor was absorbing the boron at a rapid rate. Sweet's work seemed to indicate that the tumor would have sufficiently more boron than the surrounding tissue to make irradiation therapeutically useful for about forty-five minutes. While the reactor hissed ominously due to the four hundred mph winds racing through its shielding and cooling channels, the patient lay immobile in the pit, exposed to the thermal neutron beam—for seventeen minutes in the case of the first patient, up to forty minutes for subsequent ones. The reactor was then shut down, the patient hoisted from the pit, taken back to the ambulance by orderlies, and thence to the hospital (Slatkin 1991).

The first BNCT treatment, of a fifty-one-year-old woman, gave the lab another lesson in public relations. An associate editor of *Collier's* and well-known science writer, John Lear, learned of the BNCT project from AUI head Lloyd Berkner, who happened to be a personal friend. (For an amusing anecdote of how Lear and other science journalists with an exalted view of their craft were generally regarded by scientists, see Van Allen 1997, 243.) Lear came to Brookhaven, and Van Slyke, Farr, and Sweet initially refused to speak with him. But Lear was persistent, seemed trustworthy, and persuaded the scientists to speak, on condition that he would hold the article until after the results were presented in a scientific forum; it is considered unprofessional to do otherwise, given the well-known dangers of raising false hopes about unproved treatments for terminal illnesses. The Brookhaven scientists also pointedly insisted that Lear promise to check the accuracy of the article with them beforehand, and that *Collier's* would show restraint in publicizing the article.

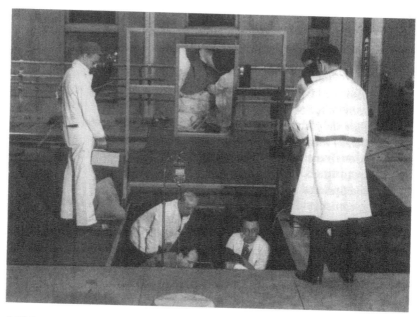

8.10 Lowering first BNCT patient into treatment facility on top of the BGRR, February 1951. Medical department head Lee Farr is in the pit at right. The mirror allowed doctors to observe the patient from the balcony of the reactor room.

Lear broke all three conditions. "ATOMIC MIRACLE" was the super-heated and misleading headline splashed across the cover of *Collier's* 21 April 1951 issue; "Science Explodes an Atom in a Woman's Brain" (Lear 1951). Lear described BNL's reactor building as a "modernistic cathedral of science" that "blazed emerald sheen." And in contrast to the six atomic explosions that the U.S. set off in the first seven weeks of 1951 in its weapons testing program, Lear described the "nuclear explosions" taking place inside the woman's head as being "as quiet as the voice of conscience," and as loosing "a gleam of hope for peaceful men of good will everywhere." Lear's article is a reminder that, once upon a time, rhetorical imagery could be used to champion the achievements of reactors. In the media, it was not a given that reactors were essentially threats to humanity, yet.

Collier's had accompanied Lear's article with an imaginative rendition of the reactor by noted illustrator Chesley Bonestell, photographs of it being restricted prior to Atoms for Peace. The illustration gave an outraged Farr the idea of charging *Collier's* with security violations, but AEC lawyers told him the illustration was too sketchy to provide the basis for a case.[40]

The fate of the patient, in fact, showed how premature the treatment was. While several days after the treatment she could speak and walk about, she quickly deteriorated and died the week the *Collier's* article was published. Although most of the nine patients who followed her experienced a temporary alleviation of their symptoms, few ultimately fared much better.

All in all, three groups of patients were treated at the BGRR: a ten-patient group from 1951 through 1953, a nine-patient group from 1954 through 1955, and a nine-patient group from 1956 through 1958. The first group received their boron dose in the form of borax, intravenously administered; though the borax made many nauseous, it had no long-term side effects, and the patients suffered only slight skin burns from the thermal neutron radiation. This raised hopes that more effective results could be obtained with stronger doses, and in the second series of patients a higher neutron exposure was used together with a new, less toxic boron-containing compound, sodium pentaborate. Also, a twenty-ton shutter was installed at the neutron port so the reactor did not have to be shut down before and after each treatment, and the port itself enlarged to 10×10 centimeters. This time, however, the exposure proved to be too high: patients in the second group suffered severe skin damage due to the high neutron dosage. In the third series, another injection method was tried so that the time of radiation exposure could be lessened while maintaining the total radiation dose; the pentaborate was injected directly into the artery that fed the brain hemisphere containing the tumor. Though this enabled the dose to be lowered significantly, some patients still suffered skin damage. Moreover, the overall results continued to disappoint. The median postoperative survival was 97 days for the first group, 147 days for the second, and 96 days for the third.

Farr continued to be enthusiastic. Still, he conveyed the risks to potential patients and their physicians, who had to sign consent forms. In light of later ethical questions about research on human subjects, it is worth quoting a paragraph from a typical letter from Farr to a physician of a prospective patient:

> It must be clearly explained to the patient's family that this is an experimental procedure for a disease, namely glioblastoma multiforme, for which at the present time no satisfactory therapy exists, that since it is an experimental procedure there are certain risks inherent in it which we believe are satisfactorily met or the patient would not be subjected to the treatment. However, as you well appreciate, absolute assurance in this regard cannot be extended. Results would suggest that there may be some prolongation of life in patients so treated. In most instances

there has been a temporary alleviation of many aspects of the symptomatology and evidence of a temporary cessation of tumor growth. Ultimately this growth has been resumed and the patient has died. Since it is an experimental procedure it is necessary also that the family be advised that an autopsy examination of the brain is expected and that the brain is to be removed for comprehensive studies to determine the degree of success of the treatment and the reason for its failure. Detailed studies done thus far on the central nervous system as yet revealed no evidence of damage to normal structures by this procedure. The decision to send a patient here must be made by the referring physician and concurred in by the family. We accept patients only from a referring physician and not directly.[41]

However, the physician, while informing the family of the patient's condition, might well follow the paternalistic doctor-patient relation of the day and not inform the patient that he or she had a malignant brain tumor. For example, the doctor might refer only to "cells suspicious of malignancy in his brain which required further attention." [42]

Meanwhile, Brookhaven scientists began to think that the neutron beam available at the BGRR was too weak for effective BNCT. One problem was that the BGRR's thermal neutrons were rapidly attenuated; each 1.8 cm into the brain, the neutron flux was cut in half. Sweet, who had collaborated on the first two series of patients at Brookhaven, began developing a BNCT center at MIT and curtailed his collaboration with Brookhaven's program. In 1955, the medical department had begun planning a Brookhaven Medical Center, to include a reactor, which would be designed principally with BNCT in mind and the center with the primary aim of supporting BNCT patients. In December 1955, AEC chairman Lewis L. Strauss officially announced that a new medical research center and medical reactor would be built at Brookhaven. Construction began the following year, the facility was completed at the end of 1958, and the Brookhaven Medical Research Reactor went critical the following March. Just as the BGRR was the first reactor designed in peacetime to be built specifically for research, the BMRR was the first designed explicitly for medical research.

When the reactor went operational, the BNCT program was transferred to the BMRR. With a substantially higher flux, the duration of exposure could be markedly reduced, and instead of seventeen to forty minutes it was twenty-three to three hundred seconds. A total of seventeen patients were ultimately treated by a standardized protocol at the new medical reactor between 1959 and 1961; several other patients were treated by individualized protocols.

But these results were even more disappointing than the earlier ones at the BGRR. Only one case looked satisfactory, a man with grave neurological symptoms who received a high BNCT radiation dose; these were completely reversed after the treatment. He lived 151 days after irradiation, and died primarily of metastasis of pancreatic cancer into the liver and abdomid lymph nodes; in his brain, at autopsy, there were no residual signs of tumor growth. He, it turns out, was the first patient ever to receive substantial clinical benefit from BNCT. But of all seventeen patients, the median post-treatment survival was only eighty-seven days. Most ominously, the four patients who were much later found to have had received the largest doses died within two weeks. Autopsies revealed acute swelling of their brains. The large doses had probably contributed to their deaths by causing the swelling (Slatkin 1991).

The ten-year optimism finally crashed, and in 1961 Brookhaven's BNCT clinical research program was terminated. At MIT, Sweet's program proved equally disappointing and that work also ceased in 1961. Thus ended, for three decades, the great hopes of those who looked for quick, therapeutic applications of atomic energy to medicine. The applications existed, but in far less dramatic (thus less newsworthy) areas than immediate cancer cures. The episode was a textbook case in the perils of exploring unproved and possibly dangerous treatments for terminal illnesses. When such treatments are successful, the researchers are generally hailed as compassionate and courageous saviors of humankind; when unsuccessful, the participants can be branded as guilty of unethical, life-threatening conduct.[43] Three decades later, Brookhaven would be sued for its early role in the BNCT effort, even as it was reviving BNCT in a much more promising form.

THE BGRR was an extremely versatile facility able to stage a wide variety of experimental performances. In addition to the groups described above, the reactor served many others, and most of Brookhaven's departments besides physics were assigned holes. Chemist Ed Sayre, to cite one example of the reactor's broad research program, used the reactor for pioneering work in art and archaeology; at one time, the Metropolitan Museum of Art in New York City sent its Rembrandts to Brookhaven for noninvasive studies of their authenticity.

The reactor department also assigned holes to off-site universities including Columbia and Pennsylvania State, and many industrial concerns, including General Electric (whose research laboratory was the first industrial laboratory to put together a full-time research program at the BGRR), Bell Telephone Laboratories (which had a spectrometer and carried out irradiations of semiconductors in a special shielding tank), Westinghouse Electric Corporation (which had a low-temperature facility), the Naval Research Laboratory (conducting a magnetic core material test irradiation), Republic Aviation Corporation (a carbon adsorption experiment), and Wright Air Development Center (capture gamma studies).[44] Esso Laboratories (Standard Oil), Phillips Petroleum, Monsanto Chemical, General Dynamics,

45. GRAPHITE OXIDATION STUDY (PHYSICS)
46. SLOW CHOPPER (PHYSICS)
55. PHASE CHANGE IN URANIUM BASE ALLOYS (WAPD)

WEST FACE

11. ION CHAMBER EVALUATION (PHYSICS)
13. FOIL SLOT
14. FOIL SLOT
16. GAMMA-RAY SPECTROMETER (AIR FORCE)
    CROSS SECTION STUDIES
    NEUTRON SPECTROMETER (PHYSICS)
23. NEUTRON DIFFRACTION SPECTROMETER (BELL TELEPHONE LABORATORIES)
24. 30°C CONTROLLED TEMPERATURE IRRADIATION FACILITY (CHEMISTRY)
25. FAST RABBIT (PHYSICS)
31. CROSS SECTION MEASUREMENT NEUTRON SPECTROMETER (COLUMBIA UNIV.)
33. NEUTRON DIFFRACTION SPECTROMETER (CHEMISTRY)
34. NEUTRON DIFFRACTION SPECTROMETER (CHEMISTRY)
35. NEUTRON DIFFRACTION SPECTROMETER (PHYSICS)
36. MAGNETIC PROPERTIES OF MATERIALS NEUTRON SPECTROMETER (PHYSICS)
42. ISOTOPE PRODUCTION TARGET CONVEYOR
45. CONDITION CONTROLLED HYDROCARBON IRRADIATION (ESSO)
51. 30°C CONTROLLED TEMPERATURE IRRADIATION FACILITY (PHYSICS)
52. NEUTRON DIFFRACTION SPECTROMETER (WESTINGHOUSE)
53. NEUTRON DIFFRACTION SPECTROMETER (GENERAL ATOMICS)
54. NEUTRON DIFFRACTION SPECTROMETER (PHYSICS)
55. NEUTRON DIFFRACTION SPECTROMETER (PHYSICS)
56. GAMMA-RAY CAPTURE SPECTROMETER (PHYSICS)

NORTH FACE

B-12-3, C-15-1. DEVELOPMENT OF FUEL ELEMENTS FOR GAS COOLED REACTOR (FORD INSTRUMENT)
C-14-0. FISSION NEUTRON IRRADIATION (PHYSICS)
C-15-14. FISSION GAS DIFFUSION STUDY (WAPD)
D-0-8. IODINE PRODUCTION FACILITY (NUCLEAR ENGINEERING)
D-11-5½. GAS IRRADIATION FACILITY NORTH CORE HOLE. FAST CHOPPER (PHYSICS)

TOP

NORTH FACE

NORTH CORE HOLE

B-12-3
C-15-1
C-14-0
D-0-8
D-11-5½
C-15-14

WEST FACE

EAST FACE

TOP

6. THERMAL COLUMN (BIOLOGY)
8. LOW ENERGY NEUTRON FACILITY (PHYSICS)
10. NEUTRON THERAPY FACILITY (MEDICAL)
17. EXPONENTIAL FACILITY (NUCLEAR ENGINEERING)
19. SHIELDING TANK (AIR FORCE)

EAST FACE

11. GRAPHITE OXIDATION STUDY (PHYSICS)
12. REACTOR CONTROL
14. U-Bi LOOP (NUCLEAR ENGINEERING)
21. CROSS SECTION STUDIES NEUTRON SPECTROMETER (PHYSICS)
22. CROSS SECTION STUDIES NEUTRON SPECTROMETER (PHYSICS)
23. RADIATION DAMAGE ON STEELS (NAVAL RESEARCH LABORATORY)
24. NEUTRON DIFFRACTION SPECTROMETER (GENERAL ELECTRIC)
25. U-Bi CORROSION LOOP (NUCLEAR ENGINEERING)
34. U-Bi STUDIES (NUCLEAR ENGINEERING)
35. U-Bi STUDIES (NUCLEAR ENGINEERING)
36. RADIATION DAMAGE ON STEELS (WAPD)
41. COLD NEUTRON SPECTROMETER (PHYSICS)
42. ISOTOPE PRODUCTION TARGET CONVEYOR
43. LIQUID NITROGEN FACILITY (PHYSICS)
44. MAGNETIC PROPERTIES OF MATERIALS NEUTRON SPECTROMETER (PENN. STATE UNIV.)

8.11 Space assignments at the BGRR, 1957.

DuPont, and Bendix also used the BGRR (see fig. 8.11). All were charged at a rate set by the AEC, which sought for the lab to "recover the equivalent of the general and administrative costs which are normally associated with one of our own scientists after taking due regard of the fact that the visitor is paying directly for shared services." [45]

The BGRR also created isotopes for hospitals and research facilities throughout the northeast. Brookhaven supplied a sodium isotope, for instance, to Harvard Medical School; samples were irradiated in the early morning, removed from the reactor at 6 A.M., taken to LaGuardia, flown to Boston by private plane, and put in use before noon. [46] The nuclear engineering department also had an isotope development program headed by Walter Tucker. [47] One of its aims was to develop "generators" for isotopes—devices containing long-lived radioactive materials that users outside Brookhaven could "milk" for isotopes with shorter half-lives, removing the need for constant express shipments of short-lived isotopes. In August 1954, Brookhaven sent its first generator (of iodine 132, generated by tellurium 132) to the Mayo Clinic in Minnesota. After Tucker's group had some difficulty purifying the ^{132}I, they discovered it was contaminated by an isomer of technetium known as ^{99m}Tc. Further research showed that ^{99m}Tc was an ideal material for a diagnostic tracer, and soon Tucker and his group began to develop a generator for it, reporting on their work at an American Nuclear Society meeting in Los Angeles in June 1958. They sought to patent the device, but the AEC balked. "The product will probably be used mostly for experimental purposes in the laboratory," it wrote in turning down the application. "On this basis, no further patent action is believed warranted." The agency misjudged. Three years later, in 1961, the Argonne Cancer Research Hospital ordered a ^{99m}Tc generator, the first of a torrent of requests as the advantages of ^{99m}Tc for research and clinical use became apparent. A Technetium Club was formed to facilitate dissemination of the generators. By 1965, several companies asked Brookhaven to cease distributing the generator so that they could take over the business, and the lab complied. ^{99m}Tc gave a huge push to nuclear medicine, and in one form or another "is the predominantly used radionuclide in the several-million nuclear-medicine procedures performed worldwide annually" (Richards et al. 1982, 793).

Service irradiations were done on everything from piston rings to seeds. One service irradiation at the BGRR must surely be the only reactor activity that may have been initiated by a maharajah. According to a story that made the rounds in the early 1950s, an Indian maharajah commissioned Harry Winston, the leading diamond merchant in New York City, to create a necklace entirely fashioned out of green diamonds. Whatever the truth of the story, a consultant to Winston named Fred Pough, former head of the mineralogy department at the American Museum of Natural History in New York, actually did contact Brookhaven about using the BGRR to irradiate diamonds. (Green diamonds are generally due to natural radiation rather than trace impurities. After an initial short-lived, fifteen-hour half-life activity decays away, irradiated diamonds are safe and cannot be distinguished from naturally occurring green diamonds.) Fox reluctantly complied when the AEC's isotopes division gave the merchants permission to "study colorations induced by irradiation in precious stones" and "observe changes in irradiated gems." But after more diamond merchants asked to use the reactor he protested to the AEC that it was unwise to use a government facility in this way ("nature ought to hold the patents," he grumbled). The AEC deliberated, but decided it could find no reason for placing gems in a different category from other service irradiations.[48] Merchants were charged on the basis of the standard schedule—pneumatic tube irradiations in a single day shift for a single client were $10 for a short irradiation, $50 for an entire day plus $25 handling charge.[49]

The Pile: 1957–1961

It was difficult to prevent the BGRR fuel elements, and those of any reactor fueled by natural uranium, from occasionally leaking. Each of the eleven hundred elements was eleven feet long and weighed about a hundred pounds. Once every ten-day cycle, when the reactor was shut down and then restarted, the element would heat up and expand, then cool and contract. If cooled suddenly, the heavy weight of uranium kept it from fully contracting, causing stress on the container. Even a tiny crack in the aluminum container could oxidize the uranium metal, widening the split, and cause volatile fission products contained in the irradiated uranium fuel to discharge into the cooling air. An innovative leak detection system, developed by Borst, bathed the uranium inside each element with helium, usually caught leaking elements quickly. Another check was radiation monitors in the exhaust air filters. The first sign of a leaking element would be a draw on the helium system, and no oxidation could take place because the continued helium pressure prevented exposure of the metal to air, giving reactor operators time to remove the element. If the leak was tiny, this could safely be postponed until the next scheduled shutdown.

Every now and then one of the "leakers," as they were known, was more urgent. The first such episode took place on 2 January 1952. Twenty minutes after a regularly scheduled startup, a drop in helium pressure signaled a leaking element. Partly due to pleas from "many research people anxious to use the reactor for the remainder of the day shift," the reactor operators decided to wait until 5:00 before shutting down the reactor to discharge the element. That proved a mistake. At 3:40, alarms in the exit filters signaled the presence of fission products in the exhaust, and the reactor was shut down. After removal, the leaking element was found to have two small cracks, each a few inches long, in which black uranium oxide was clearly visible. The removal procedure was a "dirty" operation itself, and resulted in gross contamination of tools and clothes that were inadvertently tracked to other parts of the building. The exhaust filters had caught virtually all fission products, and no sign of additional contribution to the normal background radiation appeared in the radiation monitors scattered around the site.[50]

Nonetheless, the episode prompted operators to revise their procedures for prompt removal of leaking elements and for reducing contamination during the cleanup. But two more leakers showed up in the next two months, a rate that Powell described as "somewhat ominous." After a review, the BGRR's engineers revised its operating procedures to minimize both the extent and frequency of the temperature changes to which the elements were exposed.[51] Still, a half-dozen or so serious leakers developed in the next few years. Although the released fission products contributed little to the natural background radiation, the fuel element discharge process often resulted in contamination of the entire building, which required the reactor to be shut down for several days. When discharged to the fuel storage canal, the leakers also contaminated the canal with fission products, including long-lived strontium 90 (half-life twenty-eight years) and cesium 137 (thirty-three years), some of which eventually showed up in the liquid waste discharge of the laboratory, and plumes from storage tanks, years after the old elements were replaced. Years later, too, small amounts of contamination also showed up in soil and groundwater in and around the reactor, in the vicinity of waste decontamination and concentration facilities, in sediments from the sand filter beds, and certain old landfills. The airborne amounts of strontium 90 and cesium 137 released by the lab and deposited on land were small compared to the levels due to the atmospheric weapons tests then taking place in Nevada and in other parts of the world. And the liquid radioactivity discharged into the Peconic River was a fraction (generally 2 to 15 percent) of the permissible federal limit. The principal source of radiation, above natural background levels, in the Long Island area was almost certainly atmospheric tests of nuclear weapons. Other occasional sources of accidental release into the environment included an occasional broken pneumatic tube or sample container and defective tubes in the evaporation plant.[52]

The problem of leakers was addressed indirectly, in the course of the BGRR's only important upgrade. At a visiting committee meeting of the department of nuclear engineering in January 1954, Alvin Weinberg mentioned the possibility of using enriched uranium as fuel. Enriched uranium would substantially increase the flux while allowing the reactor to be run at a lower power. Adapting the reactor would initially cost about $1 million but the savings in operating costs would be about $400,000 a year.[53] (Argonne's CP-3 had switched to enriched fuel in 1950; Holl 1997, 113.)

Haworth leapt at the opportunity, and put the department of nuclear engineering to work trying to exploit it. Modifying the fuel plate design of the Materials Testing Reactor in Idaho, which in turn was a version of the design of Oak Ridge's Low Intensity Testing Reactor, Powell devised an enriched fuel element. They were made up of thin plates of 5 percent alloy of ^{235}U and aluminum, and the plates served as their own cooling "fins," so to speak. The MTR and LITR were water cooled, and because air cooling is less efficient the BGRR plates had to be spaced further apart, and curved into an S shape to fit into the BGRR's smaller fuel channels. Fox drafted a proposal, and in July Haworth was able to ask the AEC to convert the reactor to enriched fuel elements. To avoid a long shutdown, conversion took place in several stages between December 1956 and March 1958, when loading the new elements was complete and the BGRR became the first large research reactor to use enriched uranium. The maximum central neutron flux increased fourfold, but the area in which the fuel elements were located decreased, removing the availability of a number of experimental ports.[54] While eliminating the problem of leakers—though defective elements still caused problems—the upgrade did increase the concentration of ^{41}Ar in the cooling air discharge, from about 7,000 to 20,000 curies a day. But these levels were still well within AEC limits.[55]

Over time, the acceptable emissions levels at nuclear facilities, and the standard practices for handling radioactive materials and wastes, have changed dramatically. One result of this "cultural shift," as it is often described, is that practices acceptable in the past are now considered unacceptable. Many of these practices—such as the so-called drag jobs, when highly radioactive material was put on the back of trailers in containers and driven, with escorts, to the waste management areas—would not be done today in the same manner. Another result of the changed culture is the existence, at Brookhaven and at numerous other sites with nuclear facilities, of areas of soil and groundwater containing amounts of contamination that are unacceptably high by today's standards. Brookhaven has attempted to identify, study, and remediate them where practical, though the necessity to do so sometimes has had painful consequences. In the fall of 1997, for instance, scientists discovered that rainwater infiltrating the old ducts of the

BGRR was producing pools of radioactive water, adding to large amounts there; before the ducts were finally sealed, an inch of rainfall at the lab created about 500 gallons of contaminated liquid that had to be disposed of at a cost of $15–17 a gallon. Because no budgetary provision had been set aside for this cleanup, funds for it were taken partly out of the overhead that normally would have gone into research, causing the programs of some individuals to suffer. For these scientists, thanks to the unfortunate legacy of past practices, it was literally raining on their research.

In 1989, Brookhaven became a Superfund site—less because of the to-tal amount of contamination (other sites on the Superfund list had vastly higher levels of contamination) than because of the risk involved in the lab's location above a sole source aquifer. The furious controversy about the lab's emissions that would develop toward the end of the following decade would concern the hazard, if any, posed by the residual amounts of plutonium, strontium 90, and cesium 137 left over mainly from the BGRR, and possible long-term effects of the ^{41}Ar and emissions from the leakers. These health concerns raised questions about the adequacy of the contem-porary regulations and standard practices, particularly in the possible rela-tion of any or all of these to an alleged increase of breast cancer on Long Island. But judging these requirements and standard practices by today's standards would require analyzing them in the context of present-day inter-ests, agendas, and knowledge, and hence must be deferred.

The HFBR

In 1954, with an eye to the future only shortly after completion of the BGRR, Hughes began to pressure Haworth for another reactor. Already, several reactors had equaled or bettered the BGRR for cross-section work (Chalk River, in Ontario, had a reactor with a flux several times that of the BGRR). In addition, the resolution of accelerator-based neutron sources (the one at Columbia's Nevis laboratory, for instance), was rapidly over-taking that of reactor-based sources. In August 1956, Haworth appointed a reactor study group to examine "the question of a high-flux research reactor," to cost $10 million, chaired by Hughes and including Dienes, Hastings, Gurinsky, Powell, and reactor theorist Jack Chernick. Haworth told the committee to come up with a "definitive budget proposal to be submitted in the spring of 1957 calling for construction funds in the fiscal year 1959." [57] The committee met weekly to discuss issues such as fuel loading and unloading, experimental arrangements, and so forth. The group decided early on to optimize the new reactor for intense external beams of neutrons, which were especially useful for neutron diffraction. Optimizing reactors for specific functions was a new approach. Existing research reactors were multipurpose, like the BGRR, and incorporated a number of design compromises to satisfy different groups of users. Oak Ridge, which completed a new reactor in 1957 (the Oak Ridge Research Reactor, known as the ORR), was also planning a new one to be optimized for the isotope production of transactinides, the High Flux Isotope Reactor (HFIR). Argonne's scientists, meanwhile, began planning a high-flux Argonne Advanced Research Reactor, or A^2R^2. In November 1958, members of those three labs, and scientists from the University of California, met with AEC officials for a general discussion of future high-flux research facilities (Holl 1997, 236).

Chernick was a key figure in Brookhaven's reactor study group, for he was able to translate theoretical specifications into practical design concepts. In late 1956 and early 1957, the committee reviewed two major types of reactors under consideration: a Chernick-designed, 10-megawatt D_2O-moderated reactor with small core, and a sodium-cooled, 100-megawatt reactor of slightly larger volume. It soon became clear that the first was superior for the lab's purposes, with the second costing significantly more than $10 million and involving many engineering difficulties.

Chernick's work, completed in 1957, was a breakthrough in reactor design. Apart from so-called fast reactors, reactors depend on *thermalizing* (slowing down) neutrons in a moderator to the energy where they can cause uranium atoms to fission. In conventional reactors such as the BGRR, the moderator is placed between the uranium fuel elements, which are spaced widely enough apart so that the neutrons emitted from one element have time to thermalize during the trip to the next element. Chernick's idea was to put all the fuel elements closely together and surround them by a large vat of heavy water. When first produced, most neutrons would be too energetic to cause fission and would leave the core area into the heavy water moderator. But it would take them longer to slow down in heavy water than in regular water, precisely because it's heavier: when one billiard ball collides with another of about the same mass, the two tend to split the energy between them, on the average—but when striking a much heavier object, it tends to rebound elastically, keeping its energy. A further advantage of heavy water is that the deuterium in it has a lower absorption cross-section than hydrogen, because one neutron is already tacked on to its proton. Thus the neutron flux would peak outside the core (the core is "undermoderated"), slowing down gradually in the heavy water, though enough thermalized neutrons would return to sustain the chain reaction, resulting in a remarkable flux distribution that is illustrated in figure 8.12 (Auerbach et al. 1958).

By contrast, Oak Ridge's HFIR was a "flux-trap" reactor, which traps the maximum flux in the dead center of the reactor for isotope production.

It was originally designed without beam tubes, which engineers feared might interfere with its central flux intensity, but Weinberg insisted that the reactor have beam tubes for neutron scattering, and some were installed.

Kouts and company did mock-ups of Chernick's designs, while Hastings kept looking for ways to maximize slow neutrons and minimize fast ones. Fast neutrons were a liability for neutron scatterers; not only did they contribute unwanted background, they also forced experimenters to install much extra shielding. Here Hastings came up with a design innovation: mount the beam tubes tangentially, thus reducing the high-energy neutron background. "Thermal neutrons are going in all directions," Hastings said. "So whichever way the beam tubes point, you get the same number. But if you point the tube tangentially, so they don't look at the core, you won't get the fast neutrons."[58]

The five designers—Chernick, Hastings, Joseph Hendrie, Kenneth

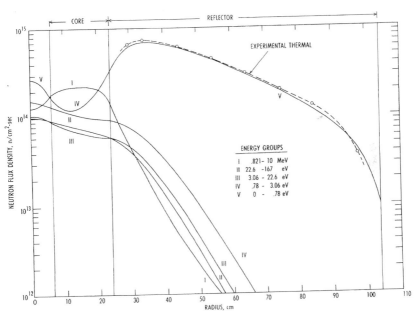

8.12 HFBR neutron profile. The calculated (and for the thermal neutrons, measured) neutron flux density in neutrons per square centimeter per second at different distances from the core, for five groups of neutrons. Group I are the most energetic neutrons, which have just come from fissioning nuclei. Groups II, III, and IV contain neutrons that have slowed down through collisions with heavy water nuclei in the moderator. Group V, the thermal neutrons, have lost all of their fission energy and are in equilibrium with the motion of the surrounding atoms. The significant feature of the HFBR was that its thermal neutron flux peaks outside rather than inside the core, so that they can be most efficiently used by experimenters.

8.13 HFBR inventors. In July 1961, Clarke Williams presents the HFBR's five coinventors with an honorarium ($25) paid to every BNL inventor on signing a patent application for the Atomic Energy Commission. Left to right: Jack Chernick, Julius Hastings, Joseph Hendrie, Williams, Kenneth Downes, and Herbert Kouts.

Downes, and Kouts—applied for and received a patent on the new system. They submitted their first proposal to the AEC in July 1957. The AEC had been on an austerity budget for two years, and the response was not encouraging. But on 4 October 1957, a tiny Soviet satellite named Sputnik altered the situation. Worried about declining American superiority in science, the U.S. Congress voted to include extra money for research in the AEC's budget. In March 1958, the AEC asked Brookhaven to resubmit its reactor proposal (another beneficiary was Oak Ridge's HFIR). The lab then contracted with General Nuclear Engineering Company for a conceptual design study of the heat transfer system, on the basis of which the lab could ask the AEC for money. General Nuclear was run by Wally Zinn, one of the principal members of the Chicago part of the Manhattan Project, father of the CP-5 at Argonne, who had gone into industry. Heat transfer was the crucial factor in determining the ultimate performance of the reactor; how fast one could get the heat out of the core determined the upper limit of the flux. The General Nuclear study concluded that the power level would have to be twenty megawatts and that the reactor could be built for

$10 million. Kouts's group conducted critical experiments on the design between July 1958 and April 1959, allowing finalization of features like core size and shape, fuel elements, and flux. Haworth struggled with the AEC for the lab to hold the prime contract, but in mid-1960 the AEC announced that it would hold the prime contract. Ground was broken in September 1961 (fig. 8.13).[59]

Donald Hughes never lived to see the groundbreaking. He was a diabetic but a hard drinker, and in the late 1950s circulatory problems began to take a toll on his health. In 1960, he died a week after suffering a heart attack. After his death, Palevsky took over the neutron physics group, and the Sigma Center was moved from the physics to the nuclear engineering department.

Although some groups continued to mount major efforts at the BGRR during the 1960s, the HFBR's impending completion caused many to spend their time preparing experiments there. After the HFBR came on line, in 1965, there was little reason to keep the old reactor running. Experiments could normally be done faster and more efficiently on the new reactor. The BGRR was put to certain specialized uses, such as testing out chemical engineering loop concepts, but there were not enough of these to justify its continued operation. It was shut down in June 1968, periodically reactivated over the next few months, and then shut down for good in 1969. Over its eighteen-year lifetime, it supported an experimental community that was exceptionally diverse in its people, auxiliary instruments, and scientific fields. Its closing heralded the end of an era for reactors; henceforth, they would be much more specialized tools.

While the graphite pile and support structures were preserved as part of a display in an exhibit center, all fuel elements were removed with the last batch shipped to the Savannah River reprocessing plant in 1972. In 1986, the BGRR was designated a nuclear historic landmark by the American Nuclear Society, and a bronze commemorative plaque can be seen on the reactor's south face.

CHAPTER 9

"For the Enlightenment and Benefit of Mankind": Research at the Cosmotron

UNLIKE THE BGRR, whose box-and-stack profile dominated the landscape from atop the highest point at the lab, the Cosmotron did not stand out visually. It was an ungainly-looking device sprawled out on the capacious, open floor of building 902; from the room's balcony, one could take in the seventy-five-foot machine with a single gaze (fig. 9.1). Still, the Cosmotron, like the BGRR, was a landmark in postwar U.S. science, as it was the first accelerator able to make cosmic ray energy particles reliably and routinely, "at home" in a laboratory environment well furnished with facilities.

While previous accelerators had been built, operated, and used by the same people, the Cosmotron's size made that no longer possible. As at the BGRR, a distinction evolved between operators and users, which Haworth formalized by creating a Cosmotron department, headed by George Collins, for the machine's operators.[1] But relations between users and operators are closer at accelerators than at reactors. While reactors produce neutrons at essentially one intensity and energy spectrum, accelerators can tune their intensity and energy, and often must to suit demands of experimenters, making for closer contact and often conflict between users and operators. At first, though, the number of operators and users at the Cosmotron was so small that both groups felt as if they belonged to a single community; everyone involved with the machine knew everyone else on it, including the janitors. Indeed, many former operators recall the Cosmotron as the last accelerator at which they had a good grasp of the physics experiments, and many former users remember it as the last accelerator at which they had a good understanding of the machine.

It was almost a family enterprise [Clifford Swartz of the Cosmotron department recalled]. I remember once walking over to building 902 to check up on an experiment on a clear summer night about 2:00 A.M. From afar I could see the doors were wide open, and soon I was happy to feel through the soles of my shoes the familiar thump! thump! of the motor-generator every five seconds telling me the machine was running. Getting closer, I began to *smell* it—the Cosmotron had a distinctive ozony-oily odor which permeated the surroundings. When I walked in the door I didn't see a soul, but I did hear music. I followed the sound through the warren of shielding blocks, and finally came upon the source. A graduate student, the only person on the Cosmotron floor—only one was needed!—was sitting with her feet up on the console, playing the flute while she watched the needles and gauges of the instruments. It was a different era in science.[2]

As this recollection suggests, each Cosmotron experiment usually was performed by a single person with an associate or two, and half a day's use of the machine at a time often sufficed to produce enough data. Collins initially proposed that everyone submit a one-page "request for experimental time" to a "Cosmotron Experimental Program Committee," which he ran, which would "consider the merits of all proposed experiments and presumably reach agreement in recommending a course of action to the Department Chair" (fig. 9.2).[3] In practice, since Collins ran the committee essentially by himself, experimenters informally approached him when they wanted machine time, and Collins managed to handle it this way, though with increasing difficulty, for several years.

It was all rather casual [Cosmotron department physicist Robert Adair recalled]. George [Collins] made the decisions, and dished out the time in an informal way and in small pieces: "Oreste [Piccioni], you'll run Monday and Tuesday; Sam [Lindenbaum], you'll run Wednesday and Thursday." That was the sensible thing to do at the time. The accelerator was not too oversubscribed, we ran it ourselves, and having the department head make all the important decisions was the style of the universities where we had been trained. Having a schedule? That would have been alien to the whole atmosphere. *Nobody* had *ever* done *that*![4]

Once a week, all the Cosmotron's experimenters gathered for a colloquium in the conference room of the Cosmotron building. This, too, was informal; the first choice of a speaker was whatever visitor happened by that day. If none were found, a Cosmotron group leader would fill in, or a group member.

It was one of the best times in my professional life [recalled physics department physicist Ralph Shutt]. We were seeing new things, week after week. At the weekly colloquia, when nobody else was available, I would often give the talk. But it wasn't hard, since each week there was something new to report, and I'd begin by saying something like, "This week's event is such and such."[5]

The Cosmotron experimental community, like that at the BGRR, was interdisciplinary, and though attendees at these meetings were mainly experimental physicists, they also included chemists interested in high-energy nuclear reactions, who worked alongside the physicists from the beginning.

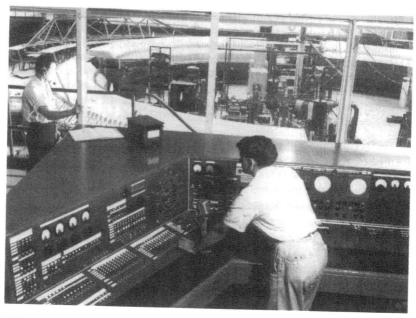

9.1 View from Cosmotron control room, overlooking the magnet ring, whose inner diameter was seventy-five feet.

Some theorists also attended, including Robert Serber from Columbia, who after 1951 spent one day a week at the lab as a consultant. Brookhaven employed several theorists, including Ernest Courant and Hartland Snyder, who pursued their own ideas, but Serber was different (fig. 9.3). Though unassuming to a degree almost unknown among physicists, Serber had a sweeping command of nearly all branches of theory and was able to see how all its different pieces fit together and thus their relevance to experiments. Said Joseph Weneser, a theorist who arrived at Brookhaven a year after Serber:

> Serber liked to tackle complex ideas and think them through until he had discovered the simple principles behind them. Then, when he'd explain them to you, he'd make it seem as easy as "What goes up must come down," and you'd wonder why there ever was a problem. And he was always available for discussions. Those kinds of people—Fermi was another—have a tremendous influence on the field, which may not come across just from papers or citations. (Serber 1998, xvi–xvii)

Serber was the most influential theorist at Brookhaven for at least a decade.

Name __M. Blau__
Request for Experimental Time   Date __July 1953__
                                 Scheduled
                                 Date _____

Purpose of Experiment: __Nuclear interactions of + 250 mesons in nuclear emulsion.__

_____

Time Requested _____ __8 hours__

Beam Requirements: Energy __2.3__ BEV, Intensity __full__
                   Repetition Rate _____ Seconds

                   Other Requirements _____

                   _____

                   _____

Target Number _____   Description: Dimensions __6" x 6"__
                                           Material __beryllium__

                                           Special Features _____

                                           _____

                              Location in Vacuum Chamber __Upstream side of E-straight__
                                                                              __section__
Fixed Target Located At __35½__ inch radius

Pulsed Moving Target: Target to begin at _____ inches radius and be at _____
                      inches radius _____ milliseconds after T_0

                      Target Travel Time Requirements, if any _____

                      _____

                      Pulsing Rate _____

Auxiliary Equipment Required: (Include those items requiring rigging)
            __Large magnet in 32° direct beam.__

_____

_____

Remarks: _____ __Other people may be able to work at same time in direct beam.__

9.2 Typical request for experimental time at the Cosmotron in the early years, by Marietta Blau, who used emulsions to study nuclear interactions (on Marietta Blau see Galison 1997, 146-60).

9.3 Several theorists (Brookhaven staff and summer visitors) in Gertrude Scharff-Goldhaber's office with her nuclear model hanging on the wall at right, 1959. From left to right: Lenard Eyges, George Trigg (with pipe), Steven Moszkowski (back to camera), Robert Serber (facing, with glasses), Ralph Behrends, Theodore Berlin, T. D. Lee (arms folded). Photograph by Gordon Parks, *Life* magazine. © Time Inc.

In 1952, at the time of the Cosmotron's dedication, about a dozen sub-atomic particles were known. The familiar proton and neutron were the building blocks of the nucleus. The electron had an antiparticle, the positron; the electron's sibling particle, the muon, had physical properties seemingly identical to the electron's except for its mass, which was two hundred times as great. The photon was the particle of light. Mesons included pions—plus (π^+), minus (π^-), and neutral (π^0)—that seemed to serve as the nuclear glue that held together the protons and neutrons in the nucleus. Especially puzzling were other types of particles known as "V" or strange particles. It was unclear how many existed or what they were, and at first they were simply called V_1, V_2, and (possibly) V_3 (Noyes et al. 1952; Pais 1952). These particles had been called V particles because they decayed into a pair of particles that traveled away, forklike, from their point of origin. They had been dubbed strange particles because, although they were produced in strong interactions (interactions involving the strong nuclear force), they did not decay with the rapidity associated with all other known strongly interacting particles. Instead, they lived a *trillion* times longer—around 10^{-10} instead of the expected 10^{-22} seconds—a time span of the sort associated with the far feebler force involved in beta decay, sometimes called the Fermi force after the physicist who described its existence. The result was a small theoretical crisis: How could particles be produced in a strong interaction but decay only weakly?

An inspired guess, by Abraham Pais of the Institute for Advanced Study and Murray Gell-Mann of the University of Chicago, was that strange particles had a previously unknown quantum number (one of a handful of numbers that describe how a particle behaves): strangeness. Strangeness was conserved in strong interactions, meaning that a particle could not lose its strangeness by decaying, as strongly interacting particles normally do, in 10^{-22} seconds. Strangeness was violated (not conserved) in Fermi-type weak interactions, letting strange particles decay into nonstrange particles after 10^{-10} seconds. If true, this idea implied, in collisions of nonstrange particles, that strange particles were created in pairs, one particle with a strangeness of $+1$, the other with a strangeness of -1, so that the net strangeness was 0 (Pais 1952).[6] Because the *associated production* of strange particles, as the phenomenon was soon called, had never been observed, physicists were eager to look for it when the Cosmotron came on line, at an energy level comfortably above that required to create strange particles (about 900 MeV).

First Discoveries: 1952–1954

Associated production was first seen by the cloud chamber group led by Ralph Shutt, in a new kind of cloud chamber whose development he and his group had pioneered.

Born in Switzerland in 1913, Shutt grew up in Berlin, where he attended the Technical University and received a Ph.D. in 1938. Under increasing danger (his mother was Jewish and would spend the war in a concentration camp), Shutt fled to Switzerland in 1939 and spent half a year scraping together the affidavits and $3,000 deposit necessary to obtain a U.S.

visa. Soon after he arrived in New York, he met Thomas Johnson, a physicist building a cloud chamber on the campus of Swarthmore College who hired Shutt as an assistant.

Cloud chambers utilize principles of cloud formation to create conditions that allow charged particles to leave contrail-like tracks of small droplets as they pass through the chamber (on cloud chambers see Galison 1987, chap. 2). These tracks can be photographed and examined for clues to the nature of the particles that created them. By surrounding the chamber with a magnet, physicists can bend the trajectories of the particles, yielding information about their momenta (from the amount of curvature) and charge (from the direction of curvature).[7] Shutt and Johnson built one two feet in diameter and six inches deep, controlled by a Geiger counter: every time a cosmic ray passed through the counter, it caused a piston in the chamber to retract. This suddenly diminished the pressure in the chamber, causing the air to become supersaturated and creating the conditions in which charged particles left visible tracks just as a camera snapped a picture. Shutt followed Johnson when the latter moved to Brookhaven National Laboratory in the summer of 1947, replacing Norman Ramsey as head of the physics department. With Johnson occupied by administration (in 1951 he would leave Brookhaven to become director of the AEC's research division), Shutt became head of a cloud chamber group, whose other members were Alan Thorndike, William Whittemore, and the brothers Earle and William Fowler.

Until the Cosmotron was completed, Shutt's group—and all others interested in high-energy particles—had no alternative but to seek out cosmic rays in the conventional ways. Thorndike sent lightweight cloud chambers aloft in balloons to study cosmic rays in the upper atmosphere; not from Long Island, where surrounding bodies of water made this risky, but from various midwest locations. Other Brookhaven physicists studied cosmic rays on airplanes or mountaintops. "In retrospect," Thorndike said, "these were things to do until the Cosmotron was in operation."[8]

Of far more long-range importance was the Shutt group's work in developing a new kind of cloud chamber. Conventional expansion chambers had serious drawbacks. One was cycling time, the time between expansions. Each expansion sent the pressurized gas swirling (the gas was pressurized to slow down the particles, making measurements more precise), and the device could not be used again until the gas settled back down, which could take up to half an hour. Because the Cosmotron produced particles every few seconds, such chambers were infuriatingly inefficient, and would be effectively idle for most of the Cosmotron's running time.

In 1950, Shutt saw an article by Argonne National Laboratory physicist Alexander Langsdorf outlining a scheme to achieve supersaturation by diffusing a vapor from a warm, nonsupersaturated region to a cooler region, where the vapor would condense.

> The basic idea was this [Shutt recalled]. Up *here* [pointing to the top of the device] you have water and/or alcohol vapor at normal temperature. Down *here* you make it cold—you have a black glass plate and you put dry ice or some refrigeration system under it. Now, since *this* is warm and *this* is cold, the vapor tends to diffuse downward, and at some point it gets so saturated that you get fog. But just before the fog, it's supersaturated and you can see particles; that is, their passage leaves tracks. And not just every few minutes—all the time.[9]

Within a few days Shutt and his group put together a small chamber of the sort Langsdorf had described, and were soon working on larger, more efficient models. The diffusion chamber was so simple in principle that one day Whittemore brought in a peanut butter jar to see if he could make one out of household ingredients. By putting a piece of felt, soaked in an alcohol mixture, at the top of the jar, and setting the jar on a piece of dry ice he was able to see cosmic rays flit through—and sometimes even cosmic ray showers, which would fill the jar with tracks. The scientists issued a press release to alert schools about the idea, which was widely publicized. Today, diffusion chamber kits for educational use can be bought commercially from outfits like Edmund Scientific Company.

Shutt's group built a large, high-pressure diffusion chamber with a two-inch sensitive layer, and late in 1952 wheeled it over to the Cosmotron. The chamber was not ideal, and was often swamped if too many tracks passed through.

> But it worked [Thorndike said], which was good because the Cosmotron was not terribly reliable in its first years, so when the beam came on you didn't want to spend two or three hours fussing with your equipment to get it to work right. It was important to be able to throw the switch and having everything on line to take data for a few hours.[10]

The chamber was installed near the south straight section. A target in that section produced swarms of secondary particles from which the group culled beams of neutrons and pions and sent them into the chamber. Thanks to tests and initial difficulties with the accelerator, the group was unable to have the first of what became known as "chamber days" until 13 March 1953. The morning was spent filling the chamber and getting it to work. Getting the Cosmotron to run smoothly was another matter, and the operators had the beam ready only that afternoon. Shutt's group then split up, with Shutt and William Whittemore remaining at the chamber, William Fowler and electrical engineer William Tuttle working the generator that powered the magnet, and Alan Thorndike stationed in the Cosmotron's control room. Shutt and Whittemore took a short sequence of pictures as a test. To their dismay, they saw that the Cosmotron's beam was flooding the chamber with particles—a problem unknown to cosmic ray researchers. Consulting via three-way telephone, the group members realized that most of the problem was created by stray particles that lingered around in the chamber long enough to get into the next picture. They had the Cosmotron crew reduce the pulse repetition rate from every five to every eight seconds, allowing more time for the chamber to clear. After another test strip showed this to be satisfactory, the run began.

Chamber days were held about once a week. During them, the Van de Graaff injector sent a bunch of particles into the Cosmotron ring every eight seconds. The motor-generator thudded, the bunch was accelerated and smashed into a target, spraying neutrons down a beam pipe into the diffusion chamber, where the camera photographed any interactions. Over these days, the group took several dozen hundred-foot rolls of thirty-five-millimeter film, twenty thousand pictures in all, which were sent to photography to develop.

To the uninitiated, a photograph of cloud chamber tracks looks as unmeaningful as the lines cut in a frozen lake by a schoolful of skaters on a winter afternoon. To the trained eye, it can be a window into a different world, one in which different types of particles collide and speed off in new directions with new identities and properties. A world appears to physicists in such photographs, observes science historian Gerald Holton, which is one reason why "some critics of science are so wrong in thinking of modern science as entirely depersonalized, cold and abstract, devoid of all personal concerns" (Holton 1973, 106). The laborious, triage-like process of sorting out the interesting photographs—by far the most time-consuming part of the experiment—is called scanning.[11] Collisions between the neutrons and hydrogen in the cloud chamber were rare, occurring only in about one out of every hundred pictures, and usually resulting in the creation of two pions. On 3 April, after one chamber day, Thorndike was leafing through a newly delivered batch of developed photographs. His eye was caught by one picture containing a fork-shaped track at almost dead center, the characteristic signature of a V_1 particle. It had been created by a collision between a neutron and a nucleus in the wall of the cloud chamber—Thorndike could not tell whether a second V had been produced—but it was the first indisputable example of an artificially produced strange particle. A few weeks later, Shutt took home a batch of pictures. After clearing off the kitchen table, he examined them with a magnifying glass. One had a strange particle produced in the cloud chamber itself, and enough information was available about the interaction to show that some momentum was missing, presumably taken away by a second neutral strange particle which had not decayed in the picture. Soon another such event showed up, and the experimenters wrote a first article on their work: "[T]he present results are consistent with the possibility of production of V together with one other heavy unstable particle" (Fowler et al. 1953a,b). Shutt and company found several more, at a rate of about one every two thousand pictures; in all, they observed about half a dozen Vs during the first set of chamber days (fig. 9.4).

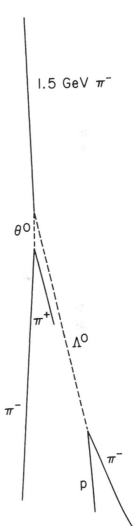

9.4 One of the first events indicating evidence of associated production of V or strange particles, showing two (a Θ^0 and a Λ^0) produced in a collision between a π^- and a proton. From Fowler et al. 1954a.

In the first two years of the Cosmotron's operation, the group took some 150,000 photographs. Over time it looked more certain that strange particles were produced in accord with the hypothesis of associated production. The group also saw other types of strange particles, and identified and measured their properties (Fowler et al. 1954a, 1955). Shutt began to farm out the exploding numbers of photographs to university groups. Yale was one. Earle Fowler, who left Brookhaven for Yale in 1952, borrowed a scanning and measuring machine from the lab for his grad students to use. They worked with Shutt's group for a few weeks to learn the ropes and returned to Yale with film, where the images were scanned by English majors. "They were available and hungry," recalled Thomas Morris, a graduate student of Fowler's in 1954. "The AEC didn't give grants to English departments." [12] Morris and his colleagues measured the interesting pictures, filled in data sheets, and mailed them to Thorndike. Soon other university physics departments followed suit, and the activity of scanning, analyzing, and interpreting the Cosmotron's cloud chamber pictures dispersed widely throughout the east coast.

Other physicists soon asked to operate cloud chambers at the Cosmotron. Among them was Leon Lederman, an energetic assistant professor at Columbia with a skillful eye for important problems. Lederman wanted to take a Columbia-built chamber out to the Cosmotron. The proposal was a perfect exemplar of the "Brookhaven concept"—that the lab's unique facilities were for experimenters from the outside—but the theory had not yet been tested. Indeed, lab researchers, especially those who had set aside their own work to help build the machine, felt entitled to priority over outsiders. In addition, they had played the role of outside users at Nevis, where Lederman and his Columbia colleagues had been insiders. As a result, Lederman recalled:

> We were not exactly encouraged. At first the inside experimental groups viewed us as competitors, but once we got set up it was all right. At the time I thought the user concept was *my* idea—but I soon learned it was what the wise Rabi had envisioned all along. [13]

Scaling up his previous chamber fivefold, Lederman built a thirty-six-inch expansion chamber at the Nevis workshops with a graduate student named Kenneth Landé. They drove out the assembly to Brookhaven in an old gray navy truck from Nevis, but discovered that the chamber plus magnet was too large to fit on the experimental floor. Lederman's group ended up using a farm catalogue to obtain a nine-foot prefabricated shed of the sort used to store corn, assembled it outside building 902, and connected it to the machine via a pipe through the wall (fig. 9.5).

Cloud chambers were biological things. You had to use magic to keep them working, and they could be easily disrupted. You were trying to take very precise measurements of how the tracks curved in the magnetic field. So you waited a few milliseconds for the tracks to grow to visible size, then popped a rubber diaphragm that's a full thirty-six inches in diameter to expand the chamber and

9.5 Rigging Columbia cloud chamber into corn crib.

cool the gas. If you didn't do it just right, the gas swirled, the tracks wiggled, and your measurement was bad. And any dirt could poison the gas. So you had to do a lot of magical things to get it to work, many of which had no scientific basis but you'd read that somebody did them so you did them, too. And you developed some witchcraft of your own and published that for the benefit of others. But then there was the moment you finally get it working and all you have to do is remember to take the lens cap off and your big flash lamps flashed and you took thousands of photographs and you felt like a scientist again.[14]

Besides studies of strange particles, another type of research at the Cosmotron involved measuring particle cross-sections. Just as a nuclear cross-section gives the likelihood of an interaction between a neutron and a nucleus, a particle cross-section gives the likelihood of an interaction between one particle and another. Brookhaven had two early counter groups measuring the pion-proton (π-p) cross-section; a group led by Seymour Lindenbaum and Luke Yuan examined the low-energy end, while one led by Oreste Piccioni and Rodney Cool studied the high-energy end. These cross-sections were especially important, for information on how pions interacted with nuclear particles was thought to be vital to understanding the nature and strength of the nuclear force. The data also shed light on a puzzle encountered by Fermi and his group the previous year.

Fermi had measured pion-proton cross-sections at the Chicago cyclotron up to the limit of its ability to produce pions. The cross-section for the pi plus was rising at the limit, and for the pi minus seemed to level off after rising steeply. An Indiana University theorist named Keith Brueckner thought the pions and protons might be behaving analogously to nuclei in "excited states," momentarily being held together as a unit by the nuclear force until shooting off again. If so, the behavior would be a "particle resonance," analogous to a Breit-Wigner nuclear resonance. But Fermi and company cautiously declined to so identify it; "Unfortunately," they wrote in 1952, "we have not been able to push our measurements to sufficiently high energies to check on this point" (Anderson et al. 1952, 936; see also Brueckner 1952). The question was important to Fermi, for it had implications for a theory according to which no yet-unobserved particles or resonances existed (Fermi 1950, 1953). But higher-energy measurements of the direct pion-proton cross-section had to await the Cosmotron.

The basic technique for measuring pion-proton cross-sections was the same as that used to measure neutron-nuclei cross-sections: one measured the intensity of an incident beam with and without a sample in it, the difference between the two measurements giving the total cross-section, or how many particles in the incident beam were scattered by the target. In 1953, Lindenbaum and Yuan discovered that the cross-sections indeed peaked at about 180 MeV, and fell off sharply thereafter. This sharp drop, the authors wrote, suggested "a resonance or near-resonance interaction" that fit Brueckner's scheme. Further cross-section work by their group, combined with the higher-energy pion-proton cross-sections measured by Piccioni and Cool, and proton-proton cross-sections measured by other groups at the Cosmotron (fig. 9.6), contributed more evidence for the resonance model (Lindenbaum and Yuan 1953, 1954; Cool, Piccioni, and Clark 1956; see also Walker 1989).

Fermi died in 1954. Today, the pion-proton peak is known as the delta particle, 3-3 resonance (spectroscopic notation referring to its spin of 3/2 and its isotopic spin of 3/2), and occasionally as the "Fermi peak." Fermi could barely have suspected how quickly "particle spectroscopy" (called that, again, by extension of the analogy with nuclei) would burgeon.

THE EXPERIMENTAL community working at the Cosmotron was interdisciplinary from the beginning. The principal user of the Cosmotron in the chemistry department, for instance, consisted of Gerhart Friedlander and a group of young nuclear chemists that joined him. Friedlander had been born in Munich in 1916, and by the time he had finished high school in 1935 Jewish students could no longer safely enroll at universities. He was fortunate enough to obtain an immigration visa to the U.S., where the Hillel Foundation at Berkeley, which had collected enough money to support one German refugee for a year, supported his enrollment. During the war, Friedlander worked at Los Alamos, and afterward went to G.E.'s Schenectady lab before coming to Brookhaven in 1948, where he had to survive an already-described security clearance process.

Friedlander's first work at Brookhaven was on nuclear decay schemes using the sixty-inch cyclotron, and nuclear reactions using the Nevis cyclotron, but he soon became interested in the study of nuclear reactions induced by the 3 BeV high-energy protons that would be available at the Cosmotron. In 1947, Robert Serber (then at Berkeley) had worked out a theory showing that a nucleus would not look opaque or even cloudy to high-energy particles, but rather like a collection of individual objects (Serber 1947). The theory described a collision between a high-energy particle and a nucleus as involving two processes. In *intranuclear cascade,* the incoming particle would strike a nucleon and knock it and maybe one or two more out in a more or less elastic collision in about 10^{-23} seconds, while in *evaporation,* the excitation energy left behind would concentrate on a nucleon and in about 10^{-20} seconds throw it off the way a kettle of boiling water throws off drops—but because little transfer of large amounts of energy would take place, this would involve at most a few nucleons. By 1952 this theory had been tested in the region covered by existing accelerators, up to a few hundred MeV.

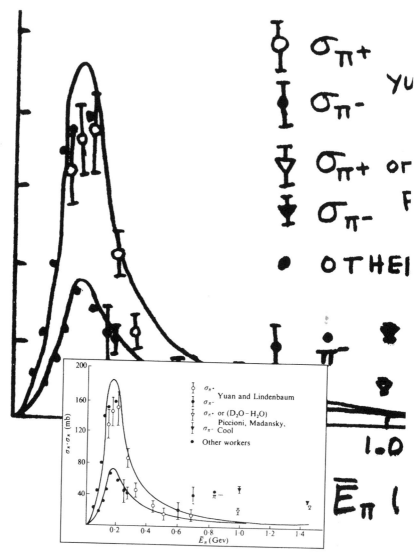

9.6 Early data from the Lindenbaum-Yuan group at the Cosmotron on pion-proton scattering, showing the "bump" indicating a resonance, and an inset showing continuation at higher energies by other Cosmotron groups.

In summer 1952, while the Cosmotron was being tested, the accelerator physicists put a lead brick in the vacuum chamber as a beam stop. After the brick served a few weeks in that role, in the course of which it was bombarded with proton beams of different energies and intensities, Friedlander persuaded the physicists to let him have it, giving them a substitute (fig. 9.7).

9.7 Gerhart Friedlander working at the Cosmotron, surrounded by shielding blocks.

Friedlander and Jack Miller, a summer visitor from Columbia, took the brick to their lab, cut off several thin slices of lead from the front of the brick with a hacksaw, and proceeded to dissolve each layer and perform chemical separations on the slices, to study what was the general pattern of nuclear reactions.

> We worked through the night, and I remember at one point doing an iodine oxidation with the purple iodine fumes rising around us like in a Hollywood movie. We knew right away we had an interesting result. In Serber's model, the nucleus looks pretty transparent at the Cosmotron's energies and the chances were very great that only one or two nucleons would evaporate. But we found immediately a huge and fantastically active fraction of rare earths, which have atomic numbers from 57 to 71. That meant that from the target nucleus lead, of atomic number 82, at least 50 or 60 nucleons were being boiled off. It was clear that we were seeing much more complex processes.[15]

They did not publish this rough-and-ready initial work because an indeterminate mix of energies and intensities had been involved in tuning the beam. Still, much larger energy transfers to the nucleus had clearly taken place than described by Serber's model, with large probabilities of hundreds of MeV being deposited in the nucleus. After further, more controlled investigations (Friedlander et al. 1954), Friedlander and a number of associates came up with the idea that the mechanism must have to do with pion production inside the nucleus. After entering the nucleus, the proton would produce a pion, which would be a much more effective avenue to transfer energy to the nucleus given its short range in nuclear matter; this, in turn, would lead to much more complicated evaporation phenomena. In collaboration with various individuals from Columbia, Los Alamos, and Chicago, the Brookhaven researchers developed a model for intranuclear cascades and evaporation in these high-energy proton-induced reactions, which has since been extended and elaborated to apply to heavy-ion reactions.

The Cosmotron was built by and operated for physicists; all others took a back seat, and often had to work in the wee hours of the morning. But Friedlander's was one of the few nonphysics groups to acquire a measure of power.

> The saving grace for us was that, for a long time, the physicists depended on us for a very crucial piece of data: the absolute calibration of the beam. We determined cross-sections for certain reactions, and once we knew the cross-section for a reaction it could be used to measure beam intensity. We would expose an aluminum foil along with an experiment; in a short irradiation we would measure the sodium 24 produced in the aluminum, and in a longer one sodium 22, to determine the intensity for the experimenters. Even more importantly, we could determine the spatial distribution of the beam. We would expose a piece of foil to the full diameter of the beam, and then cut it up into little pieces, from which

we could tell exactly what the intensity was at any given point. By being willing to do a certain amount of this service work we had a certain amount of leverage to do our own stuff.[16]

Another in Friedlander's informal group, Jim Cumming, became an expert at beam monitoring by carrying out very careful measurements of these monitoring reactions, publishing a series of papers about cross-sections consulted by accelerator users at every major accelerator into the 1980s (Cumming, Friedlander, and Swartz 1958; Cumming and Hoffman 1958; Cumming 1963).

BY FALL 1954, the Cosmotron's initially shaky performance had improved and usable beam time had steadily increased. The machine supported a large experimental program with several cloud chambers and over a dozen counter groups. But on the afternoon of 5 November 1954, smoke suddenly began issuing from one end of quadrant #2. The operator shut the machine down and helped extinguish a small fire in the insulation. The source of the fire seemed to be the inner copper coils and to reach them, all experimental equipment and shielding were removed from that quadrant and the vacuum chamber yanked. A tiny fracture in one coil had allowed a trickle of water to escape, causing a short between that bar and the next.

The coils generated the huge electromagnetic fields that held the protons in orbit. Every five seconds, the field suddenly rose, and then just as abruptly relaxed, pulling not only at the protons but also at the coils themselves. During construction, the engineers had done what they could to reduce the motion of the innermost, most active coils down to about 0.015″ per pulse. The crack had developed in the innermost coil bar, the shortest and least flexible, after 3.5 million pulses. Were all the coils growing fatigued, in which case all four quadrants would have to be repaired, or was only this one defective?[17]

Tests of the cracked coil showed it was likely weak when installed, implying an accelerator-wide overhaul was unnecessary.[18] But dismantling the entire quadrant and replacing the coils required a six-month shutdown. The Cosmotron's operators took the opportunity to introduce improvements, the most significant of which was modification of the vacuum chamber of quadrant #2 to allow an external beam to take protons out of the vacuum chamber directly to experiments (Piccioni et al., 1955; the first extracted beam was at Liverpool, Holl 1997, 209; an external beam was already proposed for the Bevatron, Wright 1954, but not installed in that machine until later).

The improvements resulted in a threefold increase in intensity, raising radiation levels in the floor and control room, and more shielding around the machine and experimental areas had to be installed. Until then, safety had been treated rather laxly, consisting mainly of a red warning light when the machine was in operation, and a bell that rang when it was turned on and off, signaling experimenters to leave the area. Willy Chinowsky, a Columbia graduate student, once had his view of the red warning light obscured; thinking the machine was off, he continued to work until he suddenly heard the bell and the words, "Cosmotron off!" Health physicists realized he had received a hefty dose of protons directly to the chest, but did not know how much because the beam had not conveniently struck the radiation badge on his lapel. Maurice Goldhaber, naturally, figured out a way to determine Chinowsky's exposure: by measuring the calcium activity that the protons had made on striking the buttons of his shirt. The incident highlighted the need for more care, and a door to the experimental area was installed that would lock automatically when the machine was on and scram the machine if opened. Still, security was taken rather lightly—no one wanted to disturb experiments in progress if they didn't have to—and Collins pretended not to notice the ladder that always leaned up against the shielding blocks, enabling one to bypass the door.

Some physicists objected vociferously to the additional shielding. When a plan surfaced to shut down the Cosmotron for about a year to install seven to eight thousand tons of concrete around the machine, Cliff Swartz argued vociferously for milder precautions in a memo entitled "A Shielding Manifesto":

> A spectre is haunting the Cosmotron—the spectre of inundation by concrete . . . The disadvantages are so serious that the whole chain of reasoning leading to such a drastic step should be described in writing and independently analyzed. . . . Several people have received doses during one week greater than tolerance, but no one has ever received over one third of the integrated 13 week tolerance. . . . [I]t is not so clear . . . why it is necessary to wrap up the whole machine in this concrete tomb. . . . We have a whole year to win. Physicists of all departments, unite! [19]

The external beam created a different safety issue. Until then, the 3 BeV protons remained inside the machine; now, stray protons might occasionally pass out of the building into the back yard of the Cosmotron. Swartz, who had set up controls to monitor the external beam (his beam measuring device was nicknamed Barrymore because it revealed "*the* profile"), was asked to find out if the protons had a long enough range to be a problem.

In the summer 1955, when the external beam was being tested, Swartz sent a health physics team armed with surveying equipment, Geiger counters, and walkie-talkies to the perimeter road encircling the laboratory, about a quarter mile away from but directly in line with the mouth of the external beam. Security guards were posted on the road to stop cars.

When the team was ready, the first two proton pulses shot through the external beam. To Swartz's amazement, the team heard clicks on their Geiger counters—the pulses had passed clean through the forest. The team moved to Route 25, several miles away on the north border of the lab. Again the Geiger counters were lined up, the Cosmotron delivered pulses, and the counters picked up a signal.

> It reminded me of a story I'd heard about Theodore van Kármán, the early aerodynamicist, who as a youth lived on the coast and would stand on the pier feeding sea gulls and watch their flight. He got so skillful that he found he could lead them into stalls; imagine, a human being, able to make sea gulls stall! Or the story of Vincent Schaefer, a General Electric scientist and inventor of cloud seeding, who when the weather was just right would go out with his fog-making machinery outside of Schenectady and bring down a fog in the entire region. Particle physicists are used to studying their phenomena in the clinical conditions of the laboratory; they only rarely encounter them outside in the world. But there they were, standing on the dirt shoulder of the deserted road, next to a line of trees, the Geiger counters picking up a tiny signal of protons, seemingly from nowhere, that had traveled several miles through the woods. It was rather romantic.[20]

An earthen berm was then built behind the Cosmotron building to catch stray protons and prevent them from going further than the back of the building.

Designing the AGS

Meanwhile, the Cosmotron's designers were hard at work on their strong focusing accelerator design. At the beginning of 1953, they received a rude shock. The idea of strong focusing had not only already been described, but patented, and a law firm had contacted the AEC claiming patent infringement.

The inventor, an American-born Greek electrical engineer named Nicolas Christofilos, had sent an unpublished manuscript, entitled "Focusing Systems for Ions and Electrons and Application in Magnetic Resonance Particle Accelerators" to Berkeley.[21] The manuscript was only superficially read, and ended up in a dusty file. Then, in late August 1952, Christofilos read of the Brookhaven discovery and recognized it as his idea. He contacted a Greek law firm with Washington associates, borrowed advance money from them, and left Athens for the U.S. in January 1953. Retrieving his manuscript from the file, red-faced Berkeley scientists saw that while the details of his scheme differed from those of the Brookhaven scientists, the principle was the same. Courant, Livingston, Snyder, and Blewett promptly acknowledged Christofilos's priority (Courant et al. 1953). The AEC followed suit, and in May a license agreement was reached between the AEC, AUI, and Christofilos granting him ten thousand dollars in exchange for the right to use the strong focusing method. Christofilos wanted to come to work in the U.S. on fusion, but while the AEC wanted him to become a part of their program, he did not have security clearance and they refused to put him in a fusion program. With the AEC's concurrence, Haworth offered him a position at Brookhaven.[22]

Newspaper accounts tended to treat the Christofilos story in terms of the popular myth of the scientist-inventor as solitary, despised, and overlooked ("Letter from 'Crazy Greek' Ignored, U.S. Misses Boat on A-Invention" was a *Washington Post* headline).[23] A fuming Green composed a response that is worth quoting for the way it expresses a scientist's perspective of such treatment.

> The cold fact is that when we needed to take another large step in energy, members of our staff sat down and made the necessary inventions and calculations without benefit of any outside inventors. A highly complex mathematical theory of the alternating gradient, or strong focusing, accelerator was developed by Courant and Snyder of this Department in cooperation with theoretical physicists all over the world. This theory forms the entire basis of the design of our machine and is quite independent of the work that Christofilos did. When it turned up that he had independently conceived the strong-focusing idea, we were intrigued and impressed. However, his mathematical work at that time was less advanced, and his calculations were scarcely suitable for the design of an operating machine. Whether the construction of this machine would have been accelerated if we had learned of his work earlier, or whether we would be on the same schedule as now, is a matter requiring a crystal ball. My own opinion is that it would have made little difference. The time had become ripe for the invention of strong-focusing and it was inevitable that someone would happen onto it. To be historically correct, the principle was really laid down in a paper by Thomas in 1938. Since we did not need to economize on magnets at low energy, neither Thomas nor the rest of the profession realized the implications of his "Butterfly Cyclotron." Scientific discoveries are always made by a large number of people who contribute various bits of information and who stimulate one another with conversation and letters concerning their ideas. The person who finally synthesizes the entire result should always have great credit for his perception, but there is an inevitable tendency to crown him with complete credit to the utter disregard of all the others. This tendency is even more marked if there are some romantic or human-interest overtones.[24]

By the time Christofilos arrived in March 1953, the working group members had made a small amount of progress. For a while, they were unable to decide on which energy goal to shoot for, with opinions ranging from 50 to 125 BeV. Nor did they have a name. George Collins suggested

"Project Proteus." John Blewett proposed "Chorotron" or "Kolotron," implying a device in which particles "dance" around a circle. Other names were "Thorotron" and "Project Procrustes." Hildred Blewett, exasperated at the inability of the working group to settle on an energy, suggested AGNES, for "Alternate Gradient Nebulous Energy Synchrotron." As no better name had emerged, for a while they called themselves the "AGNES group" before reverting to the more prosaic placeholder acronym AGS, for "Alternating Gradient Synchrotron." [25]

In March, Haworth arranged a meeting between Brookhaven staff and representatives of MIT, Harvard, and Princeton to discuss strong-focusing accelerator design. In advance, Brookhaven accelerator scientists went over plans and cost estimates for strong-focusing machines of 20, 50, and 100 BeV. These studies brought everyone back to reality, and Haworth opened the meeting by telling those assembled, "We have been finding out that the construction of machines in the 15 to 100 Bev range is neither especially easy nor particularly cheap." [26]

Two problems in particular were serious enough to threaten the very possibility of a strong-focusing machine. British scientists discovered that the beam would be extremely sensitive to minute errors in placement or design of the magnets (Adams et al. 1953). Errors as little as a thousandth of an inch could cause enough destabilizing oscillations in the beam to make it impossible to control the particles. After much discussion, Courant found what looked like a solution, involving a slight increase in aperture size. But whether the scheme would work in practice was far from clear (Courant 1953).

The second, more intractable problem, which Courant, Livingston, and Snyder had already noted in their first paper on strong focusing, was *phase transition.* The accelerating cavity of machines like the Cosmotron uses a radio frequency wave to push the bunches of particles; in effect, these particles ride a certain spot on the slope of the rf sine wave when they cross the accelerating cavity once per orbit. The particles do not arrive together in a dot, but in a small package, some ahead of others. But the package tends to stay together on the back slope of the sine wave, where the trailing particles are kicked a little more energetically and the leading ones a little less. The particles in the bunch wind up oscillating about a point about midway up the slope.

At higher energies, the relativistic limit of the speed of light comes into play. As the particles acquire more energy and approach the speed of light, they acquire mass instead of speed. The more massive and energetic particles, moving in wider orbits, now lag behind the less energetic particles, because they travel a greater distance but move at about the same speed. Here the situation is reversed, and to keep the particles together it is now necessary to give the *trailing* particles (i.e., the more energetic ones) less of a boost, and the *leading* (less energetic) ones more of a boost. This apparently paradoxical situation—as the energy goes down the speed increases—is analogous to the way the space shuttle is sometimes slowed down, dropped to a lower orbit in which it moves faster, in order to catch up with a satellite ahead of it. In the case of accelerators, the *phase* of the accelerating wave needed to keep the package stable is different, and becomes that of the front of the sine wave rather than the back. Phase is an important property that would be an ingredient of key discoveries of high-energy physics in the 1950s. In classical mechanics, the phase is the portion (usually expressed as an angle) of a periodic or cyclical motion that is traversed at a certain time. The minute hand of a clock, for instance, has a period of an hour (a 360° cycle), and its phase is what fraction of that amount it has moved through. After fifteen minutes, for instance, the hand has gone through a phase angle of 90°. In quantum mechanics, particles are described as being like waves and are said to have amplitudes, and also phase.

Courant argued that phase transition, or moving the particles from the back to the front of the sine wave, could be made at the point where the focusing forces balance, around a few billion electron volts. Aiding the physicists would be the fact that as the protons approached that energy, they tended to bunch together more tightly. Though the particle orbits would be momentarily unstable, the switch would allow them to regain stability for further acceleration. Still, many accelerator physicists feared that this would not work, that the beam would be shaken up and lost in phase transition, and that the phenomenon would put an effective limit on the energy of synchrotrons. At the end of the summer, Haworth asked the AEC for funds to test Courant's idea by building a quarter-size analogue of the AGS, which would use electrons, not protons, to provide "the maximum of orbital data with a minimum of engineering complications" to test the complex solutions of the equations of motion of charged particles in a strongly focusing environment.[27]

But Haworth was impressed enough with how confident Courant and others were about phase transition to draft a proposal. On 9 September 1953, he sent AEC director of research (and former head of the Brookhaven physics department) Thomas Johnson a proposal for construction of an accelerator of 25 BeV to be eventually upgraded to 35 BeV. It would have a 260-foot radius, cost $20 million, and with luck might be completed before the end of 1959. The vacuum chamber would be only 3 × 6 inches as opposed to Cosmotron's 30 × 7. Haworth's letter is a mere five single-spaced pages long (one of which is almost entirely a table), plus an additional five pages of supporting material. Nowadays, this letter is often cited with a mixture of nostalgia and anger by accelerator physicists; today's proposals are thousands of pages long, containing detailed documentation of technical and engineering aspects with elaborate, computer-generated graphics. But more remarkable than its brevity is the way Haworth attempted to justify the machine:

> The Cosmotron has, during its relatively short operational use, yielded much new data . . . and has even led to the observation of certain hitherto unobserved heavy meson phenomena. That extension of the available energy would yield many fruitful results seems unquestionable. . . . [F]urther extention seems highly desirable, for specific and predictable reasons as well as on the general grounds that past extensions of energy have always proved highly profitable.[28]

Rarely has the case for a scientific instrument whose aim is basic research been put so honestly and succinctly. Rather than attempt to justify the instrument based on short- or long-term, direct or indirect returns or goals, Haworth simply assumed that the AEC shared the view that achieving knowledge about the structures of the world is of value in itself, and that the proposed instrument seemed particularly promising in this respect. "I trust that the foregoing is sufficiently explanatory for your present purposes," Haworth concluded.

Meanwhile, the CERN PS was well on its way, though work was scattered all over Europe until the project was consolidated at the Geneva site in October 1953. That month, the CERN council held a public examination of the project and invited the top accelerator experts from all over world. During the meeting CERN scientists engaged in a lengthy debate over whether to build an electron analogue. Upon learning Brookhaven had decided to build one, "that news settled the question for us," J. B. Adams, who led the group constructing the PS, wrote Haworth. Quoting a remark he'd heard Livingston make about the problem, Adams said they had decided to trust that "[s]cience will find a way out and not to build one."[29]

But the Blewetts, Courant, Livingston, and the other Brookhaven accelerator physicists at CERN for the meeting thought that the analogue was unnecessary for their institution as well. Once again, as in the question of the Cosmotron's aperture, they were willing to put their faith in their calculations in the face of potentially crippling difficulties. The Blewetts (who were at CERN for several months contributing to the PS project full-time) wrote Haworth arguing heatedly against building the model; "fear of unknown effects seems to us just another argument in favor of more analysis and computation." Their concern was that it would delay the AGS significantly in the race with CERN. Thus at the same moment the Blewetts were helping CERN with the PS, they were anxiously encouraging Haworth to speed up completion of its AGS:

> Altogether the model looks to us like pretty expensive insurance. Now that Brookhaven and CERN are in the same energy range it puts Brookhaven at a terrible disadvantage. If we admit the necessity for a model we can't even freeze

the [design for the] magnet aperture until the model results are all in. By then CERN will be well into the construction phase and should be operating before us by just about the time spent on the model. That should be plenty of time to permit them to discover all of the really exciting nuclear phenomena up to 25 or 30 Bev.[30]

While contradictory from the perspective of political rivalry, the Blewetts' behavior was quite consistent with performance rivalry.

But Rabi felt strongly that "we should make our own decision as to machine size independently of what CERN might do," and Rabi's opinion again carried the day. "We are reluctant to embark on so large a venture without some experimental evidence," Haworth wrote the Blewetts.[31]

In December 1953, the AEC authorized the electron analogue, which was built in the wooden test shack behind the Cosmotron building. On 8 January 1954, Haworth called a special meeting of the accelerator development department, now thirty-six physicists strong, to announce that the AEC had just approved the AGS project itself. Having had no luck finding someone to head accelerator development, Haworth typically decided to run it himself (recall that he had done the same—twice!—during the Cosmotron's construction just a few years before). The approval gave an official stamp to Brookhaven's intense, large-scale performance rivalry with CERN. Nearly all major technical steps taken by one institution followed consultation with scientists working on similar problems at the other. The collaboration involved more than information flow, and included actual exchange of personnel.

The electron analogue, built in the test shack in back of the Cosmotron building, was one of the few major undertakings that Brookhaven did without an equivalent project at CERN.[32] It gave Brookhaven scientists greater confidence and they uncovered some problems that they would not have otherwise, though, as feared, it set back the AGS schedule by several months. In April 1955 the accelerator builders gathered in the test shack to watch it try to go through phase transition. An oscilloscope that registered the beam intensity over time would reveal whether they had succeeded or not; a sudden drop of the line at the moment of phase transition would reveal that the beam had been lost (fig. 9.8). In case beam transition turned out to be difficult, Plotkin built electronic equipment to help guide it through. But the line held, 5 MeV was reached, and phase transition achieved.

The success coincided with a visit by Vladimir Veksler, codiscoverer of the principle of phase stability, accompanied by V. Vladimirsky, coinventor of a way to avoid the need for phase transition. Veksler was fluent in English, but had to speak in Russian out of political necessity and his words had to be translated (he would occasionally correct translators, saying "Nyet, nyet!" until they had rendered his words right). He was informed

9.8 Electron analogue of Alternating Gradient Synchrotron, with control console in foreground.

about the passage through phase transition, in Russian, by Princeton chemist John Turkevitch. Veksler excitedly asked Turkevitch to repeat the news, which Turkevitch did, a second time in Russian. Veksler then demanded of Kenneth Green, in German, to hear the story once more. Finally, Veksler could not help insisting to be told the news yet a fourth time, breaking the rules by speaking directly in English! When they returned home, they revised plans for the Serpukhov machine; the negative bends in the magnet ring were replaced by regular bends, and with the same diameter ring the energy was boosted to 70 from 50 BeV.[33]

Birth of the Program Committee

By the time the Cosmotron was reactivated, in April 1955, it was no longer the most powerful machine in the world. Berkeley's Bevatron obtained its first BeV-energy beam, surpassing 5 BeV, in March 1954, and experimentation began that November. While there was plenty of work to go around, the atmosphere at the Cosmotron changed as a performance competition began between experimental teams at the two labs. Each team knew others were also out there with the same interest in fundamental questions, the same access to literature, and the same ability to build equipment. To do forefront work, one had to keep abreast of new tools and new information, all the while training new graduate students as the previous crop earned their degrees and moved on.

The more competitive atmosphere, the growing size and number of experimental teams, and the fact that the more sophisticated experiments required more beam time per experiment, also imposed the need for new organizational and management practices at accelerators. During the Cosmotron's first two years, Collins had managed the experimental program by the seat of his pants, as it were, making nearly all major decisions himself with occasional consultation. But conflicts were growing. Most of these grew out of the different interests between operators and users, and the impossibility of one person shouldering such a large operation himself the way Collins was attempting. Personality also played a small role; Collins did not deal skillfully with people in conflictual situations. He began to feel harried, especially after Haworth's creation of an accelerator development department deprived his department of its most capable senior staff (Blewett and Green). Moreover, Collins felt, Brookhaven's experimental staff did not include natural leaders to whom such decisions could be entrusted. In a memo to Haworth he complained:

> [T]here is no one here who would command national recognition were it not for the Cosmotron. . . . To make matters worse those people we have do not form a unified group which feels a responsibility for the whole program. . . . Because no responsible group with a detailed knowledge of circumstances exists I have unfortunately had to decide most questions of scientific policy and personnel changes without the benefit of discussions. The source of these difficulties or our inability to correct them lies in the *absence of clearly defined responsibility.* The undefined area of responsibility which lies between the Cosmotron and Physics Departments is the source of endless confusion and some ruffled feelings.[34]

Experimenters, meanwhile, resented that Collins, in charge of the operation of the machine, had so much control of their work. As Friedlander, Piccioni, Edward Salant, and Shutt—four powerful group leaders—wrote Haworth the following month:

> The present Chairman of the Cosmotron Department has been arbitrary and autocratic. In many cases he has not attempted to satisfy the needs of experimenters; he either did not consult them at all on questions vital to their work or disregarded their advice, often without informing them. By means of his powers as Department Chairman he tries to impose his views on the research to be carried on. Unfortunately, however, his decisions are not based on any special competence in the field of high energy physics, or on an adequate understanding of the Cosmotron, or, generally, on sound judgment. Moreover, he has created dissatisfaction and disharmony by his unfortunate way of dealing with people.[35]

In June 1955, at a general meeting on scheduling, Collins tried to calm

the waters by agreeing to schedule a full six months in advance, to appor-
tion beam time in general categories (with specific fractions to go to cloud
chamber, counter, emulsion work, etc.), and to grant more time than before
to experimenters not using the glamorous cloud chambers and counters,
who felt shortchanged.

In the meantime, Haworth had decided to form a committee to review
proposed experiments with the power to recommend that some be altered
or even turned down. Knowing the idea would be unpopular, he asked a
universally liked outsider, mild-mannered theorist Robert Serber, to break
the news. It was not welcomed.

> You can't imagine what a radical proposal this seemed at the time [Serber re-
> called]. The idea that somebody would judge whether your experiment was worse
> than someone else's—it was insulting, insufferable, and probably unconstitu-
> tional. . . . I was elected to break the news to the users, which I did and got plenty
> of verbal brickbats. (Serber 1998, 187)

Haworth established a High Energy Policy Committee, whose name and
personnel would often change, chaired by himself, which was effectively
the lab's first program committee. Experimenters would have to submit a
"Request for Experimental Time." It did not involve much paperwork, still
only a single-page form on which experimenters were given 3 lines to state
the "purpose of experiment," 5½ for "special conditions and remarks." [36]
Still, experimenters were horrified that a tribunal would be summoning
them to defend their experiments ahead of time. As Maurice Goldhaber put
it, "They felt like they were being forced to go to confession before they
had sinned." [37]

A few experimenters continued to resist these new practices as antitheti-
cal to science and to the spirit of Brookhaven, voicing strong opposition at
meetings. In May 1956, after one particularly virulent exchange, physics
department chairman Samuel Goudsmit sat down and penned an unusually
blunt and threatening memo to the complainants. The size and complexity of the Cosmotron, Goudsmit wrote, has introduced a new factor into
physics research, to which, like it or not, everyone has to adjust.

> This new factor is the need for full and extensive cooperation of a large group of
> people. The old time research worker, who was an extreme individualist, cannot
> thrive in our atmosphere since he is almost doomed to anonymity. But success
> in these new fields of experimental physics is, perhaps unfortunately, possible
> only if there exists a spirit of participation. There should be pride in the achieve-
> ments of the project as a whole, no matter whether the results are obtained by
> members of our Laboratory or from the outside. Jealousies must be suppressed,
> ambitions should be focused on accomplishments of the Laboratory and not of
> the individual. . . .
>
> It is not merely the research that is novel but also the operation and manage-

ment of the machine itself. The first lesson we have learned is that these are almost inseparable. . . .

Thus it is clear that in this new type of work experimental skill must be supplemented by personality traits which enhance and encourage the much needed cooperative loyalty. Since it is a great privilege to work with the Cosmotron, I feel that we now must deny its use to anyone whose emotional build-up might be detrimental to the cooperative spirit, no matter how good a physicist he is. From now on until my function as Chairman of the Physics Department ends, I shall reserve the right to refuse experimental work in high energy to any member of my staff whom I deem unfit for group collaboration.[38]

At the beginning of the next year, an exhausted Collins announced his desire to "get away from the Laboratory for a while" and spend a semester in Europe. The request was granted.[39]

Strange Particles Get Stranger: 1955–1957

The squabbles did not alter the Cosmotron's research program substantially. Throughout 1955 and 1956, the machine was primarily used to study strange particles and pion cross-sections (fig. 9.9). As physicists learned more about strange particles, and discovered new varieties, the nomenclature changed. They were now divided into groups (L mesons, K mesons, and Y mesons), and then given individual or "Christian" names. The V_1s, which included a number of particles of about the same mass, were referred to generically as *Y particles* or *hyperons,* and included the lambda or Λ (which decayed into a proton and pi minus). The V_2s, lighter than the V_1s, were referred to as *K particles,* and included the theta or Θ^0 (which decayed into two pions, a pi plus and a pi minus) and the tau or τ (which decayed into three pions, pi plus, pi minus, and pi zero).[40]

The Ks were particularly intriguing, and particle physicists would learn much from them, at Brookhaven and elsewhere, in the coming years. One puzzle involving Ks was the apparent ability of thetas to violate what had seemed to be a basic physical law: they seemed to be at times interchangeable with anti-thetas, defying the conservation law according to which a particle and its antiparticle were quite distinct entities. Such a brazen violation of a basic conservation law seemed unlikely, and it was unlikelier still that the theta was its own antiparticle.[41]

Seeking an explanation, Gell-Mann and Pais came up with one that can be explained by reference to polarized light. Every physics student knows that light can be polarized to vibrate in only one plane; horizontally, let us say, or vertically. And while horizontally polarized light behaves exactly like vertically polarized light (the two states are "symmetrical"), they cannot simply change into each other. But if horizontally and vertically polarized light are mixed, their crests and troughs interact in a way that gives rise to "circular polarization," where the plane of polarization corkscrews around either right or left depending on the phases of the initially polarized forms. Light that is *produced* as horizontally and vertically polarized forms thus can *appear* as though it comes in two different forms, circularly right-polarized and circularly left-polarized. Physicists speak of these two forms of circularly polarized light as being produced by the *superposition*

9.9 Cosmotron experimental program, late 1955.

of horizontally and vertically polarized light. But there's more: One can also speak with equal justice of horizontally and vertically polarized light as being produced by the superposition of the two forms of circularly polarized light. Mathematically, the two *pairs* of forms are symmetric. Each pair can be described as a superposition of the other two.

Gell-Mann and Pais thought that theta and anti-theta might be interfering in a similar way. According to the laws of quantum mechanics, the two particles can be described as acting like waves, with amplitudes analogous to the amplitudes of classical waves. The different amplitudes of the two particles might be *mixing* or interacting with one another, producing two final states different from the two initial states. The two initial particles would have opposite strangeness. Because changes in strangeness are forbidden for the strong interaction, the mixing of the two states could be caused only by the weak interaction—a very weak effect, but creating a large mixing because the particles have the same mass. And, Pais and Gell-Mann discovered, these two final states would have strikingly different decay modes: one could decay almost always into *two* pions, while the other was forbidden to decay into two pions, but could decay into a variety of more complex modes, such as *three* pions, pion-muon-neutrino, and pion-electron-neutrino. Moreover, the one that could decay into two pions could do so rapidly, in about 10^{-10} seconds, while the decay of the other would take 5×10^{-8} seconds, some 500 times longer. To an experimenter these would appear to be strikingly different particles. Gell-Mann and Pais called these particles Θ_1^0s and Θ_2^0s (which also became known as "K-longs" and "K-shorts," or K_L^0s and K_S^0s). Each pair could be described as a "mixing" or superposition of the other. This mixing, they concluded, might be responsible for the *apparent* violation of the antiparticle symmetry law. A theta that decayed into two pions was already known to exist; that was presumably the Θ_2^0 or K_S^0s. If the two theorists were right, about half of all thetas produced should also have a longer lifetime and be able to decay into three pions. As Gell-Mann and Pais wrote in a paper that appeared in March 1955:

> To sum up, our picture of the Θ^0 implies that it is a particle mixture exhibiting two distinct lifetimes, that each lifetime is associated with a different set of decay modes, and that *not more than half of all Θ^0's* can undergo the familiar decay into two pions. (Gell-Mann and Pais 1955, 1389)

It was, a theorist later remarked, "one of the most far reaching ideas ever proposed in elementary particle physics" (Sakurai 1964, 269).

Just before the paper appeared, Pais came to Brookhaven. As was customary, the visitor was asked to give the informal colloquium talk. At first he modestly professed to have nothing to say, but after amiable arm-twisting agreed to discuss particle mixing. His audience immediately understood that Gell-Mann and Pais had done far more than simply predict the existence of yet another particle. While some physicists still considered quantum mechanics little more than a statistical calculational procedure that provided only a ghostly, unsatisfactory, and perhaps ultimately incorrect picture of the subatomic world, here was a predicted quantum mechanical effect that was direct and fleshy. Physicist Robert Adair, who was present, recalled:

> I think everybody, even the good theorists, tend to think of a particle as a tiny sphere; I always think of a little steel ball bearing myself! Now here was an effect caused by the quantum mechanical nature of a particle in which a particle would *not* be acting like a steel ball at all. Not only that, the difference was dramatic— it was a matter of a whole different lifetime and decay mode! It was one of those great ideas about which you feel, "I would never have thought about it," and yet because you know quantum mechanics you understand it *instantly*.[42]

Several individuals in the audience, including Oreste Piccioni, recalled that passage of particles through matter shifts their phase (analogous to the way, for instance, the phase of light shifts when it passes through glass), and began to talk excitedly about the possible implications for particle mixing. When Pais, who had been reluctant to speak about the idea in the first place, tried to make an exit, saying he had a train to catch, the experimenters forced him to stay and continue the discussion. Not long thereafter, Piccioni and Pais had a number of lengthy discussions in the course of which they realized that this phase shift would affect the particle mixture so as to transform or "regenerate" some Θ_1^0s into Θ_2^0s or vice versa. With theoretical input from Serber, Pais and Piccioni published a paper outlining what would become known as the Pais-Piccioni regeneration effect (Pais and Piccioni 1955).

Chinowsky brought up Pais's idea with Leon Lederman; Lederman, meanwhile, had heard Pais give the same talk at Columbia's weekly colloquium. Lederman's corn crib cloud chamber, they realized, was well positioned to look for the Θ_1^0. It was placed a fair distance from the Cosmotron, giving them a shot at seeing the decays of longer-lived particles that would pass straight through closer cloud chambers. From Gell-Mann and Pais's paper, they knew the approximate lifetime of the predicted particle, and could calculate the probability of seeing a decay in the chamber, which was some fraction of the total events. Lederman, Landé, and Chinowsky created a new beam for the chamber, made by smashing a pion beam into a target and passing it through a "sweeping" magnet that removed all charged particles. The result, a beam of neutral particles that would include a number of thetas, was directed into the corn crib and its thirty-six-inch cloud chamber.

Because the cloud chamber's cycling time was about a minute, Lederman's group had to let a dozen or so of the Cosmotron's pulses go by between pictures. Thirty-three hours—some twenty-four thousand pulses—of running time gave them only twelve hundred pictures. The laborious scanning process began at the end of 1955. By the next summer, Lederman's group found about two dozen events where a theta had clearly decayed into three particles, presumed to be pions. In July 1956, the group sent a letter to the *Physical Review* announcing observation of "the disintegration of a long-lived neutral particle" (Landé et al. 1956, 1903).

Was the new object the predicted Θ_1^0/K_L? More detailed information was needed. Lederman and his colleagues petitioned Collins for more running time, and despite the huge number of wasted pulses, Collins believed the experiment was important enough to warrant it. By the end of 1956, they had run off another five thousand pictures, in one hundred of which were three-pion, pion-muon-neutrino, or pion-electron-neutrino decays. Early the next year, they published another paper concluding that the additional events plus careful analysis gave additional support for particle mixing (Landé et al. 1957).

As philosophers of science have pointed out, the discovery process in science often involves not a search for something that one already understands and is hard to find, but a coming-to-understand some phenomenon that seems a little mysterious—even unreal—when its face or "profile" first crops up in the laboratory. Scientists often find they must see how a new phenomenon appears in different profiles under various conditions before they acquire confidence that they can recognize the phenomenon for what it is; they ask, in effect, "Can it be performed in different ways"? (Crease 1993, chap. 6). The idea of mixing seemed so fantastic that some members of the team still felt a twinge of doubt, and only attained more confidence when the phenomenon revealed itself indirectly in other experiments. Chinowsky was one:

> [Columbia professor Jack] Steinberger's group was studying pi minus p making $K^0 + \Lambda_0$, and finding only half the K^0s that one would expect. I felt relieved when I heard that, because I realized it meant the longer-lived K^0s were passing clear through their chamber and decaying outside. That effectively amounted to a confirmation of our results. People don't mention it, and it's not in the literature, but for us that was important. Speaking for myself, when I heard those results of Steinberger, that was the first moment I was completely convinced that we had indeed found the K_L.[43]

For some, the discovery of the K_L even had the effect of reinforcing their faith in quantum mechanics, whose peculiar features had led theorists like David Bohm to question its ultimate validity. A few years later, writing about the K_L and K_S in a book chapter entitled "Some Consequences of Strangeness," Richard Feynman wrote:

> This is one of the greatest achievements of theoretical physics. It is not based on an elegant mathematical hocus-pocus such as the general theory of relativity yet the predictions are just as important as, say, the prediction of positrons. Especially interesting is the fact that we have taken the principle of superposition to its ultimately logical conclusion. Bohm and co-workers thought that the principles of quantum mechanics were only temporary and would eventually fail to explain new phenomena. But it works. It does not prove it right, but for my money, the principle of superposition is here to stay! (Feynman 1961, 50)[44]

The Lederman team's discovery of the K_L was almost the last hurrah for cloud chambers. Early in 1952, a twenty-five-year-old instructor in physics at the University of Michigan named Donald Glaser, frustrated by the slow rate of accumulation of strange particles in cloud chambers, turned his attention to superheated liquids like pentane. Analogous to the way the slightest disturbance of supersaturated air causes condensation, the slightest disturbance of a superheated liquid causes bubble formation; if properly adjusted, Glaser found, a charged particle passing through the liquid is sufficient. Because the liquids were three orders of magnitude denser than the gases in cloud chambers, many more events would occur in liquid-filled chambers. In addition, the denser medium would slow down secondary particles, allowing more to decay in the chamber, providing more information to experimenters. After hearing of Glaser's work, experimenters at Berkeley and Brookhaven seized on bubble chambers as the future detector of choice and aggressively pursued the technology (on bubble chambers see Galison 1997, chap. 5).[45]

The pursuit of bubble chambers at Berkeley and Brookhaven allows another comparison of the styles at the two institutions. At Berkeley, where Ernest Lawrence still ruled his laboratory with the aid of a few trusted associates, the bubble chamber effort was led by Luis Alvarez. Alvarez was an intensely ambitious experimenter who—in true Berkeley style—always kept uppermost in mind that the long-range goal was to build the biggest chamber possible, and used that goal to guide even short-term decisions. The first bubble chambers, for instance, like cloud chambers, used pistons to increase and suddenly reduce the pressure. Looking to the future, Alvarez did not believe pistons would work with large chambers because of the huge mechanical forces involved and the difficulty of operating pistons at extremely low temperatures. Thus even in small chambers of a few inches he began superheating the liquid by putting a column of gas on top of it, then releasing the pressure on the gas. During one early visit to Brookhaven, Alvarez looked scornfully at one of the lab's piston-operated chambers and said triumphantly, as if the remark sealed the doom of an antiquated technology, "What will you do when the piston breaks off?"

At Brookhaven, where (initially) independent bubble chamber groups competed, nobody felt the pressing need to stop using pistons. If the pistons were big enough with respect to the chamber, their velocity did not have to be high and local turbulence was kept to a minimum. As Adair put it, "With our naïveté or stupidity, or perhaps with more wisdom than we knew we had, we went ahead and continued to build piston-operated bubble chambers, and they worked very well."[46]

Ralph Shutt first learned about cloud chambers from Alvarez, on one of his visits to the lab, and invited Glaser to Brookhaven. Glaser came for a visit in February of 1954, and spoke about a six-inch pentane chamber he was constructing. While some members of Shutt's cloud chamber group found it "a pretty small gadget" with which to study strange particles, they were impressed with the principle and invited him back.[47] Happy to have the opportunity to test his devices on the most powerful accelerator in operation, Glaser returned in September, bringing his completed chamber and a graduate student, David Rahm. They took pictures with the Cosmotron's beam, demonstrating that their device could cycle fast enough to keep up with the Cosmotron's pulse rate of one every five seconds, and Rahm stayed on at Brookhaven with instructions to do what he could to perfect it.

By the beginning of 1955, Shutt, like Alvarez, knew he had to phase out cloud chambers and build bubble chambers. His first, six-inch model used liquid hydrogen—advantageous because its nuclei are protons (which were at once target and part of the detector) and all the particle interactions would be protons on protons, but disadvantageous because liquid hydrogen is highly explosive and required cryogenics (the technology for handling extremely low temperatures, with which few high-energy physicists had any experience). By June 1955, Shutt was confident enough to propose building a twenty-inch chamber, and began adding people, including Rahm, to his soon-to-be-called bubble chamber group.[48]

A second bubble chamber group was led by Adair. Although Adair had been an infantry sergeant in Patton's army, he was generous with ideas, credit, and jokes, and boisterous laughter seemed to fill the room when he was around. Many are the motivations for careers in physics; in Adair's case, the main reason seemed to be that he could not think of anything more fun. Fascinated by the ways physics phenomena manifest themselves in the world, Adair once wrote a book on the physics of baseball. Few physicists regarded their careers as more of a public calling than Adair; he played a major role in the debate over the alleged dangers in low-level electromagnetic radiation.

As a graduate student at the University of Wisconsin, Adair had worked on neutron cross-sections for his Ph.D., which he received in 1951. At the time, nuclear and elementary particle physics were becoming separate disciplines, and while the bulk of his training lay in the former field, Adair decided his true calling was the latter and landed a job at Brookhaven in the Cosmotron department. He found the transition to elementary particle physics more difficult than anticipated. "Graduate students knew more about cloud chamber techniques than I did," he later said. But the recent invention of bubble chambers gave him an opportunity to start on a level playing field with everyone else. In the fall of 1954, he and a technician began to build a six-inch hydrogen bubble chamber, finished in early 1955. Though tiny, it was Brookhaven's first hydrogen chamber. "I still remember expanding it for the first time," Adair said. "The lights flashed, and I suddenly saw *real tracks!* It was wonderful! [Laughs.] That was one of the great thrills of my life!" [49]

Adair then put together an experimental group. His first recruit was Lawrence Leipuner, a graduate student from the Carnegie Institute of Technology. "One time the chamber stopped expanding," Leipuner says. "We looked inside, and there we saw our little piston, its stem snapped off, lying motionless on the bottom of the chamber. We took one look at each other, remembered Alvarez's admonition, and burst out laughing!" [50] Fixing the piston was not a problem, and the team put the chamber in a pion beam and began taking pictures. Later, they built a fourteen-inch chamber.

A third Brookhaven bubble chamber group was led by Columbia professor Jack Steinberger. Steinberger was born in Germany in 1921 but fled in 1934 for the U.S. He received his Ph.D. from the University of Chicago in 1948 and went to Berkeley the following year. After refusing to sign the controversial loyalty oath imposed by the regents of the University of California, he was forced to leave in June 1950. He moved to Columbia that fall, and began to work at the Nevis cyclotron with the intention of moving to the Cosmotron when it was completed.

Steinberger was a demanding individual to work for. Aggressive, bright, critical, and short-tempered, he envisioned ambitious experiments and then impatiently drove himself and his subordinates to telescope the intermediate technological steps in the shortest time possible. This made him a terrible thesis advisor for needy students, but a terrific one for those who weren't. In 1953, he found three graduate students with the requisite stamina and thick skin: Jack Leitner, Nicholas P. Samios, and Melvin Schwartz.

Leitner would get his Ph.D. in 1956, whereupon he worked at Brookhaven for several years before moving on to Duke and Syracuse. During those years, he continued to work on his own and in collaboration with individuals at Brookhaven. He died, unexpectedly and suddenly, in 1968.

Nicholas Samios, who became the lab's director three decades later, was born on Manhattan's East Side in 1932, the child of two Greek immigrants. A "jock" as a kid, he had a particular fondness for stickball, a baseball-like game played with a bouncy rubber ball and a broom handle whose rules varied from street to street. On 40th Street, where the Samioses lived, a hit ball that stayed on a roof in fair territory was a home run; if it got stuck on a fire escape it was out, but playable if it bounced off. Before the game began, each side would bet an equal amount of money on the outcome; at the end of the day, the winners would go to a show, while the losers would hang out on stoops. On weekends, young Nick played the game enthusiastically from sunup to sundown.

Nick's father ran a successful midtown Manhattan seafood restaurant and in other circumstances it would have been natural for the son to follow in the father's footsteps. This path was diverted by the New York City public school system. After a first-class science education at Stuyvesant High, Samios was admitted to Columbia and enrolled in a special course sequence in mathematics for science majors—along with classmates Leitner and Schwartz. He majored in physics, came to Brookhaven as a summer student in 1952 (as an extra set of hands completing the electrostatic injector), graduated in 1953, and returned to Columbia as a graduate student. Attracted by Steinberger's dynamism and enthusiasm, Samios sought him as a thesis advisor.

Mel Schwartz was an impatient, ambitious youth who became a Nobel laureate in 1988 and would also serve as Brookhaven's associate director. He came close to quitting physics as a cocky Columbia undergraduate when he found himself "bored as hell" by a series of elderly physics professors who basically retold what he viewed as dry material from textbooks. At the beginning of his junior year (1951), Schwartz decided to give physics one last shot and jumped ahead to a fourth-year course, "Atomic Physics and Introductory Quantum Mechanics."

On the first day the teacher walked in. All my other teachers had come in wearing suits and ties, but this one had on sandals and an open shirt. He was sleepy and bleary eyed, could hardly stand up, and mumbled as an apology that he had been running an experiment all night at Nevis. When he began to stumble through his lecture, he made mistakes that even I, a bright, fresh kid, could pick out; technically, it was the worst lecture I'd heard in my life.

Schwartz was enthralled.

> This was the first young, working physicist I had ever met. It was an unpolished human appearance of somebody who *really* was involved in physics. I had the sense he was enjoying his work. He wasn't just a teacher; he was a real researcher who was *learning* things! I knew right away that this was the guy I wanted to work with.[51]

The teacher was Steinberger. His interest in physics reinvigorated, Schwartz stayed on as Steinberger's graduate student.

In June 1953, the AEC informed Haworth that it was denying Steinberger clearance; evidently, the AEC had some testimony against him in its files that it refused to divulge.[52] Though he did not need clearance to get past the lab gate, he was technically prohibited from working there even as a day visitor. While nobody at the lab cared, Schwartz recalls that during one surprise visit by an AEC official it was thought prudent to send Steinberger, one of the lab's most eminent experimenters, to a back room. Steinberger finally solved the problem by applying for clearance at the Hudson labs, a secret lab run by Columbia on a navy contract, whose clearance investigations were open. The evidence against him turned out to be an anonymous statement informant that Steinberger was able to disprove. He received clearance from the navy, which made AEC clearance routine.

In July 1954, Steinberger attended an international conference on particle physics in Glasgow at which his interest was captivated by two things: a talk on bubble chambers, and a talk by Gell-Mann and Pais on strange particles. Steinberger abruptly switched his focus to strange particles, and his instruments to bubble chambers. He and his team of graduate students became known as the Columbia bubblers.

Samios and Schwartz built, assembled, and tested a six-inch bubble chamber at the Nevis machine shop. Since Nevis was relatively remote from both Columbia and Brookhaven, the students depended on Schwartz's Plymouth. Steinberger had his students try several innovations to get the chamber running as rapidly as possible. One was to recompress the liquid a few milliseconds after expansion, right after snapping the picture, to prevent the bubbles from growing further and rising to the top of the liquid. This cut the recycling period to several seconds. Constantly tinkering, the Columbia group rebuilt its chamber many times before assembling a viable chamber, a ten-inch device of aluminum with glass windows.

On the Cosmotron floor, the Columbia group built a new beam line to collect negative pions from a target in the Cosmotron and lead them to the bubble chamber. When they got beam they discovered to their horror that it was contaminated by hordes of other particles. The pion beam, they found, was hitting a metal post inside the Cosmotron, collisions with which were producing secondary particles. Schwartz redesigned the beam to skirt the post, and the experiment was run all over again.

In November 1955, Steinberger's group was granted forty hours of beam time—five eight-hour periods on the midnight shift. With Steinberger required to teach during the day, the three graduate students ran the entire experiment.

> The chamber was piston operated, and the piston had O-rings which had to be lubricated constantly. The lubricant kept slowly leaking into the hot propane, clouding it. Cleaning it was a laborious task, for it required dumping and refilling all the liquid in the chamber. At the end of each run, we'd have to dump the liquid and clean the chamber, get it ready for the next day, have breakfast, go to sleep in the lab dorms, and come back about 8 P.M. to do it all over again.[53]

The experiment went off without a hitch, producing some thirty thousand pictures. Then they took the film to Nevis where it was scanned, leaving them with some twelve hundred pictures of two-pronged events.

> Analysis meant we'd measure on the pictures the angles between where the tracks stopped and where the V's started, and the angles between the legs of the V's. That's called doing the kinematics. For each event we had two pictures from two different projections. By knowing the positions of the cameras we could then work out the positions of these angles in space. That's called doing the geometry. After that we could figure out whether a particle was a lambda or theta, say, and what its energy was.[54]

Then they separated the events into three categories. One was elastic scattering—cases in which the pion had simply rebounded off the proton—which Leitner wrote up for his thesis. The inelastic events, in which the collision had changed the identities of the particles, included some fifty-five examples of associated production—roughly five times the number Shutt had seen in his two years of using cloud chambers, a testimony to the superiority of the bubble chamber. These fifty-five events were divided in turn into two groups: those where lambdas and thetas were produced, and those in which charged sigmas and K's were produced. For their theses, Samios wrote up the first group, Schwartz the second. The team also published a joint paper on their work, "Properties of Heavy Unstable Particles Produced by 1.3-Bev π^- Mesons," the first publication of a high-energy physics experiment using bubble chambers (Budde et al. 1956).

After earning their Ph.D.s from Columbia, Leitner left for Duke and then Syracuse, while Samios and Schwartz stayed on in Steinberger's group and built a twelve-inch propane chamber with eight times the volume of the first, and then a twelve-inch hydrogen chamber.

As science historian Peter Galison has observed in his seminal article, "Bubble Chambers and the Experimental Workplace," bubble chambers had a far-reaching impact on the practice of experimental research, crossing a threshold into new levels of complexity (Galison 1985). Hitherto, experimental collaborations had been ad hoc affairs, consisting of loosely affiliated people with similar interests. Bubble chambers—complex, time-consuming, and costly to construct—brought, among other changes mentioned by Galison, a new inertia to research groups.

> By the time you had what you needed for one experiment [Thorndike recalled], you'd invested a lot of time and effort in building and understanding the equipment and in learning the techniques. And the pressure was intense: If you saw an interesting experiment to do, and had to take the time to stop and build the equipment from scratch, someone else likely would get there first. That's why groups would generally improve the apparatus they already had; they'd use one device to do physics, and simultaneously build the next-generation device. New physics was coming up all the time, and everybody was in a hurry.[55]

This development led to a change in the Brookhaven concept. With state-of-the-art detectors rapidly becoming impossible for university groups to support, the original goal of allowing outside groups with their own equipment to compete on a par with inside groups for use of the Cosmotron was becoming obsolete. Shutt's bubble chamber group, the largest and best supported, began to dominate bubble chamber work at the Cosmotron. Worried about the increasing disparity between outside teams and his own, Shutt decided to let university groups submit proposals to use his new twenty-inch chamber to the program committee, competing with proposals from his own group. In practice, however, outsiders fared best if they worked with people from Shutt's group. A new pattern thus evolved. Outside teams would recruit members of Shutt's group before submitting proposals to the program committee. If the experiment was approved, Shutt's group prepared and operated the chamber, while the university groups made arrangements with the accelerator operators and afterward took charge of scanning and analyzing the film (fig. 9.10). In the process, Shutt's group grew into a permanent institution at Brookhaven's high-energy physics facilities. In this fashion, the Brookhaven concept was rescued (and the user concept of bubble chambers soon extended to other labs; Holl 1997, 199).

The emergence of a single dominant bubble chamber group at Brookhaven had another effect; it crystallized the difference between the experimental programs at Berkeley and Brookhaven in a dramatic contrast in style, expressed in the characters of the leaders of each institution's bubble chamber group: Luis Alvarez and Ralph Shutt. Alvarez, whose hobby was

9.10 The twenty-inch bubble chamber at the Cosmotron, 1960.

flying airplanes, was a daring physicist-engineer who always kept in mind that the purpose of the devices was to do physics, and entrusted the engineering of his projects to a topflight engineering staff. His was a dominating personality that did not take well to being contradicted, and he tended to view individuals as for or against him. He brooked no rivals; if an individual wanted to work on bubble chambers at Berkeley, it would be in one of Alvarez's projects and on work of which Alvarez approved. Fortunately, Alvarez was a brilliant leader; he had a fertile imagination and was a relentless risk taker.

Ralph Shutt, an accomplished musician whose tastes ran to Beethoven's last sonatas, was an engineer-physicist who loved the microdetail of project engineering. He was uninterested in dominating the experimental program; while Berkeley's bubble chamber group was called the Alvarez group, its equivalent at Brookhaven was generally called simply the bubble chamber group. Shutt not only did not mind outsiders working on his equipment, he encouraged it. A perfection seeker, Shutt brooded over the construction of his instruments. He fought with others at the lab—not about control, but about support, for he tended to view other departments at Brookhaven as service departments.

Mel Schwartz, who worked with both men, recalls:

Alvarez's basic instinct was to put someone else in charge of the engineering of the bubble chamber, make sure it was guided by the best support staff he could get, and then do the experiments himself. Shutt's basic instinct was to engineer something and let others worry about the experiments. Also, Lawrence and the princes of his realm ran the Berkeley laboratory with a strong hand, while at Brookhaven under Haworth and Goldhaber the experimental program was pretty much a function of individual initiative. That made Berkeley a little more systematic place than Brookhaven. [Brookhaven] could think of experiments that were cleverer, but [Berkeley was] better at mainstream physics.[56]

Oligarchy versus democracy: this was the contrast in styles in the performance competition between the Berkeley and Brookhaven bubble chamber programs the 1950s and early 1960s, as the chambers grew steadily in size: Berkeley's 10″ in 1956, Brookhaven's 20″ in 1958, Berkeley's 15″ in 1957, Brookhaven's 30″ in 1960, Berkeley's 72″ in 1959, Brookhaven's 80″ in 1962. Meanwhile, too, CERN was in the process of joining the performance rivalry, and sent Bernard Gregory to Brookhaven for a year to work with Shutt's group studying the design of its 20″ in preparation for building a scaled-up version—the first large hydrogen bubble chamber in Europe— for use at CERN's proton synchrotron (to whose completion Brookhaven scientists were also enthusiastically contributing).

As BUBBLE chambers increased in size, the tally of strange particles mounted, too. Once again, a new theoretical crisis involving K particles appeared. These had been thought to come in different types depending on their decay modes, and assigned a number of different names accordingly. The theta and the tau, for instance, were similar in mass and lifetime, but were treated as separate particles because they decayed differently: the theta was able to decay into two pions while the tau was able to decay into three pions (in their neutral versions they were the neutral K particles that underwent the particle mixing described by Gell-Mann and Pais). As experimenters measured their physical properties ever more precisely, it became clear that the two particles were identical in every aspect but one: parity.

In ordinary language, parity simply means equality: two things are on a par or have parity if they are equal in some way. In the vocabulary of physicists, parity (P) is a specific kind of equality that arises from a basic operation that can be performed on the wave functions of a set of particles in an interaction. It involves reversing all the spatial variables—flipping the X's, Y's, and Z's of position from negative to positive—to produce the mirror image of the event, so to speak. If an interaction is spatially reversed in this way, the parity is said to be even if the sign of the wave function remains the same and odd if it changes.[57]

The parity of the theta was easily determined—that it decayed into two pions implied that it had even parity—whereas the parity of the tau was harder to determine. But British theorist Richard Dalitz devised a clever technique to do so (on the Dalitz plot see Galison 1997, 218–24). By plotting the kinetic energy of the pi plus and pi minus in each tau decay, Dalitz discovered, the data points would all be contained within a boundary line determined by considerations of conservation of energy and angular momentum. If the tau had odd parity, the data points would be found near that boundary line; if it had even parity, they would cluster in the middle of the figure circumscribed by the line (Dalitz 1954). When only a few tau decays were known the situation was unclear, but as experiments logged more and more tau decays, many were discovered near or at the boundary, meaning the tau had odd parity. Since parity was a fundamental conservation principle, the tau and theta were different particles.[58]

Or were they? The curiosity grew into a puzzle as more and more evidence turned up that, otherwise, the theta and tau were identical. Experimenters at the Cosmotron and the Bevatron produced thetas and taus and ran them through different kinds of material, under the assumption that two different particles would interact differently with targets. But no difference appeared.

Physicists were between a rock and a hard place. The rock was to treat tau and theta as different particles. Because they differed by just one property, this seemed as absurd as treating a person with two coats as two different people. The hard place was to treat tau and theta as simply two different operational names given to the same particle that had different decay modes but didn't conserve parity. Because parity was a fundamental property, *this* seemed as absurd as allowing that, just because one had heard that two similar-looking people had been at widely separated places at the same time, they could be treated as the *same* person.

The Cosmotron, like the BGRR, attracted several theorists who liked to follow experimental work closely. Two particularly intrigued by the tau-theta puzzle were C. N. Yang, a physicist at the Institute for Advanced Study, and T. D. Lee of Columbia University. In 1953, Collins had invited Yang to Brookhaven to advise physicists about experiments; because Yang was not a citizen, special permission had to be arranged. During Yang's year at Brookhaven, he and office-mate Robert Mills had developed a new theoretical structure, called gauge symmetry, that turned out to be one of the most important theoretical breakthroughs in high-energy physics of the second half of the twentieth century, though this was not appreciated for a decade.[59] In April 1956, Yang returned to Brookhaven for the summer after the end of the spring session of the Institute for Advanced Study, and began to work on the tau-theta puzzle with his long-time friend and collaborator T. D. Lee. The two had worked together on and off from the time they were both graduate students at the University of Chicago right after the war. At first, they entertained the half-serious thought that this and only this tau-theta particle violated a fundamental space-time symmetry. If so, there was hardly more to say—it would be a phenomenon with a unique profile—and all relevant information was already available. But one could hardly give up there: if so, why *that* particle?

One morning in late April or early May, Yang made the long drive in from Brookhaven to meet Lee at Columbia, and they retired to a cafe to await the opening of a restaurant for lunch. While mulling over the tau-theta problem, they discussed whether parity could be conserved in the strong interaction by which the particles were produced, but violated in the weak interaction by which they decayed. That seemed implausible, given that beta decay had been extensively studied without any sign of parity violation. Or had it? Just to be sure, they reanalyzed beta-decay data and found that, incredibly, none of the hundreds of experiments on beta decay had tested directly for parity violation. They then set out to determine what other testable profiles parity violation in the weak interaction might have.

In June, Columbia's term ended and Lee moved to Brookhaven for the summer. It was a busy time at the lab. Keith Brueckner, Richard Feynman, and Robert Oppenheimer all visited. Lunch conversation ranged from the usual physics topics to the sinking of the *Andrea Doria* to the periodic hurricanes in the Atlantic that created high-amplitude, long-wavelength waves that sometimes made the traditional picnics at Westhampton beach hazardous. Meanwhile, Yang and Lee carried on a constant dialogue over parity in a mixture of Chinese and English that both amused and bewildered their colleagues. They refused to let the subject go, even at Westhampton beach, where they used sticks to trace out equations in the sand. On 22 June, the two completed a Brookhaven report, BNL 2819, outlining their discovery. Entitled "Is Parity Violated in Weak Interactions?" the paper did not claim that parity could actually be violated. All it claimed was that there was a surprising lacuna in knowledge on this point. Still, for many physicists, this was like saying that there was no hard evidence that no dinosaurs do not still live on the bottom of the ocean—it was not something one dropped everything to test. That same day, they sent the report to the *Physical Review*. Its editor, Samuel Goudsmit, loathed question marks in titles and maintained the dignity of his journal by retitling it "Question of Parity Conservation in Weak Interactions" (Lee and Yang 1957).

But the authors were unwilling to let the matter rest. Examining a fundamental but untested law of nature was important regardless of how it turned out. Yang and Lee tried to recruit experimenters at Brookhaven and Columbia whose work involved weak interactions to look for parity violation.

To Adair, they suggested: Use your bubble chamber to observe what happens when a pion decays (weakly) into a muon plus neutrino, and the muon then decays (weakly) into an electron and two neutrinos. Parity symmetry in the weak interaction would lead to the electrons shooting off independently (or *isotropically*), not exhibiting a preferred direction with respect to the muon's direction. Parity violation would have the electrons going off *anisotropically,* in preferred directions. To Steinberger, they suggested: Use your bubble chamber to observe what happens when a lambda that stops inside the chamber decays (weakly). Parity symmetry would mean that the decays would be isotropic, parity violation that the decays would be anisotropic. To Lederman, they suggested: Use your counters at your muon experiment at Nevis to examine whether the electrons are emitted isotropically or anisotropically.

Recalled Adair, "We all said something to humor him like, 'That's interesting, T. D.; we'll get around to it after we do this next experiment!' "[60] Lederman joked to Yang that he would do it "when he found a very bright graduate student to be a slave" (Yang 1983, 31). Steinberger *had* several bright graduate students (Leitner, Samios, and Schwartz), who in fact were looking at lambdas for their theses. When they plotted the angular distribution of the decays in the paper they finished in June 1956 they noted a *slight* asymmetry (a two–standard deviation effect). They quite correctly said their statistics were too poor to make a judgment—slight asymmetries happen all the time—but getting enough (they would have to triple their data) would take much more time. Ronald Rau, who joined Shutt's group in February 1956, recalled:

> There was a lot of important science that we just didn't get around to doing, that we might have done, and that we'd even talked about doing—and those parity decay experiments were a good example. What we really needed was a red-hot, mature physicist who knew which problems *should* be done, had the authority to compel others to do them, and could push them until it was done right. The Cosmotron didn't have a Donald Hughes. Life at the Cosmotron was, as a result, something of a laissez-faire operation. It took a while to figure out how to develop a systematic accelerator program. Later we got much better at this.[61]

One who *was* willing to drop everything was Chien-Shiung Wu at Columbia, enough so that she canceled a long-planned visit to the Far East with her husband, Luke Yuan, on the twentieth anniversary of their exodus from China. Hoping to get a jump start on the experiment, she joined collaborators weeks before Lee and Yang submitted the paper. (At the same time, Valentine Telegdi, another experimenter in Chicago, also began an experiment.) The experiment involved the beta decay of cobalt 60, one place where parity violation might show its face. By New Year's Day, 1957, the experimenters were confident that they had seen evidence of parity violation and soon submitted a paper saying so. Parity was violated not just in the tau-theta decay, but elsewhere in the weak interaction (Wu et al. 1957).

It is difficult now to imagine just how revolutionary this result appeared. Parity was not a mere property like strangeness, but seemed to be one of the most fundamental of all features of things; spatial equality, as it were, in the universe. Physicists now raced to explore this marvelous new phenomenon of nature. It was not a question of *replicating* Wu's experiment— what was the point of doing that?—but to evoke, in the laboratory, different profiles of the phenomenon wherever it might be found.[62] Parity violation was one of the few utterly unexpected scientific phenomena (X rays and fission are two other examples) of which, once it was known to exist, scientists could see hundreds of other profiles in just a few days in their labs, making the discovery process a particularly dramatic event. First out of the block was Lederman's group at Nevis. After hearing of Wu's results, they sat up most of the night figuring out how to do the experiment quickly, threw it together the next morning, and had results by noon.

On the Cosmotron floor, everyone with a stack of bubble chamber photographs of lambdas began rifling through them, looking for evidence of anisotropic decay. But most bubble chambers by then were equipped with magnetic fields, which aided in particle identification for most experiments but also had the effect of ruining this one by introducing its own anisotropy into the decays. It would be easier and quicker simply to take more data. Indeed, Adair and others realized that it might even be possible to measure something that Wu and Lederman hadn't examined. The Lee-Yang theory contained a parameter that described the "amount" of parity violation. That is, all one needed to explain the tau-theta puzzle, and even Wu's result, was a little bit of parity violation. But it could be violated by different amounts, with the maximum theoretical value of the parameter being 1.

The Cosmotron experimenters set out to measure this parameter. No approval was sought or needed. Individuals from many different groups volunteered to help. Adair and Leipuner, whose hydrogen chamber would be the simplest to use, took charge of preparing and operating it. Group leader Rod Cool, who had built and operated a pion beam for other purposes, adapted it for this one. Rahm built coils to block out the earth's magnetic field so there would be as little induced anisotropy as possible.

Within a few days they photographed almost a thousand events. While the film was being developed, Leipuner set up projection equipment and added a drafting machine to read off the data on the angles and positions of the particles. These data were written down by hand and transferred to punch cards, which Leipuner ran through the lab's rudimentary computer, the Remington-Rand 409-2R payroll machine. The verdict: the parameter was "nearly equal to its maximum possible value of one" (Abashian et al. 1957, 1928). Parity was being violated as much as it could be.

At the end of January, the annual New York meeting of the American Physical Society began at the New Yorker Hotel. The news about parity came too late for a regularly scheduled session, and a special session was added on Saturday afternoon, 2 February. The four main speakers were Wu, Lederman, Yang, and Telegdi (who had finished his own parity experiment), followed by a parade of other speakers, including a representative of the Brookhaven effort. None of the experiments "replicated" any other, but rather exhibited different profiles of the same phenomenon. To present their work, the group members selected Ronald Rau. Perhaps awestruck by those who had preceded him, and fearing he would be perceived as daring to rank the Brookhaven experiment with theirs, Rau downplayed the results and even seemed to apologize for getting up to speak about it at all. Adair recalled:

> It is part of our culture that it's considered reasonable and proper to *mildly* exaggerate [laughs] the importance of what you are doing. You always like to get as much credit as you deserve, and maybe a little more. You try not to make an absolute fool out of yourself, but you want to make it seem as though *you* think it's important. So here we are at the APS meeting, we'd worked hard, done the experiment at the Cosmotron—and then Ronnie gets up there to speak for us and says something like, "I don't even know why we are here, all these other people have done the real work, but we did this anyway" [laughter]. Oh, Christ, we could have shot him![63]

The impact of a discovery like parity violation cannot be neatly summarized. To those in the field, such a discovery is many things at once: an instructional lesson, an episode in a story, and a tool, and more. Parity violation was an instructional lesson in that it contributed a piece of outstanding information to a puzzle. It also carried forward a story—the story of particle physics—with its own excitement, drama, and moral lessons ("See how blinded we can become by untested assumptions?"). Finally, the phenomenon of parity violation was a tool in that it could be used in turn to carry out hitherto impossible experiments. Discoveries such as that of parity violation thus become deeply woven into the science, equipment, kinds of experiments, and even the personal relations of the physics community.

The Brookhaven parity experimenters were lucky. Shortly after they finished, the Cosmotron short-circuited again, for exactly the same reason as before, though at the other end of the quadrant. Once again, the machine was shut down for about six months while the coil was repaired. The Cosmotron's operators thought it likely that fatigue cracks were appearing in the copper coils of other quadrants, but gambled that improved clamping would forestall more short circuits, for a few years at least. In the summer of 1957, the Cosmotron came back on line—but that fall suffered yet another breakdown. "We lost the gamble," wrote Lyle Smith (who had replaced Collins as head of the Cosmotron department during the latter's sabbatical).[64] After what was now a total of 7.7 million pulses, an innermost coil bar had once again fractured at the right-angle junction, this time in quadrant #1. An investigation now concluded, "The present design of the coil is satisfactory electrically, but is mechanically inadequate. We know of no way to repair the present end connections to produce a truly reliable coil."[65]

Haworth called the AEC in Washington, and found them not eager to spring for another expensive repair of the Cosmotron, given that the Bevatron was on line at twice the Cosmotron's energy and the AGS was on the way at ten times better. For a while the possibility loomed that the machine would be shut down entirely; failing that, the AEC might go for another local fix rather than a full-scale repair of the entire coil. The AEC asked Haworth to stop work on the machine until the agency had reached a decision, and he passed the order to Smith.

In a classic conflict of interests between users and operators, a discussion ensued about whether to patch just that spot, as they had in January and three years previously, or to fix the design error, which would involve dismantling the Cosmotron for at least a year to rewind the entire coil assembly. Users tended to favor the stopgap approach. Piccioni, the senior experimental physicist on the floor, was in the middle of an experiment and desperately wanted more data. Adair, too, was unhappy about the prospect of putting his own work on hold for so long. "We were all young and ferociously ambitious and thought we had great experiments going," he recalled.[66]

The Cosmotron's operators were adamantly opposed to a stopgap measure. There were sixteen places in the machine where the innermost coil loop took a ninety-degree turn, and fatigue fractures had occurred at three of them; it was only a matter of time, the operators argued, before the others would break, too. Smith therefore ordered his technicians to take Skil saws to the magnet coils, chopping them into pieces, making a local repair impossible. When Haworth learned about the action an hour or so later, he was livid at Smith's blatant disobedience. But now he had no choice but to go after a full repair.

Thus the argument over the machine was settled, and control restored to its makers, in this anti-Solomonic method, by cutting up the baby. Nonetheless, the wisdom of Smith's action was realized as soon as the other coils were disassembled. Some of the thick ($3'' \times 1\frac{1}{2}''$) copper bars at the inner coils were so weak from metal fatigue that engineers were able to break them apart with their bare hands.

PARITY VIOLATION threw a sudden and dramatic spotlight on the weak interaction—or, rather, whether such a thing as *the* weak interaction existed. At the Rochester conference in April 1957, Wu described a still unpublished experiment, by her students Brice Rustad and Stanley Ruby, at the BGRR suggesting that beta decay, the prototypical weak interaction, might have different forms and therefore could actually be the product of several forces. One piece of the evidence for this had to do with the spin of the neutrinos emitted in beta decay. More exactly, it had to do with what is called the *helicity* of the neutrinos, or their spin relative to their direction of motion. When a particle is spinning clockwise while moving away from an observer, its spin is said to be pointing in its direction of motion, and the particle is said to have positive helicity; if counterclockwise, it has negative helicity. Imagine extending your hand and turning a doorknob to the right; that's positive helicity. The experiment described by Wu suggested beta-decay neutrinos could be right-handed, which implied in turn that beta decay itself had different forms and was not one force.

When Leonard Schiff, chairman of the physics department at Stanford, began organizing the program for an APS meeting to be held just before Christmas 1957, he cast about for someone to survey the beta-decay problem. Maurice Goldhaber's name leapt to mind. Goldhaber had a thorough grasp of nuclear problems, and, as mentioned, with his wife had carried out the definitive experiment confirming that the electrons in beta decay and orbital electrons were, in fact, identical. And Goldhaber had a reputation for settling controversial scientific issues. Goldhaber recalled:

> My first inclination was to say no, because I had not worked on beta spectra. My second inclination was that it would do me a lot of good to read these contradictory papers. I accepted, got the key reprints together, and stayed home one Friday morning in mid-October to read them. Before I finished the first paper I thought, "There must be a better way to do this." Twenty minutes later, I had thought of one.[67]

In those twenty minutes, Goldhaber rifled through his extensive mental file of nuclear transitions to find one (an isomer of europium called ^{152m}Eu) in which the momenta and angular momenta of every other particle in the decay *but* the neutrino could be established. [152m]Eu undergoes K-capture, a kind of inverse beta decay in which a nucleus absorbs an electron and emits a neutrino, transforming itself momentarily into an excited state of samarium, [152]Sm, which then emits a gamma ray to go to the ground state. If one could measure the momentum and spin of the gamma ray by which the [152]Sm excited state returned to the ground state, one could figure out the momentum and spin of the neutrino that the [152m]Eu emitted when turning into [152]Sm; these two quantities would indicate the neutrino's helicity. Goldhaber said, "On Monday morning, I rushed to the lab and said, 'Boys, drop everything!' And they did." [68] The "boys" were Lee Grodzins, a postdoc who had been studying the decay of [152m]Eu, and Andy Sunyar, who had worked with Goldhaber on isomers for a decade. By the end of the day the three had started preparing the experiment. A [152m]Eu source was made by irradiating europium in the BGRR, then placed inside a magnet (whose direction of magnetization was periodically changed) perched atop a conical piece of lead. When K-capture occurs, the nucleus shoots off a neutrino, transforming itself into an excited state of [152]Sm and recoiling in the process (step 1). This excited samarium nucleus then emits a gamma ray going to the ground state (step 2). Which way had the neutrino gone? The team could determine that by a trick called "resonance scattering." Europium had been chosen because the Goldhaber team found that the emitted neutrino in step 1 and the emitted gamma in step 2 in this case are of about the same energy. Therefore, the gamma ray emitted opposite to the neutrino has the same helicity as the neutrino, and has the correct energy to be resonantly scattered in a ring of samarium placed around the sodium iodide detector, so that it can be recorded by the detector. (The lead cone between the source and detector prevents gammas from directly reaching the detector without being scattered first.) The helicity of the resonance gamma (and thus of the neutrino) follows from the difference in scattering it experiences as the electron spin direction changes in the magnet. The experiment took about ten days to complete, followed by a month of checks and rechecks; the team sent off a paper to the *Physical Review* on 11 December. Goldhaber and company found that the neutrinos were left-handed just like the electrons (Goldhaber, Grodzins, and Sunyar 1958). Nine days later, Goldhaber stood before an attentive audience at Stanford and told them about the work. "I felt good," he said, "because I knew the answer and didn't have to talk about all these contradictory experiments" (fig. 9.11).

The episode also illustrates the crucial role experience can play in devising an experiment. Of the thousands of known nuclear transitions, the one involving [152m]Eu was, and still is, the only one known to work. Moreover, the experiment relied on a number of clever techniques that might easily have been impossible. So ingenious was this experiment that it is safe to say that most physicists at the time did not realize that it was even possible in principle. Asked how he came up with the idea, Maurice Goldhaber likes to answer: "You work for twenty years on isomers, and think for twenty minutes." [69]

In the aftermath of that and other experiments over the next few months, the weak force was firmly established as a single fundamental interaction, opening the door for further exploration. For decades thereafter, moderators often introduced Goldhaber at conferences by saying, "Goldhaber's the one who—," then gesturing as if twisting a doorknob to open a door, instinctively rotating their hands clockwise in the process. The ever precise Goldhaber then would begin his talk by gently reminding the speaker that the neutrino is *left*-handed, and repeat the gesture the other way.

9.11 The Goldhaber team's equipment for studying neutrino helicity.

Completion of the AGS, and Overhaul of the Cosmotron

Meanwhile, construction was proceeding feverishly on the AGS, and on CERN's PS. (Argonne, too, was building an accelerator, the 12 BeV Zero Gradient Accelerator or ZGS, but it was universally recognized to be the result of a "quintessential political decision" to stay ahead of the Russians in BeV until the AGS came on line; instead, it became operational four years after the AGS; Holl 1977, 169, 328.) In constructing the AGS, as the Cosmotron, the scientists had to commit themselves to the machine assuming that they would be able to find solutions to yet-unsolved technological problems, which is what makes construction of a forefront research instrument intrinsically different from an industrial project. As CERN's J. B. Adams wrote of his institution's proton synchrotron:

> When one orders a railway bridge, one can with complete confidence write a specification, ask for tenders, adjudicate, place the order and sit back and wait for the bridge to appear. It is not necessary to employ a team of bridge building experts to check the calculations of the chosen firm; all the work including any development can be confidently left to the firm. There are almost no component parts of the Proton Synchrotron that we can treat in this way. Every single piece has to be worked on in the Division; even such conventional sounding components as flywheel generator plant and rectifier/inverter banks are outside the range of European firms.[70]

Brookhaven's accelerator department was designing and constructing the machine itself, while Stone & Webster designed and supervised construction of the tunnel and complementary equipment. Digging the half-mile-long trench for the AGS began in January 1956. Stephen Weinberg, then a young assistant professor at Columbia, recalled:

> Until that point, whenever I'd seen an accelerator, they were always things you could look *at*. They were large, yes, but they were *in* buildings, and you could look at the building and then go inside and see the accelerator. The AGS was the first accelerator I'd seen that was part of the landscape.[71]

Thanks in part to the expanding cost of experimental support facilities, the AGS price tag had slowly crept upward from the original estimate of $20 million in 1953 to a final figure of $31 million (fig. 9.12).[72]

The CERN PS came on line before the AGS, circulating a proton beam of 24 BeV on 24 November 1959. The AGS was not far behind. Brookhaven finished building the 240 synchrotron magnets in June 1959. Throughout the summer and fall, the magnets were installed on their support girders, their locations were surveyed, and they were leveled into position. On 13 April 1960, the linear accelerator produced a 50 MeV beam, setting the minds of the accelerator builders at ease about that major component. On

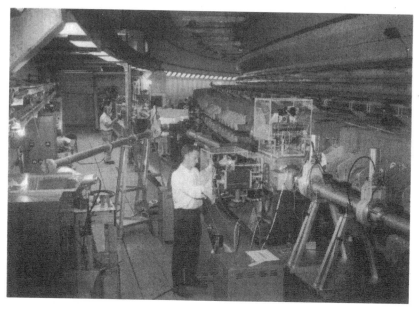

9.12 AGS injection system.

26 May 1960, the linear accelerator injected a beam into the AGS and the physicists were able to keep the beam circling a hundred times around the whole machine. On 22 July, the beam was accelerated through phase transition energy—another worry laid to rest!—to about 7 BeV, but the control system was behaving erratically. The accelerator builders spent a week tinkering with the rf system and becoming increasingly frustrated; Green later recalled "screaming in black rage" until discovering that "a machine does not respond to screaming." After laboriously examining the rf stations one by one, on 28 July 1960, a Thursday evening, the group concluded among themselves that "tomorrow it will work." [73]

On the morning of Friday, 29 July 1960, the accelerator control room was jammed. At first, operators kept losing the beam during the delicate phase transition. Though tests indicated that the switching equipment was functioning correctly, someone on a hunch reversed a transition switch. The beam promptly shot to 24 BeV and a few minutes later to 31 BeV at an intensity of 8×10^9 protons per pulse, exceeding design specifications (30 BeV and 10^9 protons per pulse). Haworth sent out dozens of cables saying, "AGS OPERATED ABOVE THIRTY BEV TODAY," and was promptly deluged by congratulatory replies. "NICE TO HAVE COMPANY," CERN officials cabled back, with just a touch of self-congratulation.

BY THE standards of 1960 the Cosmotron was a crude, hand-operated machine. All its major systems were separately run, and the beam lacked feedback control. Its safety mechanisms were easily circumvented by impatient experimenters. The design flaw in its coils had kept it out of operation for almost a third of its eight years, a poor record for a major experimental facility. On the other hand, it was built inexpensively (at least in comparison with the Bevatron), and played an important role in the maturation of high-energy research. The Cosmotron brought accelerator physicists into contact with the new social difficulties that would be involved in large-scale experimental programs, providing early models for such important accelerator practices as scanning, shielding, program committees, use of bubble chambers as user facilities, and university-lab collaborations.

When the Cosmotron came on again, in mid-1959, its performance was disappointing at first. In October 1959, the beam intensity was about 3×10^{10} particles per pulse, down by a factor of 3 or 4 compared to 1957. Haworth told the executive committee that the Cosmotron "unquestionably suffered by premature diversion of most of the people active in its construction to the design and construction of the Alternating Gradient Synchrotron."[74] He was only partly right. Early the next year, a more fundamental reason surfaced: the new coil configuration had some undesirable side effects. The old coils in front of the magnet had served to constrain the field; with their removal the field got into the ferrite in the rf system. A steel box had to be built around the ferrite in the north straight section in place of the usual aluminum box. But shielding the rf cavity with a steel box now introduced a magnetic asymmetry into the machine, making the beam want to go around in an egg-shaped path instead of a circle. The machine was shut down again, and steel boxes were put around the *other* three straight sections, replacing the existing aluminum ones and making the machine symmetrical.

As the problems multiplied, Haworth became convinced that Collins was also part of the problem. In the past few years, he had come to lose confidence in his old friend. Collins had often let the Cosmotron department run out of money, which outraged the fiscally conservative lab director. And he did not seem to ride herd on the Cosmotron's problem solvers enough to suit the lab director's taste. In 1958, when Collins returned from his six-month leave, Haworth asked Charles Falk, a junior administrator, to oversee the Cosmotron department; henceforth, Collins could not purchase anything without Falk's signature. But Collins continued to act in ways Haworth considered irresponsible, and the two essentially ceased speaking, which was painfully felt by both sides. "It has been a long time since we have had a talk and this breakdown in communications is causing serious difficulties at the Cosmotron," wrote Collins in January 1960. A month later, he angrily complained that the reports he submitted to Falk on the Cosmotron's performance were subject to being changed without his knowledge or permission.[75]

Worse, the Cosmotron's mechanical problems continued. At the end of January 1960, faulty welds caused some retaining bolts to break, limiting operation of the machine for several more weeks. On April Fool's Day an alternator failure shut down the machine for another month. Haworth ran out of patience. On 4 April, he telegrammed Lyle Smith, away in Paris:

COLLINS AND I HAVE MUTUALLY AGREED THAT HE WILL BE RELIEVED OF COS-
MOTRON RESPONSIBILITY AS SOON AS YOU RETURN. YOU WILL BE IN CHARGE.
. . . UNDER CIRCUMSTANCES, IMPOSSIBLE TO GRANT EXTENSION OF LEAVE.[76]

Haworth soon undertook a major reorganization. During most of the 1950s, Collins had helmed the Cosmotron department while Haworth led a separate accelerator development department. Haworth now dissolved the Cosmotron department (Collins went back to physics), put all its experimenters in an enlarged physics department, and created an accelerator department to operate the Cosmotron and the AGS, headed by Green with deputy chairman Smith in charge of the Cosmotron. Chairmanship of the physics department, held for a long time by Samuel Goudsmit, was assumed by Maurice Goldhaber in May 1960, and when Goldhaber became lab director the next year the chairmanship was assumed by G. H. Vineyard (who would also succeed Goldhaber as lab director a decade later).[77]

By the summer of 1960, the Cosmotron almost miraculously got back on its feet, and was operating much better than it ever had. Its facilities were also greatly improved: the experimental area, which had been 5,000 square feet, was expanded to over 25,000 square feet, and two more external beams were added. And its running time shot up: while in March 1953 it operated ten shifts a week (two eight-hour shifts five days a week), and in April 1955, fifteen shifts per week, in 1961 the Cosmotron began running twenty-one shifts a week, or around the clock.

But by then the AGS was running, and it soon took over most of the lab's high-energy work. Of the major high-energy research teams, only Adair and Leipuner remained behind at the Cosmotron, for the moment at least, working with a fourteen-inch chamber they had built, which was the biggest bubble chamber in building 902. While the Cosmotron's life as a high-energy physics instrument was virtually over, it continued to operate for several more years serving medium- and low-energy physicists, nuclear chemists, and biologists. In the early 1960s, Donald Hughes's lieutenant Harry Palevsky created high-energy neutrons for cross-section research at the Cosmotron using a method called *charge-exchange:* a proton whizzing by a neutron sometimes swaps electric charge with it, continuing on its way as a neutron while leaving the proton behind. Palevsky would turn 1 to 3 BeV protons into neutrons, scatter them, and then charge-exchange them back into protons so that they could be detected. In the mid-1960s, Palevsky became interested in scattering the protons themselves off nuclei, pioneering a field later called intermediate-energy physics.

With the exception of a major generator failure in April 1965, which kept the Cosmotron off line until the end of August, the machine ran better in the 1960s than it ever had, at new levels of reliability and intensity. Control room log book entries sound contented and even cheery in ways unknown back in the 1950s: "hunky-dory," "Very," "Very very," "no perspiration" "problems none, smooth run," and even a (slightly inaccurate) Cole Porter reference ("just one of those fabulous nights:—a ride to the moon on gossamer wings").

But it is an irony of advanced basic research equipment that they are usually working better than ever at the moment they become obsolete. In 1965, the AEC delivered to President Lyndon Johnson a "Policy for National Action in the Field of High Energy Physics." The agency was feeling pinched by expensive new projects that included the Stanford Linear Accelerator Center (SLAC), the Los Alamos Meson Physics Facility (LAMPF), and a 200 BeV accelerator then still on the drawing board. The report named four accelerators nearing the end of maximum usefulness: the Cosmotron, the Bevatron, the Cambridge Electron Accelerator in Massachusetts, and the Princeton-Pennsylvania accelerator in Princeton. Tentative shutdown dates were scheduled (U.S. A.E.C. 1965).

On the night of 30 December 1966, two experimental groups, Palevsky's and Adair's, were still taking data on the Cosmotron floor. The machine was scheduled for permanent shutoff at midnight, but Adair's group pressed for more time, and the Cosmotron got a stay of execution until eight the next morning. Just before midnight, John Blewett turned up, as did Collins. Blewett, who together with Green had formed the backbone of the team that had built the Cosmotron, wanted to be around the moment when it met its end. Discovering the shutoff had been postponed a few hours, he decided to hang around, passing the time by playing cards with Leipuner and other members of Adair's group, growing progressively lachrymose and not a little inebriated along the way; "I made a lot of money," said Leipuner. At 0700, Blewett mumbled a few inchoate words, and switched off the machine. The final log entry:

Machine ran well till final shutdown at 0700.
FINI

The Cosmotron, like the BGRR, supported an experimental community that by the time of its shutdown belonged to the past. Henceforth, accelerator experimenters were increasingly divided into subcultures (Galison 1997) organized around their ever larger and more complex detectors.

After its shutdown, the Cosmotron was disassembled. Many parts were recycled. The injector was converted for biomedical research purposes at Columbia. Some groups took a few C magnets for experiments, while the Smithsonian Institution added some to its historical collection; one still stands outside the AGS building as a monument. Building 902 is now a magnet facility for Brookhaven's newest accelerator, the Relativistic Heavy Ion Collider. Its floor is littered with hundreds of magnets a mere ten inches in diameter, a fraction the size of the Cosmotron's, and the magnet builders keep track of them by branding each with a bar code. A line on the floor, noticeable only to those who know to look for it, is the only sign left of the Cosmotron, which has otherwise vanished into the oblivion with which laboratories condemn bygone forefront research instruments.

Goldhaber's Directorship

LELAND HAWORTH had overseen the lab's development from a still mud-splattered army camp to a forefront institution with world-class instruments. Dedicated to the lab, he turned down, in 1953, an invitation to become deputy assistant secretary of defense, and, in 1958, an offer to become head of the AEC's reactor division.[1] He took a personal interest not only in the lab's scientific activities but in its social life as well; he marched at the heads of parades, tossed the first pitch at the opening day of Brookhaven's fast-pitch softball season, lent money to those who needed it, and participated in theatrical events (he was the first body exhumed from the cellar in a production of *Arsenic and Old Lace*). At Christmas in the early years, junior scientists were flabbergasted when the lab director came around personally to their offices to thank them individually for their service.

Haworth understood that, while exploiting one set of machines, a laboratory had to plan and build the next generation to survive. He pressed the lab staff to work on such machines and participated closely in the technology development of these machines. In the long run these projects moved forward more rapidly because of Haworth's knowledge and intimate involvement. His own honesty and integrity were beyond question, even by the ordinarily suspicious AEC. As Kouts recalled:

> We all profited in those days from a relationship that Haworth had developed with the AEC, which was based on complete frankness and honesty with them. It's always common in enterprises where you deal with a funding agency to inflate the estimate of what you need. You say, "I need more," because you expect them to cut you back. If you do it right, you get cut back to what you really wanted in the first place. Leland never did that. He gave an honest account of what he expected to do and what he expected to require in order to do it. And this was done so clearly, with such obvious character to it, that nobody would cut what Leland asked for. That relationship was a very important one in the molding of Brookhaven, and the reputation Brookhaven had with the Atomic Energy Commission and with the outside world.[2]

During Haworth's directorship, Brookhaven's peer laboratories multiplied. The number of AEC laboratories grew to twenty, of which eight— including Brookhaven, Argonne, and Oak Ridge—were multipurpose national labs. Each of them offered a unique staging ground for projects, and was able to secure an adequate supply of work while leaving enough for others. Competition thrived, with each lab having areas of specialty that it wielded to land new facilities and projects. They were managed through administrative or Goco contracts, though the labs fought a running battle with the AEC over their interpretation, with the labs fighting to stave off what they felt was stifling bureaucracy, as well as the agency's growing tendency to treat the labs, not as communities in themselves, but as places to carry out independent research programs supported in Washington. Haworth and other lab directors fought these tendencies, and in 1957, at a meeting in Yosemite National Park in California, participated in what he and Oak Ridge's Alvin Weinberg optimistically called a Yosemite revolution against the AEC's oppressive hand.[3] The revolution failed to materialize, not because the revolutionaries lacked vision and energy nor their case merit, but because a tiny Soviet satellite named Sputnik turned massive amounts of federal attention and resources toward the labs, relieving some of the pressure. By the end of the decade, their future looked rosy again. In 1960, the congressional Joint Committee on Atomic Energy issued a report that said:

> The laboratories are an effective means for integrating the scientific and technical capabilities of the nation through their complementary position with respect to the universities on the one hand and industry on the other. They are indispensable to achieving the basic objectives set forth in the Atomic Energy Act: Development and use of atomic energy for (1) the common defense and security; (2) promotion of world peace; (3) improvement of the general welfare; (4) increasing the standard of living; and (5) strengthening free competition in private enterprise.

The labs, the report continued, concentrate varied disciplines and facilities that "support, augment, and stimulate" one another to create "reservoirs of organized and interrelated talent and experience." Scientists and engineers from universities and industry can participate, creating an "effective tool for attacking the complexities of science and technology," the results of which exceed what could be expected by distributing resources otherwise. These "national assets," the report concluded, must be kept "vigorous." The labs would be "fully occupied" filling "national needs" for at least the next decade.[4]

Haworth's last year and a half at Brookhaven were marked by several personal troubles. He was discovered to have colon cancer, for which he had an operation in November 1959. The prospect left him terribly depressed; "I don't know whether I'll ever see you again," this normally op-

timistic man told an associate the night before. Moreover, his wife, Barbara
Mottier, was also severely ill and wheelchair bound; she died on 6 February 1961.

Ten days later, on 16 February 1961, Haworth attended a meeting at the
National Science Foundation on AUI's plan to build a National Radio Astronomy Observatory at Green Bank, West Virginia.[5] After the meeting,
Haworth called on Glenn Seaborg, a physicist from the University of California at Berkeley, winner of the Nobel Prize in chemistry for 1951, and
the man whom the newly inaugurated president John F. Kennedy had appointed AEC chairman. A vacancy had arisen among the five AEC commissioners, and Kennedy had specifically asked Seaborg to fill it with a
"young scientist" whose experience on the commission could serve later
as background for another position in government where scientific expertise was important. Kennedy was also concerned about rebuilding the
AEC's declining reputation and influence in Congress. The AEC was now
facing looming battles with Congress involving the atomic energy program
and projects involving large accelerators. Seaborg immediately thought of
Haworth, though at fifty-six he was hardly a young scientist.

"He had the combination that I wanted," Seaborg recalled. "He was
an effective administrator, a soundly based scientist, and had an effective
personality—the very qualities that had made him so effective as director
of Brookhaven."[6] Haworth was intrigued by the challenge of joining an
agency that had such an impact on research throughout the United States,
even though it meant a substantial cut in salary. He eventually accepted,
resigning as director of the lab effective 1 April 1961. Pending the outcome
of a search, the position of director was assumed by Haworth's deputy,
Gerald Tape. Haworth served on the commission for two years, and did
exactly what Kennedy had in mind, moving on to become director of the
National Science Foundation in 1963.

In a resolution thanking Haworth for his services to the lab, the AUI
trustees recalled the famous epitaph of Sir Christopher Wren: "If you
would seek his monument, look around."[7] Brookhaven was a peculiar kind
of monument. Other monuments may stand out by their grandeur or the
materials with which they are made. Brookhaven, instead, stood out by its
achievements; its scientists had managed to build a set of instruments and
tease from them performances that disclosed new features of the world.
Under Haworth, the lab had completed its first round of major facilities,
and had all but completed the second; the AGS was already in operation
and about to be dedicated, while the HFBR was approved and about to have
its groundbreaking. Thanks to Haworth, Brookhaven's vitality as a staging
area was guaranteed well into the next decade.

10.1 Maurice Goldhaber as lab director (1961–1972).

Haworth's deputy, Gerald Tape, succeeded him as acting director while a search committee considered candidates for a permanent replacement. Though I. I. Rabi was technically not on the committee, he played the principal role in seeking and questioning candidates. Rabi felt it was crucial for the director to have stature as a scientist, and the three outside names that kept cropping up were Norman Ramsey, a Harvard physics professor who had been one of the lab's founders and the first head of the physics department; Edward Purcell, also a Harvard professor, who won the 1952 Nobel Prize in physics for the discovery of the nuclear resonance method of measuring nuclear properties; and Robert Bacher, another physicist, BNL founder, and former AEC commissioner. All three, when informally approached, were uninterested. Internal candidates included Tape, Maurice Goldhaber, and chemistry department head Richard Dodson. Some at the laboratory felt that scientific prominence was less important than management ability and knowledge of the lab's operations.[8] Rabi, as usual, prevailed, and the decision was made to seek a prominent scientist. Maurice Goldhaber was selected; he had been at Brookhaven since 1950, and chairman of the physics department since 1960 (fig. 10.1).

The Goldhaber Years, 1961–1965

Goldhaber was still a dashing-looking fifty despite a set of heavy, horn-rimmed glasses that obscured his clear features. He spoke carefully, in a voice with a musical lilt and the soft timbre of a tenor clarinet, and in phrases that gave off the impression of an experienced, if somewhat detached, intelligence. An occasional twinkle in the eye and slight upturn at the corners of his mouth were the only physical manifestations of an impish streak that often manifested itself in witty pronouncements.

He brought an entirely different style to the lab directorship. If Haworth ran the lab like a president who sought to participate actively—too actively, for some tastes—in shaping legislation and in steering lab projects into being, Goldhaber governed more like a royal figure, remaining aloof from day-to-day operations and leaving operations to the initiative of deputies. To make things happen, Goldhaber relied less on personal intervention and more on the authority of the office.

In Haworth's conservative, plain style, the director's office was small and spare, with concrete block walls, table and desk, no rug on the floor. Not only was there no air conditioner, a hot air duct ran through the ceiling that warmed the office even in summer. Under Haworth, every spare cent had gone toward research, and about a hundred buildings from Camp Upton, barracks thrown up to house troops on a temporary basis, were still in use. The lab was still relying on the obsolete Camp Upton street-lighting system, telephone system, water tank, and sewage system. One of Goldhaber's first acts was to have the director's office enlarged and renovated. The pipe was removed, an air conditioner installed, and the floor covered with a carpet. Many buildings were renovated, and the site began to look less like a converted army camp and more like a university.

Haworth had found it next to impossible to delegate: when a job came along which he could not handle by himself, he would appoint a committee with himself as head. Goldhaber recalled:

> Take the way he handled tenure cases. For each person, he would appoint an ad hoc committee of four or five, and would himself write personally to individuals to solicit letters from outside. I remember once receiving a phone call from Haworth on a Saturday just before Christmas. "Say, Goldhaber, whom should I get to write a letter of recommendation for Mr. X? I'm proposing him for tenure at the January meeting of the board and need a letter before then." It was a clumsy system, and resulted in his getting overburdened—but it was typical of the way he ran things, with a heavy personal touch.[9]

Goldhaber appointed associate directors—one for high-energy physics, another for life sciences and chemistry—to handle problems formerly dealt with by the director, and a panel called the Brookhaven Council to handle tenure cases (it soon came to advise the director in other areas as well). Whereas Haworth appointed chairs with no term limit, trusting that he had chosen the best person for the job, Goldhaber announced that, henceforth, chairs would be appointed for a specific period.

People said of Goldhaber that, as director, "he spent more time in seminars than the director's office"; the remark was intended as praise by supporters and criticism by detractors. For good or bad, he was reluctant to engage in political conflict. This was a marvelous attribute in a scientist and in a leader of scientists, but could prove unfortunate for an institution that relied so heavily on government resources at a time when the price tag of new accelerators and detectors was skyrocketing and federal enthusiasm for basic research soon to plummet.

Most of all, Goldhaber carried to the directorship a calm and lofty view of scientific inquiry. For him, the integrity of science made it worth doing in and of itself, wholly apart from whatever uses it might have. If one just took care of healthy science, as he put it later, "the applications will fall off like ripe plums from a tree." [10] The gentlemanly class that Goldhaber brought to the lab extended even to its series of elegant Christmas cards. Goldhaber would choose a cover depicting an important discovery or development, emblematic of the lab, that had taken place there that year. These cards reveal where his heart lay: in the scientific accomplishments of the institution.

For the first half dozen years of his directorship—"the sixties before The Sixties," in David A. Hollinger's witty phrase—the lab benefited from continuing federal largesse toward science, partly reflecting what Hollinger has identified as a crest in the hope and respect that American society and in particular its intellectuals placed in science (Hollinger 1996). At the beginning of the 1960s, the U.S. led all other nations in spending on research and development as a percentage of GNP (Smith 1990, 71). In the immediate wake of Sputnik, the science-government-industry partnership forged during World War II was revitalized. Congress earmarked plenty of funding for programs and facilities, and devoted much enthusiasm for basic research; good leadership meant encouraging high standards of scientific integrity.

These were heady days for Brookhaven, In 1963, a President's Science Advisory Committee/General Advisory Committee (PSAC/GAC) panel, headed by Norman Ramsey (now at Harvard), ratified the "two-step" approach in which Berkeley and Brookhaven leapfrogged each other in building forefront accelerators, recommending that the next machine, in the 200 GeV range (this was about the time the nomenclature changed from "BeV" to "GeV," the "G" standing for giga), be built at Berkeley, while

Brookhaven should proceed with a design study for a 600 to 1,000 GeV machine. But it also recommended that Brookhaven pursue a program to construct and use storage rings: two circular rings to store beams of protons, injected from an accelerator, orbiting in opposite directions. At intersection points between these beams, head-on collisions would occur between particles in the two beams, much more violent than collisions between such particles and fixed targets. For a number of reasons, Brookhaven scientists rejected this recommendation, deciding instead to pursue an upgrade of the AGS while they confidently worked on plans for the next large accelerator to be built in their soil.[11]

Premonitions

But there were omens of more difficult times. An incident that followed a fatal accident, in January 1961, at an AEC reactor in Idaho hinted at new troubles in the lab's relationship with that agency. The AEC began a study of operation and maintenance procedures for 131 reactors that it owned or licensed. Although an AEC review team found no violations of rules or regulations at the BGRR, the agency ordered the reactor shut down until it implemented some procedural changes. Infuriated lab officials felt that the AEC was using the lab to appear hard-nosed about safety; a few called friends in the AEC, some of whom apologized. As lab officials feared, newspapers erroneously assumed that the reactor itself had been deemed unsafe, and prominently featured the story. The episode heralded the AEC's more cautious approach to its contractors. In the first decade of the existence of the national laboratories, they viewed themselves, and were encouraged to view themselves, as involved in a collaborative and mutually supportive relationship with the AEC. The episode signaled a new wariness on the part of the AEC to become too closely involved with contractors.

An omen of another sort was delivered by Derek de Solla Price, a historian of science from Yale, during his Pegram lectures, given in June 1962 and published the next year as *Little Science, Big Science* (de Solla Price 1963). The phrase "big science," referring to the large-scale character of modern science, quickly became part of the everyday lexicon of science.[12] De Solla Price was enthusiastic about big science, but also pointed out that the exponential growth of postwar science would inevitably trail off. His overall optimism masked the more ominous message that the pace of scientific funding was due to slacken.

Some physicists were listening—among them Luke Yuan. By 1963, Yuan had noticed a trend in the fundamental problems of his field: the energies at which they were resolvable were climbing fast. He also realized that the cost of the accelerators needed to reach those energies would climb proportionately, and would tie up a larger fraction of the science budget in a single project. Yuan was not at all sure the necessary consensus for such projects existed, either among the general public or the scientific community. Hoping to inspire greater ardor among his colleagues, and to shore up the consensus in Congress and the general public, he hatched the scheme of publishing a small book on the goals and values of high-energy physics.

> At the time, there was no universal theory, quarks had only just been proposed, and there was the feeling that it was very complicated. The AGS had just been completed, people were talking about the 200 GeV machine that became Fermi-lab, and at Brookhaven we were studying a 600 to 1,000 GeV machine. But I thought it might take much bigger machines, much more money, much more commitment to get to the bottom of it, to the foundations. I thought the book would be a way of generating the enthusiasm. It had a political purpose that hadn't been required before.[13]

Yuan broached the idea to Oppenheimer, who was enthusiastic, as were several others. He then contacted about thirty scientists, asking them to set down their views on the importance of high-energy physics, and its problems and implications. The articles were published late in 1964 as a BNL report, *Nature of Matter: Purposes of High Energy Physics* (Yuan 1964).

The 150-page document is a remarkable snapshot of the hopes and expectations of the key figures of the field. It was an ambitious attempt to shore up the faltering partnership between science, government, and industry in the specific area of accelerators by providing leaders in the physics community with a mix of scientific arguments and popular rhetoric and imagery that they might use to persuade others. In the foreword, Oppenheimer quickly described what was known about particles and fields, adding that many important questions about the nature of matter remained unresolved. "It may be asked," Oppenheimer admitted, "why this state of affairs is important enough to warrant the effort, the expense, the public support needed to enter the domain of much higher energy physics." The answer is neither simply that the construction of such state-of-the-art devices inevitably creates new techniques valuable and applicable to other sciences, nor that such devices sometimes lead to unexpected discoveries of importance to technology or human welfare.

> It is also this: the last centuries of science have been marked by an unabating struggle to describe and comprehend the nature of matter, its regularities, its laws, and the language that makes it intelligible. The successes in this struggle, from the Sixteenth Century until our own day, have inspired the whole scientific enterprise, and lighted the world of technology, and the whole of man's life. They have informed the education and the devotion of young people. They have played an ineluctable part in the growth, the health, the spirit, and the nature of science.

We are now, despite tempting and brilliant topical successes, deep in the agony of this struggle. This volume attests the conviction of those who are in it that, without further penetration into the realm of the very small, the agony may this time not end in a triumph of human reason. That is what is at stake; that is why this book is written.

The book ends with an essay by Yuan on the prospective 600 GeV machine. Between Oppenheimer's preface and Yuan's conclusion are twenty-six articles varying widely in content and style. Many authors seemed unsure of what they were being asked to do. Some advanced new theories or gave technical descriptions in language addressed to colleagues, while others offered justifications for their field in popular language using an assortment of different kinds of images. While some of the imagery was eloquent, some was awkward and even far-fetched, revealing how unfamiliar an experience it was for these scientists to justify their research.

Each human society excels at a small number of the many activities that people carry out. Our own society is preeminent at large-scale technological and scientific projects, such as the building of high energy accelerators. It is therefore an expression of the highest spirit of our culture to carry on with the task which we have begun, the exploration of nature to all its limits. Indeed, it may well be judged that this spirit is our greatest contribution to the human outlook. High energy physics is clearly one of the subjects on the frontiers of such exploration. If we cut back on it for reasons of budgetary limitation, or political squabbling, I think we will have seriously damaged the best single element we have contributed to human culture. (Feinberg)

A great society is ultimately known for the monuments it leaves for later generations . . . [S]uch a machine, which is on the scale of a national effort, will without question be a source of inspiration for new science and a monument to our days. (Pais)

[R]esearch in experimental high energy physics, precisely because of its financial magnitude and its eventual broad distribution of expended funds, can play a modest but not entirely negligible role in maintaining economic stability. (Primakoff)

One should not overlook how fateful a decision to curtail the continued development of an essential element of the society can be. By the Fifteenth Century, the Chinese had developed a mastery of ocean voyaging far beyond anything existing in Europe. Then, in an abrupt change of intellectual climate, the insular party at court took control. The great ships were burnt and the crews disbanded. It was in those years that small Portuguese ships rounded the Cape of Good Hope. (Schwinger)

We find strong support today for space technology, which may allow us to explore the unknown parts of the solar system. Exploration of the unknown was always a strong component of human endeavor in our modern civilization. But it must go together, as it always did, with an equally strong component: the explanation of the unknown in whatever form it faces us. In the beginning of the Sixteenth Century, when the scientific era began, Magellan performed the first trip around the earth. But also in the same period Copernicus published his work on the motion of the planets. (Weisskopf)

Instead of feuding with one another for public favor, it would be fitting for scientists to think of themselves as members of an expedition sent to explore an unfamiliar but civilized commonwealth whose laws and customs are dimly understood. However exciting and profitable it may be to establish themselves in the rich coastal cities of biochemistry and solid state physics, it would be tragic to cut off support to the parties already working their way up river, past the portages of particle physics and cosmology, toward the mysterious inland capital where the laws are made. (Weinberg)

But even as Yuan was completing and distributing the book, the fears that had motivated it were being confirmed in the unraveling of Berkeley's confident plans to win approval of its 200 GeV machine. The reasons for Berkeley's troubles were many (Westfall 1988; Jachim 1967). One was a backlash in the wake of the rejection, in 1964, of the proposal by the Midwest University Research Association (MURA) to build a new accelerator in the midwest after initial support, which led midwestern scientists to complain more loudly than ever about inequitable distribution of resources (Westfall 1988, 172). Another was the hostility of many scientists to Berkeley, a lab well known for unfriendliness to outsiders, leading to a new lack of consensus within the high-energy physics community that, while not the kind Yuan feared, was just as corrosive to the community's prospects. In addition, the sheer cost of the $350 million Berkeley project meant a new visibility and vulnerability to political attack. An article in *Nation's Business* not long before Yuan's article suggested that the federal role in supporting the national laboratories might be harming science in the long run.[14] *The Evening Star* of Washington, D.C., remarked that with Uncle Sam having run out of dam sites, the new accelerator was "federal pork in the grand old tradition."[15]

The unexpected controversy surrounding the Berkeley accelerator made the AEC especially careful about its site selection process. In April 1965, it asked the National Academy of Sciences to form a site selection committee, which then issued an open invitation for site proposals. Goldhaber and deputy director Clarke Williams did not want Brookhaven to join the competition. They believed that Berkeley's troubles would soon blow over, that it would wind up with the machine, and that the leapfrogging would then continue as in the past, giving Brookhaven the next leap forward. But others sensed that things were up for grabs in an unprecedented way and urged lab officials to compete aggressively.

In May, Van Horn encouraged Goldhaber to make a proposal for the 200 GeV machine. Goldhaber was reluctant, and argued that joining the fray might threaten the AGS upgrade. Also, he was temperamentally disinclined to bring upon himself the political conflicts that would be involved. "As long as a national policy of construction of a 200 Bev machine to be followed by a 600–1000 Bev machine is adhered to," Goldhaber wrote back, "Brookhaven would concentrate on the second machine." [16]

Goldhaber then received a barrage of pleas to be more aggressive in courting the machine. Some pleas came from East Coast accelerator physicists, who sensed that the rules of the game in their field were about to be entirely rewritten, while others came from Long Island organizations such as the Nassau County Planning Commission, who were proud of the internationally renowned lab in their midst and aware of the additional jobs the project would bring. [17] Goldhaber reluctantly agreed to compete, writing a somewhat apologetic letter to Berkeley's director Edwin McMillan for so doing (fig. 10.2). Still, he declared, the lab's official position was that it would not be disappointed if the machine went elsewhere. [18]

In October, two additional monkey wrenches were thrown into the works. One, tossed by Columbia's Samuel Devons, was the argument that a less adventurous machine should be built; the other, thrown with brutal force by Robert Wilson of Cornell, was the argument that a more adventurous one could be constructed at much lower cost.

Devons—whom many saw as acting as the stalking-horse for Rabi, who was outraged by Goldhaber's failure to be aggressive—circulated a paper entitled "Comments on the High Energy Physics Program in the United States." Devons claimed that much money could be saved by reducing the target energy of the next accelerator to 150 GeV and building it at Brookhaven, so that the AGS could serve as an injector. [19] With Goldhaber on the sidelines, AUI president T. Keith Glennan began to take the initiative for promoting this idea, though some derisively dubbed the proposed machine the "Devatron." Friction developed between Goldhaber and Glennan, though over more than this one plan; Glennan tended to view himself as the lab's chief executive officer, and Goldhaber as its chief operating officer.

Wilson was more daring. In language harsh even by physics standards, he called the Berkeley machine "lacking in imagination" and saddled with an overly conservative design that needlessly drove up cost and time of construction. "[H]ad Ernest Lawrence proceeded in this way in 1930," Wilson wrote to McMillan, "he would have connected dry cells together to get a million volts." [20] Wilson offered a redesigned plan for a 200 GeV machine, based on synchrotrons that had been built at Cornell, that he said could be built at a new site for under $100 million in under seven years. [21]

By the fall of 1965, Berkeley had clearly lost control of the process, and both the site and design of the new accelerator were up in the air. Many Brookhaven administrators now changed their tunes. The situation threatened the lab, Williams realized, for if Berkeley did not get the 200 GeV machine Brookhaven would not be the automatic choice for the one there-

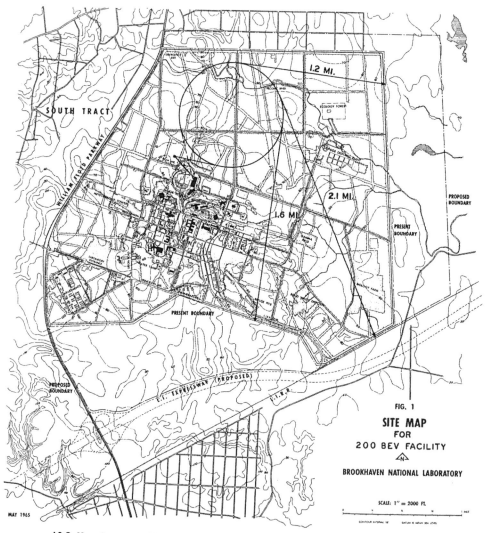

10.2 How the proposed 200 GeV (then BeV) accelerator might have fit on the Brookhaven site. The machine was ultimately sited in Illinois, at the present Fermilab facility.

after.[22] Letters began to pour in to Goldhaber expressing concern that the lab's vitality would be hurt if the machine were located elsewhere. "Those of our Government," wrote one correspondent, "who guide the fortunes of Science should be asked, why force a going, vigorous laboratory into decline while, at the same time, trying to build on a vacant site another laboratory to perform the same function"?[23] But Goldhaber refused to be converted. He continued to play the role of a gentleman who was not about to break a promise; he was also convinced, privately, that waiting for the next generation machine was the wiser scientific course.

Others—including congressmen and especially midwestern congressmen—were fighting the political battles Goldhaber shunned. In January 1966, Brookhaven sponsored a meeting at the Hotel Biltmore, in New York City, of some 150 high-energy accelerator users and designers and some AEC staff. At the meeting, later characterized by Wilson as a "sort of advertised rumble," Devons and Wilson argued for their schemes, and the idea of Berkeley's managing the next big accelerator met with open hostility (Westfall 1988, 248). Others, angling for a site near their own institutions, argued just as heatedly against Brookhaven. One was Edwin Goldwasser of the University of Illinois. Elaborating on remarks he had made at the Biltmore meeting, Goldwasser wrote Paul McDaniel, the AEC's director of research, that the AGS is "an old machine saddled with many obsolete components," and that to use it as the injector for a 200 GeV machine would involve reducing the scope of the project in a way that would be "crippling," involving the "real risk" that "the remarkable progress made during the past two decades would slow down drastically and possibly even come to a halt," in which case leadership in high-energy physics "would almost certainly pass over to Western Europe and to the Soviet Union."[24]

In March 1966, the NAS site evaluation committee named six finalists. Brookhaven was among them. Governor Nelson Rockefeller sent Glennan (whom he correctly perceived to be the accelerator's principal champion at Brookhaven) a congratulatory telegram, saying that the "full resources of the government of the state of New York" are behind the site. The Long Island Association of Commerce & Industry issued an "Action Program for the 200 Bev Accelerator."[25] But even though Glennan went all-out to bring the accelerator to Brookhaven, the Weston site was chosen later that year.[26] The town of Weston went out of existence, and construction began on the National Accelerator Laboratory, eventually known as Fermilab. To manage the laboratory, an AUI-like consortium of institutions called Universities Research Association was formed, which in turn selected Robert Wilson as the lab's director.

Thus ended the Berkeley-Brookhaven domination of high-energy accelerators. Brookhaven was left with its AGS upgrade project (for which the AEC budgeted $45.8 million in 1966), and a promissory note, of dubious value, for a larger machine someday.[27] The event indicated, too, the increasing role of political considerations in realizing large scientific projects, meaning that laboratory leaders would have to be more aggressive and learn how to read and capitalize on political openings. "The dynamics of the political economy of high energy physics in this period," wrote science historian Zuoyue Wang in an article about the origins of the Stanford Linear Accelerator Facility, "hinged on the pull of international politics and the push of the scientists" (Wang 1995, 354).

The Fourteen-Foot Bubble Chamber

At the same time, Brookhaven was also competing for a detector. In the era of big science, detectors, too, had grown in size to the point where the AEC felt the need to consolidate their construction programs. Here Brookhaven was handed a major defeat, although clever machinations recovered a share of what was lost.

Since the first working bubble chamber was built in 1952, bubble chambers had jumped in size. The larger size had many advantages: accuracy in momentum measurements was proportional to the square of track length, and longer track lengths also aided in identifying particles by secondary interactions. Large chambers were also crucial to high-energy neutrino research, for they had a greater effective interaction region—and in the mid-1960s, neutrino interactions in hydrogen were proving invaluable for understanding fundamental weak interactions. But the cost of a bubble chamber was proportional to the inner dimensions of the chamber squared.[28]

Until the mid-1960s, however, a design limitation involving the windows placed an effective cap on the cost of big chambers. Two windows were required; one for a light source, the other for the camera. These windows had to be thin enough for the camera to see through without distortion but thick enough to withstand tremendous pressures. Alvarez improved the situation by putting both light source and camera in one window and placing retrodirectors in the back of the chamber to redirect the light, but the limits imposed by requirements of strength and nondistortion remained. Bubble chamber builders at CERN, at work on a huge chamber called Gargamelle, partly solved the big window problem by placing eight cameras at an equal number of small windows each containing a tiny fisheye lens. Because smaller windows were much stronger, the size of the chamber could be much larger. And while no one camera could see more than a part of the chamber, all parts of the chamber could be seen by at least two. Brookhaven's Robert Palmer then made a significant improvement by utilizing nondistorting wide-angle lenses, recently available, able to see all of the chamber. Bubble chamber builders at several institutions then worked on designs for large bubble chambers based on Palmer's ideas: a 14′ to 16′ at Brookhaven, and a 20′ to 25′ at Fermilab (fig. 10.3).[29]

Using Palmer's ideas, Shutt's group submitted a plan in mid-1964 for a fourteen-foot hydrogen bubble chamber to the lab management.[30] But Goldhaber did not push the idea as hard as Shutt thought he should. To Shutt, the new bubble chamber was an integral part of the AGS upgrade. "Brookhaven's ability to do significant research and to compete successfully in high energy physics will have been cut considerably" if the new bubble chamber is delayed, Shutt complained to Goldhaber. "It seems necessary to fight for our whole high energy physics program and not just for a part of it." Shutt added, "I am sorry to have to feel that for the first time we are not being given effective support in a project that seems very important to many of us."[31]

Goldhaber was as disinclined to fight for the fourteen-foot detector as

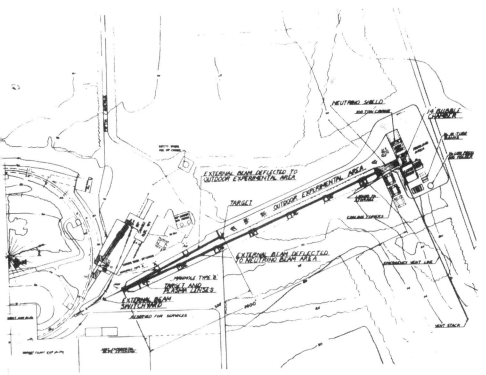

10.3 Schematic of proposed fourteen-foot bubble chamber as it would be attached to AGS.

he was for the 200 GeV accelerator, and for the same reasons. In communicating the request to the AEC, Goldhaber described it as of lesser importance than the upgrade; "we do not wish to take any steps which would interfere with the chances for acceptance, now or later, of the AGS conversion program and we therefore indicate a comparatively lesser priority for this bubble chamber proposal." [32] And indeed, the AEC turned down Brookhaven's requests for a fourteen-foot chamber—for the fourth and final time in 1968 for FY 1969—the first time that one of the lab's bubble chamber projects was not approved. The AEC, meanwhile, approved an Argonne proposal for a twelve-foot chamber, an attempt to placate the midwestern physicists for canceling a new accelerator they had proposed, though a large bubble chamber at their existing low-power Zero Gradient Synchrotron (ZGS) would be far less effective than at the AGS. [33]

Not only had the lab's bubble chamber group lost the fourteen-foot chamber to Argonne, but Fermilab was also exerting mounting pressure on Brookhaven. Bubble chamber builders there were turning against Brookhaven the very arguments Brookhaven had used against Argonne: that the lab with the most powerful accelerator was owed the best detector support. Fermilab officials attempted to convince Brookhaven to move its existing eighty-inch chamber to their laboratory, and hired away several key members of Shutt's group, at one point even seeking to get Shutt himself. In August 1968, Shutt angrily reacted. Goldwasser's reply, in its invocation of the best interests of the field, was based on principles that might have been uttered by Shutt in his battle against Argonne.

> I think that you are the first to imply that I am ruthless in the pursuit of my goals, and I assure you that, in fact, ruthlessness is not my style. . . . On the other hand, we also recognize that all participants in the national high energy physics program have an enormous investment in our Laboratory. The sooner we can become an operating research laboratory and begin to make noteworthy discoveries, the better it will be for the field as a whole. Thus we are breaking our necks to get the accelerator on the air as soon as possible and also to bring into being a complement of equipment that will enable us to make our first exploratory observations at the earliest possible date. [34]

Persevering, Shutt's bubble chamber group continued to develop bubble chamber technology with AEC research and development funds, and in 1969 obtained AEC permission to construct a seven-foot "test facility" on which to explore their ideas. In 1971, the ingeniously cobbled-together test facility was modified to serve as an operational bubble chamber, surrounded by a superconducting magnet and installed at the AGS. With seven times the volume of the eighty-inch chamber, the seven-foot facility became the principal detector used by Brookhaven's neutrino physicists for the next decade (fig. 10.4).

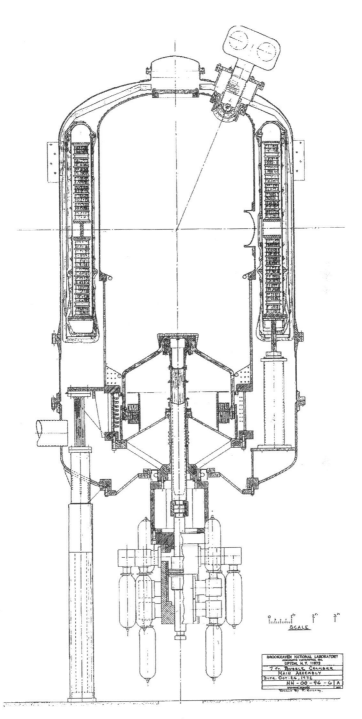

10.4 Diagram of seven-foot bubble chamber.

The New Political Spotlight

Goldhaber's tenure also coincided with the awakening of new environment concerns and a new sensitivity to the hazards of radiation. In 1960, the AEC responded to protests by stopping sea disposal of radioactive wastes from its national labs, Brookhaven included (Weart 1988, 297–98). The agency also mandated regular public reports by its labs on the amount of radioactivity at their boundaries, after which Brookhaven began releasing quarterly reports to Long Island newspapers on its contributions to environmental levels. And as issues involving radioactivity and the environment moved more to the fore of public and political concern, the lab sometimes found itself inadvertently drawn into the public spotlight for research that had been uncontroversial when first carried out.

One example will suffice, involving research to evaluate the potential costs of a disaster at a commercial nuclear power plant. In the early 1950s, the AEC had envisioned the development of a nuclear power industry, and was anxious to encourage private ownership and operation of nuclear power plants. Utilities, however, were not complying. They were wary of the astronomical liabilities potentially involved in the case of what was then known as a "runaway," a nuclear accident in which the reaction got out of hand and took off by itself. (In 1954, Argonne conducted an experiment in which one of its reactors, BORAX-1, was deliberately allowed to "run away," whereupon it exploded; Holl 1997, 118–21.) Proponents of nuclear power felt that the only solution was for the government to step in and cover the liability.[35] Following passage of the 1954 Atomic Energy Act, proponents of nuclear power began to tackle the liability issue. Early in 1956, Senator Clinton Anderson of New Mexico, chairman of the joint committee and a former insurance executive, and Congressman Melvin Price of Illinois, also of the joint committee, began preparing appropriate legislation, and July 1956, Anderson asked the AEC for a comprehensive report "on estimated damage caused by nuclear reactor runaways."[36]

On 29 August, the AEC sent Haworth a teletype:

> The JCAE has requested that a comprehensive study be made of the damage which would be caused by runaway nuclear reactors, . . . We feel that BNL is the logical choice. . . . The proposed report would resemble studies of the quote maximum creditable [or credible] accident unquote now generally made for each reactor in obtaining approval from the advisory committee on reactor safeguards.[37]

"Maximum credible accident" was a trademark phrase of the AEC's reactor safeguard committee, whose thinking about nuclear safety was oriented toward conceiving and attempting to forestall the worst disaster imaginable. The agency's request to Haworth to quantify it effectively institutionalized this orientation and would also mold the public's orientation toward nuclear safety as well (Weart 1988, 284–85, 292–93). Haworth handed the job over to Herbert Kouts, now of the nuclear engineering department; Kouts in turn named his assistant Kenneth Downes project manager. Downes assembled a team that included members from the reactor engineering (to determine what and how much radioactive material could be released), meteorology (to determine factors affecting distribution of this material), and health physics and medical departments (to determine how much injury the material could cause).

Conflicts soon arose between the scientific research agenda of the BNL team and the AEC's political agenda. One became known as the "25 r" problem. Downes asked the AEC to provide him with a figure for radiation exposure below which individuals would not be considered injured and therefore would not receive compensation. He suggested twenty-five roentgens, a figure that had been mentioned for years but that he had never seen written down. Not wanting to commit itself in print to a number, the AEC balked. Downes refused to proceed without such a number, and the AEC finally gave in and agreed on 25 r (the figure is still in use). Another conflict arose over "fallout," an emotionally charged word in the mid-1950s. The AEC wanted a substitute term used, a ruse some BNL scientists denounced as "inane." The AEC won this round, and Downes had to use terms like "deposition" instead. The most serious dispute, however, came over AEC pressure to state the probability of a worst-case scenario. In the absence of a definite probability, the actuarial problem, which after all was the ultimate motivation for the study, could not be completely resolved. The Brookhaven team, while deciding the probability was "extremely low," found no scientific basis for pinning it to a number, and refused to commit itself in the final document, which would bring down heated criticism from both supporters and opponents of nuclear power. This conflict was exacerbated by the AEC's refused to specify how the accident was supposed to happen—which meant that every possibility, no matter how far-fetched and fantastical, had to be taken into account.

> We several times tried to pin the AEC down to an assumption of how the accident happened, because we needed some idea of the mechanism in order to make reasonable assumptions about particle distributions. All we could get from the AEC by way of reply was to take the worst case. Only at the very end, when it was very clear that the worst case would be *very bad* indeed, did the AEC people start to worry about the possibilities of occurrence.[38]

The study was entitled "Theoretical Consequences of Major Accidents in Large Nuclear Power Plants," but is almost universally known by its

AEC publication number, WASH-740. It contemplates the possibility of an accident to about a two hundred megawatt reactor (what was then considered large), located thirty miles upwind of a major city. The report considered three kinds of accidents: In the first or "contained" case, all the radioactive materials escape from the core but none escape the containment building. In the second or "volatile release" case, all volatile fission products and 1 percent of the strontium are released from the containment building. In the third or "worst" case, which they admitted was highly improbable, 50 percent of all fission products in the reactor are released. In their deliberately pessimistic scenario, as many as 3,400 people could be killed and 43,000 injured, with possible property damage as high as $7 billion. Nevertheless, the authors state, "this study does not set an upper limit for the potential damages; there is no known way at present to do this."[39]

In the course of their work, the BNL scientists realized that preventing runaways was a serious though manageable problem, but that a worse danger existed in the form of a "fuel meltdown." The fission process inside each operating reactor generates a large amount of heat, which must be dissipated. If the cooling system fails, the fuel might melt and burn its way through the containment vessel to the ground, possibly leading to an enormous release of radioactive materials to the environment. (In later years, this problem would be called the "China syndrome" for the direction the molten core would take in burning through the earth was headed.) Indeed, the authors state that of all possible mechanisms of fission product release a meltdown is the "most likely to occur." While many nuclear engineers knew of this potential problem, for many others less familiar with reactors, the Brookhaven report underscored for the first time this kind of threat.[40]

The Brookhaven team, placed under fantastic time pressure by the AEC, had a draft report ready by 1 January 1957. In March, the AEC sent it over to the joint committee, which had already begun a new push for a bill providing federal liability insurance for the nuclear power industry. The Price-Anderson Act of 1957, as the bill was known, did not seek to cover fully the worst-case scenario of the WASH-740 report (which might have been too costly for Congress), but only more modest accidents.[41] Despite the gap between the worst-case estimates and available coverage, the knowledge that a scientific study existed helped the bill sail through Congress. When the joint committee voted to report the bill to Congress on 9 May, the only opponent was Chet Holifield of California, who objected to the extravagant and unusual subsidy provided the nuclear power industry, and who made liberal use of WASH-740 in his criticisms. Nevertheless, the Price-Anderson Act passed the House by voice vote and the Senate without debate, and was signed into law on 2 September 1957. Thus was laid the groundwork for the nuclear power industry. Few outside the industry realized its significance; virtually the only national periodical to remark on passage of Price-Anderson was, appropriately enough, *Business Week*.[42] Hardly anyone had any inkling of the controversy this report would stir just a few years later, when the bloom wore off the rose of nuclear power and critics started examining its underpinnings.

In July 1964, the AEC informed Goldhaber it intended to have Brookhaven reexamine and update WASH-740. The agency hoped that new safety measures in reactors would lower damage estimates, making Price-Anderson less controversial by its expiration date in 1967.[43] Again, Downes was project manager. But when an AEC steering committee met with him to hear the preliminary results, it did not like what it heard. The new safety features, Downes said, made no difference to the worst-case scenario; in fact, the potential estimated damage had scaled up because large reactors in 1964 were about five times as large as those of 1956. Far from lowering the numbers, Downes said, in the updated study "the results were frightening."[44] The AEC decided to discontinue work on the draft submitted by the Brookhaven team, which later led to accusations of a cover-up on the part of the commission, and, by association, Brookhaven.

For years, the original WASH-740 and the subsequent draft of an update have been exploited and attacked by supporters as well as opponents of nuclear power, who extracted portions and put them to different ideological uses.[45]

DESPITE THE fact that important changes in science were under way, the early and mid-1960s were a golden era for Brookhaven. The lab was able to provide ample support not only to its physicists but to members of its other departments as well. Some of the lab's most significant research programs in the life sciences, for instance, took place during this time: Panayotis Katsoyannis's synthesis of human insulin, the first human protein ever synthesized; George Cotzias's discovery of L-dopa as a treatment for Parkinson's, the first therapy ever discovered for a degenerative neurological disease (fig. 10.5); and Lou Dahl's quantitative studies of the link between hypertension and salt intake and his development of salt-sensitive and salt-resistant strains of rats (which have blood pressure characteristics similar to human beings), as well as his role in a heated controversy about the high concentration of salt in commercial baby foods (figs. 10.6 and 10.7). Cold Spring Harbor biologist Barbara McClintock, who won the 1983 Nobel Prize in medicine, was a guest geneticist at Brookhaven between 1953 and 1972, and grew a summer crop of corn for her studies at

A PARKINSONIAN PATIENT BEFORE AND DURING ADMINISTRATION OF L-Dopo

10.5 George Cotzias of the medical department, discoverer of L-dopa as a therapy for Parkinson's disease. L-dopa was the first therapy ever discovered for a degenerative neurological disease, and it revolutionized the treatment of such diseases, for it showed that certain specific chemical substances may directly affect their course.

10.6 Lou Dahl, who made quantitative studies of the link between hypertension and salt, and also developed salt-sensitive and salt-resistant strains of rats (which have blood pressure characteristics similar to human beings). Dahl acquired a high public profile when he reported that commercial baby foods had induced severe hypertension in his salt-sensitive strains of rats.

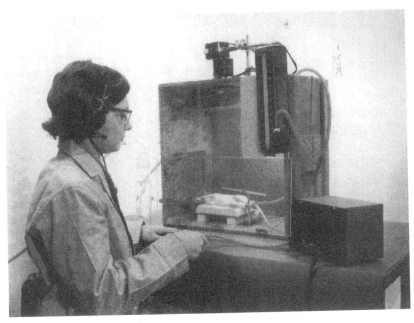

10.7 One of Dahl's assistants taking a mouse's blood pressure through device attached to tail, 1961.

the lab for many years. Raymond Davis of the chemistry department discovered an important discrepancy between theory and experiment in the flux of neutrinos from the Sun (a story discussed in Pinch 1982, 1986).

Although efforts to land construction projects for major new facilities at the lab were stalled, lab physicists were able to make good use of their new recently completed machines. The increasing scope of theory (especially in high-energy physics) had begun to change the field somewhat. Experimental groups often focused on making single big discoveries, at the expense of the inquisitive exploration of energy regions that had characterized accelerator work during the lab's early years, or exhaustive surveys of particular areas through thorough data collection. Berkeley excelled at the latter, in a way that even some at Brookhaven came to regard as providing a superior example of how to conduct research. As Ronald Rau, a keen observer of institutional styles, wrote to Nicholas Samios in 1963, shortly after arriving in CERN as Brookhaven's liaison to that laboratory:

> From a distance, I am able to see more objectively the difference between what Berkeley has been doing & what we and CERN have been doing. I was impressed *again*, but more strongly, with the impressive statistics which Berkeley *now* can bring to bear on a problem [using the 6 GeV Bevatron, at a time when BNL was operating its 33 GeV AGS]. I really feel that we must do experiments that have plenty of data. These carry the weight with the physics community. I hope you will argue and will try to push this in our group. Fewer experiments—better physics seems preferable to me.[46]

Nevertheless, the relatively informal and flexible Brookhaven experimental community, which allowed researchers to change directions rapidly to capitalize on changing events, rewarded particularly inventive and aggressive groups and allowed them to come in ahead of possibly better equipped groups elsewhere, as illustrated in particular by the stories of the key discoveries at the AGS during its early years.

Research at the Alternating Gradient Synchrotron

"IT SEEMS very strange to be here as a guest," remarked Atomic Energy Commissioner Leland Haworth in mid-September of 1961, after an introduction by I. I. Rabi at the dedication of the Alternating Gradient Synchrotron (AGS).

Haworth's audience was a crowd of about 650, including many Brookhaven founders there for the occasion. But Haworth had been the most responsible of them all for overseeing the machine's successful completion. In his last weeks as director he had even started to plan the AGS dedication and worked hard to persuade Rabi to deliver the dedicatory address. Haworth had finally prevailed and had intended to introduce Rabi, but events changed rapidly. In the first half of 1961, Haworth became an AEC commissioner and Rabi AUI president. By a strange quirk, their roles were reversed, and after Rabi introduced him Haworth said:

> I feel somewhat as might the father of a newborn child who suddenly finds that the minister has possession of the baby, and that he, the father, has been asked to do the christening. (Haworth 1961, 23)

New and important changes were at work in high-energy physics, Haworth continued. The cost, size, and management of accelerators and experimental teams had shot up. Detectors at the AGS dwarfed those at the Cosmotron, and their construction and operation required teams of specialists. "Bigness," Haworth told his audience, "is the price we pay to have the opportunity of making progress in the field."

Bigness had profoundly altered the communities that worked around accelerators. Each community was dependent on numerous others. "One no longer builds accelerators as such," Ken Green wrote in a memo in which he outlined many of the changes, "but as a component of a complete high energy facility." The project staff of such a facility, Green continued, typically consists of "roughly equal numbers of physicists, electrical engineers and mechanical engineers," and "is backed by technicians, machinists, and clerical personnel numbering 2 to 3 times the scientific staff." Finally, Green wrote, "an accelerator is never finished," but is constantly in the process of being upgraded.[1]

Bigness also meant the need for large organizational structures, and Berkeley, CERN, and Brookhaven had adopted different organizational strategies. Berkeley had never encouraged outside users, and did not have or need a strong experimental support organization. An experimenter working at Berkeley learned the ropes by collaborating with an existing group, and only then might be allowed to mount a separate experiment. CERN's proton synchrotron (PS) laboratory organization, meanwhile, had little specifically delegated experimental support; it, too, did not encourage visiting groups to arrive with their equipment, run an experiment, and leave, but usually required such a group to work with experimenters on site. Brookhaven, by contrast, attempted to aid outside experimenters. It had a physics department and a separate accelerator department in charge of operating, maintaining, and improving the machine and auxiliary experimental equipment and assisting with experimental design and layout. The result was a large experimental support group able to supply visiting researchers with virtually everything. While a few groups brought their own bubble chambers to the AGS, Brookhaven was unique in that it had a number of chambers available to visitors, including the Shutt group's 20″ (brought over from the Cosmotron and soon to be upgraded to 31″), a 30″ (built by Steinberger's group but now operated by the Shutt group), and an 80″ under way. Thus, Green wrote (with, admittedly, some bias), "it makes relatively little difference whether the experimental team comes from BNL, from an outside institutions, or is a mixture of the two."[2]

Bigness had also magnified the differences between the use of counter chambers and bubble chambers to the point where they were now, in effect, associated with two different experimental cultures (a point emphasized in Galison 1997). Bubble chamber users were typically interested in finding and imaging single events with complicated final states, such as were typical of hadron and strong-interaction physics; one thus needed a high probability of an event in the collision between the input particle and the target protons in the liquid hydrogen if one hoped to photograph it. Counter chamber users were typically interested in leptons and weak-interaction physics. Counter chambers electronically recorded certain kinds of signals, and were useful in studying weak interactions, which typically involved a small number of bodies—three or at most four rather than, say, twenty. They were useful for looking at total cross-sections of easily identifiable particles like leptons, and at picking out of a tangle of other extremely rare events typical of the weak interaction. Counter chambers were generally not user facilities: experimenters designed and built their own instruments and tailored them to register (by a "trigger") only the types of decays they were interested in, which they then collected in massive quantities. Rather than looking for a needle in a haystack, counter chamber users were usually interested in measuring the hay. A final difference between the two cultures was size: building and operating a bubble chamber was a huge enterprise requiring a substantial collaboration, while counter experiments were basically small and simple and could be handled by a few people. These two cultures coexisted on the experimental floor; only very rarely did people move from one to the other (Galison 1997).

Bigness also magnified the rivalries between laboratories. They were still performance rivalries: once the AGS and PS started up, Brookhaven and CERN began an exchange, each lab sending a prominent physicist to the other to facilitate the exchange of information and technologies. Rod Cool was Brookhaven's first representative, in 1962–63, followed by Ronald Rau in 1963–64. But important quarries were rarer and the stakes higher, intensifying the rivalry and sometimes giving rise to a peculiar neurosis that one CERN physicist called a "big discovery" complex. In that more intense atmosphere, Brookhaven's informality—created in part by Goldhaber's hands-off approach to administration—could prove advantageous and even decisive to the lab's more aggressive groups.

The Weak Interaction, Leptons, and Spark Chambers

Of the hundreds of proposals received by the High Energy Advisory Committee (HEAC) in the first few years of the AGS's operation, #28, by Mel Schwartz, Leon Lederman, and Jack Steinberger, was significant for several reasons. Its authors had crossed the cultural divide, from bubble chambers to spark (a kind of counter) chambers. It was an undertaking on a scale new in experimental physics: the first "big science" experiment. It had a very specific quarry. And it was an examination of a fundamental issue in the weak interaction.

In the fall of 1958, the weak interaction had suddenly become a hot topic among physicists. Once physicists knew that the weak force involved only one form of interaction, it was firmly established as a distinctive type of fundamental interaction, and the finishing touches were put on a theory. But development of a successful theory in science is rarely a resting point; it opens the way for further exploration and tends to inspire more sophisticated questions about the theory's shortcomings and validity. And the freshly secured theory of the weak interaction was already known to possess one large limitation and one glaring inconsistency, which now became a focus of attention.

The limitation was that the theory broke down at a certain energy. The original theory pictured weak interactions as taking place between point-like particles. It is a peculiarity of such theories that as the energy of inter-action rises, so does the cross-section. To use a crude analogy (of the sort that scientists tend to loathe), it would be as if the harder one threw a dart at a dartboard across the room, the more often one tended to hit it. But according to a fundamental principle of physics, the higher the energy of a particle the smaller its wavelength or size (the smaller the size of both dart and dartboard, so to speak) and thus the *less* likely an interaction will take place. At some point (about 300 GeV, physicists calculated), the two re-quirements collided, turning the theory into gibberish. Several theorists had an idea which would save the theory (e.g., Feinberg 1958). If the force were transmitted via a particle known as an intermediate vector boson (a W^{+-}), the interaction region would not be pointlike but have a finite size deter-mined by the mass of the boson; the limitation would be overcome. But in 1958, this was merely speculation.

The inconsistency was that experimenters did not see a reaction pre-dicted by the theory. The theory described different kinds of leptons as related somewhat as are ground states and excited states of atoms, which have the same basic parts but possess a different amount of total energy. The muon, for instance, acted like a plump version of the electron; it had the same properties but was 207 times as heavy. If the muon were a kind of excited state of the electron, there was no reason why it couldn't decay into an electron plus a gamma ray ($\mu \rightarrow e + \gamma$, or "mu to e gamma"). And as Columbia theorist Gerald Feinberg pointed out, in a paper written while on sabbatical at Brookhaven, theories involving intermediate vector bosons *had* to admit such decays, once every 10,000 normal mu decays. By 1960, some 100 million decaying muons had been observed with no mu to e gamma. According to a "totalitarian" principle of physics attributed to Gell-Mann, if something *can* happen, it *must*. Failure to observe this sup-posedly possible reaction therefore grew from a curiosity into an anomaly into a crisis. Some physical principle, theorists thought, must be interven-ing to prevent mu from going to e gamma. In a cryptic footnote to his article, Feinberg buried an astute observation: the $\mu \rightarrow e + \gamma$ process was expected because the neutrino could be emitted by the muon and absorbed by the e in accordance with the Fermi theory of beta decay, but only if the same kind of neutrino was associated with both the muon and the e. That is, if one supposed that two different types of neutrinos existed, one linked or "coupled" with the muon, the other coupled with the electron, it would explain why mu never went to e gamma. But this, too, was just speculation (Feinberg 1958).

At the beginning of the 1960s, therefore, the weak interaction had re-cently matured as a promising branch of particle physics with fundamental problems crying out for study. But a formidable obstacle stood in the way. The interaction was not misnamed: it has 10^{-10} the strength of the electromagnetic and 10^{-12} the strength of the strong interaction, meaning that in an experiment its effects would be swamped by those of the other two. Trying to pick out its effects from among the others would be like trying to hear a whisper across Grand Central Terminal at rush hour. Weak-interaction physics at high energies seemed hopeless, unless some new method could be found to study it.

The weak interaction was avidly discussed at Columbia, thanks to the presence of physicists like Feinberg and Tsung-Dao Lee, who made it the focus of their work. It was a hot topic at the coffee breaks attended by most faculty members. Mel Schwartz recalled:

> One Tuesday afternoon in November, 1959, I happened to arrive late at coffee to find a lively group around T. D. (which is what we called T. D. Lee) with a conversation ensuing as to the best way of measuring weak interactions at high energies. A large number of possible reactions were on the board making use of all the hitherto known beam particles—electrons, protons, neutrons. None of them seemed at all reasonable. In each case the very rare weak interactions were completely obscured by the vast number of strong and electromagnetic interactions. Indeed, as the coffee hour ended, it was on a note of hopelessness; there seemed to be no decent way of exploring the terribly small cross-sections characteristic of weak interactions. (Schwartz 1972, 82)

On his way home, Schwartz pondered what he had heard. That evening it suddenly became clear to him that the right way to study the weak interaction was with neutrinos, for they experience only the weak interaction and are oblivious to the strong and electromagnetic interactions. Excited, Schwartz called Lee at home later that evening, and Lee was encouraging. The next morning, Schwartz pushed into Lee's office, interrupting a conversation between Lee and his then-frequent collaborator C. N. Yang saying, "I know how to do the experiment!" The three began to do preparatory work, mainly talking and thinking. Lee and Yang emphasized that the first and most important task of high-energy weak-interaction physics was to learn whether two types of neutrinos existed.

Schwartz then wrote an article, "Feasibility of Using High-Energy Neutrinos to Study the Weak Interactions," that outlined how neutrinos could be used to investigate the behavior of weak interactions at high energies. (As Schwartz would discover, the idea had already been proposed independently the year before by Soviet physicist Bruno Pontecorvo.) One would need a proton accelerator to create a beam of high-energy pions, which would be allowed to travel unimpeded for a few dozen feet. In that time, a certain number would decay into muons and neutrinos, and a high fraction of the neutrinos would be emitted in the forward direction. Then the beam would be run smack into a thick wall of iron shielding, behind which would be a detector. If it were thick enough, the shielding would stop all secondary and tertiary particles like pions and muons that were strongly and electromagnetically interacting. The only particles to reach the detector would be neutrinos, which would sail right through the steel. Any interactions they had with particles inside the detector would *have* to be weak interactions. These interactions could be studied easily because it would not be necessary to select them from among the large numbers of other types of interactions.

Making certain assumptions about the energy and intensity of the pion beam and about detector size, Schwartz calculated the neutrino flux. The experiment was feasible, he declared, though "outside the capabilities of existing machines by one or two orders of magnitude" (Schwartz 1960, 307). He didn't bother to mention BNL's AGS, which at that time was planned to have an intensity of 10^9 protons per second (too low for the proposed experiment) and instead mentioned the possibility of using two proposed machines—neither ultimately built—whose planned intensity was 10^{15}.

Schwartz submitted the article to *Physical Review Letters,* which would not ordinarily publish sketchy proposals for experiments. The article's prospects were aided greatly by the fact that Lee and Yang (two recent Nobel Prize winners) submitted an accompanying paper, entitled "Theoretical Discussions on Possible High-Energy Neutrino Experiments" (Lee and Yang 1960). A key paper in the history of the weak interaction, Lee and Yang's article listed the principal issues in high-energy weak interactions, or what physicists might expect to learn from the kind of experiment Schwartz had proposed. One was the possible existence of more than one type of neutrino; another was the possible existence of a W^{+-}. Together, the two papers demonstrated that high-energy neutrino physics was not only possible but of fundamental interest, and the papers were printed back-to-back.

Meanwhile, Leon Lederman had recently returned from a sabbatical at CERN, where he had worked on plans for a key experiment that bore on the structure of muons, and at Brookhaven had already begun to design another experiment, a counter experiment, to scatter muons off protons (AGS experiment #4). At Berkeley, Robert Hofstadter had begun to scatter electrons off protons, yielding information that would shed light on the structure of protons. Lederman's plan was to do a similar experiment with muons and protons. In conjunction with Hofstadter's work, this experiment might yield information about the difference between electrons and muons, a continuing puzzle to physicists. (Richard Feynman wrote on the blackboard of his Caltech office, underneath a sign that said DO NOT ERASE, "Why does a muon weigh?") When Lederman learned of the possibility of a truly high-energy neutrino experiment, Lederman was enthusiastic and began to collaborate with Schwartz.

Early in 1960, the AGS went on line at about a factor ten higher intensity than expected, about 10^{10} protons per second, putting the neutrino experiment just within the bounds of possibility at that machine. Schwartz and Lederman planned to use a bubble chamber rather than a counter or spark chamber because they needed a detector with a large amount of matter with which the neutrinos could interact. "The big challenge," Lederman says, "was this: How do you make a ten-ton device sensitive all throughout its mass which you can see inside easily and which will give you a few of the collisions in detail?"[3] Jack Steinberger was also interested, as well as graduate student Jean-Marc Gaillard, an experienced researcher who would write up the experiment as his thesis at the École Polytechnique in Paris. Accelerator department chairman Ken Green was contacted to initiate planning. In May 1960, three days after Haworth announced the creation of the High Energy Advisory Committee, the four wrote proposal #28, for a neutrino experiment to look for two types of neutrinos and for intermediate vector bosons W^{+-}, using Steinberger's thirty-inch propane bubble chamber.[4] But Steinberger soon left the group, deciding to mount his own effort at CERN, where he would have a substantial head start given that its PS had come on line six months before the AGS.

One night while working late at the Cosmotron, Schwartz heard about a new kind of spark chamber that James Cronin was developing at Princeton. While proportional counters measure the amount of ionization, and hence give information about particle charge and velocity, spark chambers produce only a spark, but give superior spatial resolution—and while a proportional counter produces an electronic pulse, Cronin's chamber produced a visible spark that could be photographed. The next day, Schwartz, Lederman, and Gaillard piled into Schwartz's Plymouth and drove to New Jersey to see the device. Cronin's chamber involved flat pieces of metal, attached to a high voltage, positioned between and parallel to two others, with the space in between filled with gas. A charged particle flying through will disturb the gas atoms, causing a spark to jump between the plates. A spark chamber seemed superior to a bubble chamber for the purpose, because it could be made very heavy (providing more material with which the neutrinos could interact), and had good spatial resolution (allowing precise determination of the particles' locations and thus identities).

Cronin's experimental model needed more development before it would be useful for physics. One major task was to simplify the overall structure. Cronin had put his plates in a fancy housing and separated them with special insulators. Schwartz, who had acquired Steinberger's instinct for economy and for designing things that could be manufactured in a hurry, simply built a lucite frame around his plates—nine in each module, alternating four charged plates and five grounded ones—and clamped them together with screws, so the frame sealed in the neon but left the sparks visible from the outside. The particle trajectories would glow with the familiar red light of ionized neon and be recorded by automatically triggered cameras. Schwartz's detector consisted of ten such modules, one group of five stacked on the other, and weighed ten tons overall. Schwartz's design involved simple machinery and engineering, and could be done on a large scale.

In the experiment, a pulse of protons from the AGS would smash into a target to create a hail of pions and other particles, which would fly freely for about seventy feet. In that time, about a tenth of the pions would decay and create more particles, including many neutrinos. This tangle of about 10^{11} particles would then run into a forty-two-foot-thick pile of steel, which would stop everything but the neutrinos. On the other side of the steel was the well-shielded spark chamber, equipped with cameras to photograph the trails of sparks in the gas that would be created by neutrino collisions.

If a neutrino collided with something in the aluminum of the chamber, the signature would be either a muon or electron starting in the chamber, the product of a collision whose momentum pointed away from the target. If an equal mixture of electrons and muons were produced, it would indicate that one type of neutrino existed, able to couple alike to each of the two kinds of particles. If only muons were produced, this would indicate that the neutrinos created by the decay of pions to muons could couple only to muons, and that *two* types of neutrinos existed. But the question whether they would be able to see any neutrino collisions at all was still open.

In September 1960, Lederman and Schwartz reworked their proposal to include a spark chamber. "There is a lot of hard work in it," they wrote in an accompanying letter to the High Energy Advisory Committee, "but with strong support, there is the excitement in possible discovery of really fundamental phenomena. This would be a nice way to launch the AGS."[5] The HEAC gave formal approval in November. In another description of the proposal, written up for a conference that fall and presented by Lederman. Lederman said, "By the grace of the AEC, BNL, God, Green, and Haworth (alphabetical order), we should see neutrinos" (Lederman et al. 1960, 201).

Only one other facility in the world could stage such an ambitious experiment: CERN's PS. Naturally, several groups of experimenters there were interested. Steinberger planned to build a large bubble chamber for the job. So did French physicist André Lagarrigue, in an independent experiment. And a rival spark chamber group contained some of Europe's most eminent researchers, including cosmic ray physicists Giuseppe Cocconi and Gilberto Bernardini, as well as a younger colleague, Helmut Faissner. All three groups were to share one beam line. The CERN spark chamber group vacillated for a while when it came to actually deciding whether to go ahead. But several of the Europeans who attended the 1960 Rochester conference at the end of August (where Bernardini gave a talk on CERN's neutrino experiments) were impressed both by recent theoretical ideas on the neutrino and recent developments in spark chambers, and on their return decided to give the neutrino experiment highest priority. Faissner, who in the meantime had prepared a detailed proposal, was ordered to have his equipment ready for installation at the PS on 15 January 1961 (Jungk 1968, chap. 4).

Back at Brookhaven, Lederman and Schwartz began to arrange running time on the AGS and space on the experimental floor. Shutt's group was scheduled to run its twenty-inch bubble chamber, but Schwartz and Lederman brashly argued that their work was sufficiently urgent to warrant priority over Shutt. Shutt adamantly refused. Rod Cool, secretary of the High Energy Advisory Committee, was caught in the middle, but sufficiently impressed by Schwartz's arguments to call Goldhaber, who was away in Scandinavia on a lecture tour, to ask him to pull rank, override Shutt, and assign the beam to the neutrino experimenters.

> As much as I sympathized [Goldhaber recalled], I felt that I had to save them from their youthful enthusiasm. I was skeptical of their promise to finish this important experiment in six weeks, and insisted that they would have to wait for their own beam so that they would do their work in whatever time it took. I therefore assigned a beam for their sole use.[6]

The team set themselves up at another straight section a few dozen feet further down the AGS ring. Since there was no necessity to carry protons out of the AGS, they installed a three-inch beryllium target in the vacuum chamber, letting the secondary pions drift in the open air toward the shielding and, beyond that, the detector. The tracks had to be taken off at a slight angle, seven degrees, to avoid the magnetic field of the next magnet.

Meanwhile, the CERN group had worked through Christmas to meet its deadline. They began to run, months ahead of the Brookhaven team, but had problems coping with the "background," or events in the detector caused by stray particles that were making their way through the shielding.

> Everyone worked to the point of collapse [Faissner recalls]. I suggested several times that we take it a little easier, but that was rejected completely. We had to

observe neutrinos by this June at all costs. Otherwise, Brookhaven's machine might steal this important discovery right out from under our noses. A kind of neurosis seized us: a "big discovery" complex. (Jungk 1968, 111)

The Brookhaven team was also experiencing difficulty with background. With each pulse, the AGS was sending 10^{11} particles in the direction of the detector, which had to be reduced essentially to zero, and the team struggled to find enough shielding, stacking huge piles in front of and all around the detector. Eventually, five thousand tons of scrap armor plate from several decommissioned navy cruisers was shipped down the Hudson to Nevis, and thence to Brookhaven, where it was piled on the floor of the AGS experimental hall. The team took maximum advantage of every opportunity offered by Brookhaven's flexible and informal atmosphere. With the background still a serious problem despite all the steel shielding, the team concocted a scheme to put lead bricks in the AGS itself, on both sides of the vacuum chamber—in effect, packing in the magnets (fig. 11.1). Green and other operators refused to allow it, fearing that the weight of the bricks would cause the machine to settle and that the machine's magnetic fields would create eddy currents in the lead (a conductor), which in turn would adversely affect the operation of the machine. The result was a standoff—until one weekend, when Green and his principal staff members were absent, the team went ahead and installed the lead bricks, then showed that the machine could work successfully despite their presence.

Still, the Brookhaven group was far behind its CERN rivals. CERN was "clearly ahead of us," Shutt wrote in June 1961, and Brookhaven would get important results from the neutrino experiment "only if the corresponding CERN experiment fails." [7]

Remarkably, it did. Just a few weeks later, the CERN team met with unexpected disaster and folded. Guy von Dardel, a Swedish physicist, discovered an error in the calculations involving the straight section, or segment of pipe between the ring of the accelerator and the target. While Brookhaven's team had been working with a ten-foot straight section at the AGS, the CERN groups had been assigned a station with a five-foot straight section at the PS. Von Dardel found this was too short; the magnets would defocus the pion beam, diminishing the neutrino intensity by about a factor of fifty, sufficient to make the experiment unfeasible. The CERN management decided not to move the experiment to another port on the PS with a longer straight section, which would have involved bumping other experiments (Crease and Mann 1996, 288–89).

Faissner and a few others, reluctant to give up so readily on all the work they had invested in the detector, proposed studying even bigger spark chambers, but CERN's management had lost faith in its neutrino group and

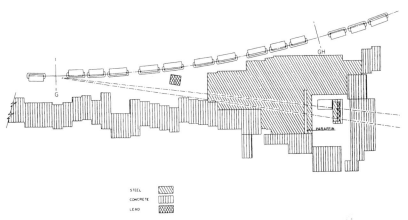

STEEL

CONCRETE

LEAD

11.1 Neutrino experiment floor layout, showing massive amounts of shielding involved.

effectively put them on hold. Steinberger, depressed, left CERN and returned to Brookhaven. A diplomatic intermediary (T. D. Lee) undertook the delicate business of securing a place for Steinberger on the Brookhaven experiment, which by this time was far along. In October 1961, the team's senior members were once again Lederman, Schwartz, and Steinberger, widely regarded as three of Brookhaven's most determined and aggressive experimenters, each with a deserved reputation as someone not easy to cross. One day the trio was walking across the floor of the AGS together past another group of experimenters; Lederman recollects that one of the latter said, in a deliberately audible sotto voce, "There goes 'Murder, Incorporated!' " using a name that had been applied to a Jewish-led gang of the 1930s notorious for its ruthless extermination of adversaries.[8]

Preliminary runs were made at the end of 1961, and the Brookhaven experimenters promptly uncovered problems. The most serious was an unusually large number of neutrons still making their way under, over, and around the forty-two-foot-long pile of steel, creating several events a day in the detector that threatened to mask the still rarer neutrino events. The experimenters then reinforced the shielding with lead. The neutron events seemed to be occurring low in the chamber, so the group figured they were coming through the floor. Schwartz crawled into the culverts underneath the electrical cable conduits in the floor and, sure enough, found another hole and stacked more lead bricks down there. The group had the AGS proton beam energy dropped to 15 GeV, which reduced the muon leakage through the shielding. Extra running time was requested.

Early in 1962, during the first few weeks of the experiment, Faissner visited Brookhaven and was welcomed as a colleague. The experimenters were still having difficulty reducing the neutron background; Faissner did some of the scanning himself and was able to appreciate the problem first-hand. He dashed off an encouraging telex to his CERN colleagues saying, "No neutrinos, only neutrons. Press on." Faissner put the telex on the Brookhaven group's account, and a copy was eventually returned to Schwartz's mailbox. Though it was normal for Faissner to keep his CERN colleagues abreast of developments, the Brookhaven group was highly sensitive about their continuing neutron problem and somewhat irritated that the tone of Faissner's telegram was more one of spy than colleague. Faissner was promptly dispatched back to CERN. After dropping off Faissner at Idlewild Airport, Schwartz drove back to Columbia.

> As I walked into Leon's office, he had just gotten a call from [colleague Dino] Goulianos—they had scanned the first large batch of film and found the first neutrino event. It was a beautiful event—a long muon was produced in a two-prong star and traveled through forty inches of aluminum before leaving the chamber. From this point on it was almost anti-climactic. We ran a total of 800 hours and it soon became clear that we were not producing electrons with our neutrinos, only muons. (Schwartz 1972, 85)

They had run a total of about eight months and several times had to request additional running time (a total of eight hundred actual operating hours), thus confirming Goldhaber's wisdom about their requiring their own beam. In that time, 100 trillion neutrinos passed through the forty-two-foot steel wall into the ten-ton detector. Out of these 100 trillion, a mere 50 interacted in the chamber to make energetic events. Of these, 29 showed only a single energetic muon produced, while the rest showed muons produced with other particles. In no event was a single energetic electron produced (fig. 11.2).

Schwartz carried their paper, "Observation of High-Energy Neutrino Reactions and the Existence of Two Kinds of Neutrinos," into the *Physical Review Letters* office at Brookhaven on 15 June 1962, and it was published two weeks later.

> In the course of an experiment at the Brookhaven AGS, we have observed the interaction of high-energy neutrinos with matter. These neutrinos were produced primarily as the result of the decay of the pion . . . . [T]he neutrinos we have used produce mesons but do not produce electrons, and hence are very likely different from the neutrinos produced in μ decay (Danby et al. 1962, 36).

In the next two weeks, the authors gave a number of reports on the discovery to colleagues. To head off an already growing conflict among the chieftains of Murder, Incorporated, the three tossed coins to determine who

11.2 Mel Schwartz with the neutrino detector.

would give which talks. The winner, Schwartz, elected to report on the discovery at the 1962 Rochester conference, to be held in early July that year at CERN in a gesture of international collaboration. This was the plum, for virtually all the high-energy physics community would be present, and the talk would be the first major presentation of the work.

The first crisis in the weak interaction was now over; it was clear why mu never went to e gamma. The other crisis, that the theory broke down around 300 GeV, persisted. This crisis would be resolved if an intermediate vector boson (W^{+-}) existed, which became the next prominent target of weak-interaction physicists. The neutrino group began to hunt for it, but no candidate events showed up. The group succeeded only in setting a lower limit of 2 GeV for the W mass.[9]

But, by the grace of the AEC, BNL, God, Green, and Haworth, the bubble-turned-spark chamber group had at least seen neutrinos, proving the existence of two species thereof. In the process, they had inaugurated the age of large-scale (on the order of ten-ton) detectors, as well as the new field of neutrino physics, still an important field at high-energy accelerator laboratories today. In 1988, Lederman, Schwartz, and Steinberger shared the Nobel Prize, in part for their discovery, and in part for demonstrating the viability of weak interaction experiments at high energy. Indeed a nice way to launch the AGS.

ONE REASON why the weak interaction fascinated physicists was that it had the peculiar property of violating parity. In the wake of the discovery of parity violation in the weak interaction, theorists had attempted to rescue symmetry in nature by postulating the existence of the deeper-lying CP symmetry. But the weak interaction was about to grow still stranger.

At the time the AGS was turned on in 1960, Robert Adair was still working on the Cosmotron, preferring the familiar routine and quieter pace there to the competitive scramble at its newer and more powerful sibling. He, Leipuner, and Rich Larsen had finally planned an experiment on the 14″ they had built and had decided to study K_Ls. They put the bubble chamber in a beam of neutral particles and shielded it in a concrete cavernous structure. Members of his group, as good-natured as their leader, had gaily decorated it like a prehistoric cave, with drawings of aurochs and saber-toothed tigers, and an occasional more modern slogan ("Yanqui go home!"). In early 1961, the group took a total of ninety thousand pictures, which were given to a scanning team headed by Eileen O'Donnell (though Adair spent much time scanning himself). The information was then processed in the Remington-Rand payroll computer, programmed by Leipuner and processed by O'Donnell. As computer time was rationed, O'Donnell had to cope creatively, and sometimes signed in as O'Adair and O'Leipuner.

Adair noticed the familiar pion-to-muon or -electron decays and the three-body decays that Lederman and company had seen in discovering the K_L. But every once in a while, he saw what *looked* like a K_L decay into *two* pions. Impossible, according to CP symmetry.

These events vexed Adair. He, like the rest of the physics community, simply believed in CP symmetry. Two previous experiments that looked at K_L^0 decays (at Brookhaven and Dubna in the Soviet Union) had produced a total of 411 decays without a single CP-violating decay into two pi (Bardon, Landé, and Lederman 1958; Neagu et al. 1961). Adair knew that "regeneration" effects (involving retransformation of a K_L^0 back into a K_S^0) might be responsible for the two-pi decays, but he was convinced these effects were not large enough to account for the number of two-pi decays that he was finding. He spent more time scanning the bubble chamber film himself, searching for clues. Colleagues remember him showing up to lunch saying excitedly, "I found another one!" almost daily. Adair would pull out a picture of a K_L into two pions and pass it around. Fingers would scratch heads: One of the tracks was *surely* not a pion. But how could that be?

In what he later called a "real intellectual jump," Adair then made an explicit decision to pull out and analyze all events that looked to be K_L to two pi. There were twice as many as could be accounted for by chance. Where did they come from?

> One possibility [Adair recalls] was CP violation, but there was *no way* I was going to believe in CP violation. I believed that God wouldn't do such a God-damned dumb thing—still feel that way a little! [Laughs.] So I made the mistake of trying to outguess God. I came to the conclusion that this fundamental mechanism would work if there was a force that would not be seen in any other way, but would be enough to make this regeneration.[10]

Adair proposed, in short, the existence of a *fifth* force of nature to explain the puzzling decays. The explanation was dissatisfying, for it assumed that the force appeared only in this one place—but the alternative, that CP was violated, was unthinkable. Adair and his collaborators then submitted a paper with this interpretation to *Physical Review* in March 1963, though it was not published until the fall (Leipuner et al. 1963). Most physicists thought that Adair's K_L results were sensational and his fifth force interpretation preposterous, but trusted his group enough to take the matter seriously. In the K system, again, something was out of joint.

James Cronin and Val Fitch, two Princeton experimenters working at Brookhaven on separate projects shared an office next door to Adair and Leipuner. Cronin and Fitch happened to be well positioned to look at the matter quickly. Cronin had the equipment: he had been running an experiment at the Cosmotron using a pair of spark chambers, improved versions of the one Schwartz and Lederman had examined at Princeton two years earlier. The experiment was not succeeding in getting a result, and Cronin was only too happy to end it and move the equipment to the AGS for more interesting work. Fitch, meanwhile, was also interested. Spark chambers were far superior to bubble chambers for looking at the apparent CP-violating effect because of their speed and ability to trigger rare events. Lederman encouraged Fitch and Cronin from the sidelines. He had published a negative result on K_L into two pi and would be a little embarrassed if Adair's results held up. Sure that Fitch and Cronin would discover whatever Adair had missed, Lederman began referring to the Fitch-Cronin work as the "demolish Adair" experiment.

In April 1963, Fitch and Cronin, together with René Turlay, a visitor from Saclay, sent their "Proposal for K^0_2 Decay and Interaction Experiment" to the HEAC. It began:

> The present proposal was largely stimulated by the recent anomalous results of Adair et al., on the coherent regeneration of K^0_1 mesons. It is the purpose of this experiment to check these results with a precision far transcending that attained in the previous experiment.[11]

The proposal was approved the next month, with a rapidity illustrating, again, the flexibility of the experimental program at Brookhaven. The experiment was scheduled to run between the end of May and the middle of July, sandwiched between two scheduled AGS shutdowns. Fitch and Cronin had trouble finding space at such quick notice on the crowded AGS experimental floor but managed to fit themselves on the inside of the ring, a little-used area Green had baptized Inner Mongolia. They set up and ran the experiment in June 1963 and exhaustively analyzed and reanalyzed the results over the course of a year. In mid-1964, they concluded that in a small but significant number of occasions K_Ls do indeed decay into *two* pions (fig. 11.3). In short, CP symmetry was sometimes violated (Christenson et al. 1964).

For most physicists, it seemed impossible. Adair recalls that, upon first hearing that Fitch and Cronin had seen enough K_L to two pi decays to convince them of CP violation, he told the bearer of the news (Abraham Pais) that, while the two Princetonians were good physicists, they had to be wrong, for CP *must* be invariant.[12] The effect was tiny—the forbidden decay was a fraction of a percent of the K decays—which would ordinarily lead physicists to suspect that experimental error was involved. But Fitch and Cronin were well respected, which immediately turned the focus of the community on the details of what had taken place, on the performance of the equipment, to see if they had anticipated every little possible tweak. When Cronin presented the results at the 1964 International Conference on High Energy Physics (yet another in the Rochester conference series, held at Dubna in the Soviet Union), a Soviet colleague suggested that a fly trapped in the apparatus might have produced enough regeneration to account for the effect. A quick calculation showed that the fly would have had to be much denser than uranium: *that* tweak could not have accounted for the result. Few descriptions could be more distorting than social constructivist ones that would treat the acceptance of results such as those of

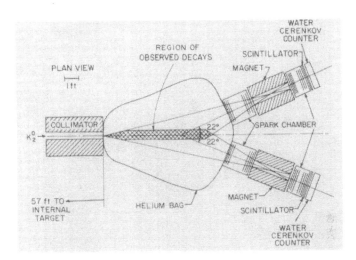

11.3 Diagram and picture of CP violation experiment and picture of the equipment in "Inner Mongolia." The two counters looked for simultaneous pions (one plus and one minus) that had resulted from K_2^0 decays. From the angles and velocities, one could determine whether these had been the *only* byproducts of the decay. The helium bag was an inexpensive substitute for a vacuum.

CP violation as a case of "negotiation" between different camps of physicists, akin to arguments between lawyers and politicians (see Eger 1997).

CP violation is still unexplained. (A simultaneous T violation preserves the combined CPT symmetry.) At the end of his acceptance speech for the Nobel Prize, in 1980, Fitch recalled Lewis Thomas's axiom: "You measure the quality of the work by the intensity of the astonishment." "After 16 years," Fitch said, "the world of physics is still astonished by CP and T invariance" (Fitch 1981, 92). The remark is still valid.

And what of Adair's results? He once posed the question to himself: "Was it simply some kind of experimental error, a gross statistical fluctuation, or was the 'Old One' generous enough to provide a hint of CP-invariance violation for they who would listen?" [13] Subsequent analysis by a member of Adair's group indicated that, of all the K_L to two pi events his group had reported, CP violation had contributed one of three pieces, with the others due to a pair of regeneration effects (Hawkins 1967). Adair's experiment exemplifies part of one path to discovery (which often happens in science, though not usually as dramatically as here) in which the discoverers do not know ahead of time what they are looking for, and must learn to recognize—have confidence that they can recognize—a new phenomenon. As in the case of the K_L, acquiring such confidence is not an immediate process; learning to recognize takes time, and one speaks of "dim" and "dawning" recognition (Crease 1993, chap. 6). Adair, rather than making an error, had a dim recognition that some new phenomenon was appearing in the performance of his equipment, without achieving a full apprehension of it. [14]

Typically, Adair himself accepted with equanimity the fact that, for the second time, he had had the opportunity to make a fundamental discovery within his grasp, and on at least one occasion at which he was asked to speak about the matter recalled a biblical verse that he had recited at his confirmation in the Lutheran Church (Luke 9:62): "No man having put his hand to the plow, and looking back, is fit for the Kingdom of God." [15]

MEANWHILE, THREE independent proposals were made to look for the W^{+-} at Brookhaven, by groups led by Cool, Piccioni, and Schwartz. All wanted to produce W's by pi-p or p-p, and look for them from leptonic decay. The advisory committee recommended against the proposals, and Goldhaber made the final decision, on the grounds that the cross-section was not large enough.

Schwartz grew weary of the hunt for the W and in 1966 left Brookhaven for Stanford to do other types of weak-interaction experiments. Lederman, however, was still determined to hunt the W, and led a team that tried to use AGS 30 GeV protons to create W's directly in collisions, turning off the neutrinos by not allowing the pions a chance to decay. If a W were created, the signature would be a decay into a muon with a high transverse momentum. This managed to set a lower limit of 2 GeV on the W mass (Burns et al. 1965). Many other searches for the W, at Brookhaven and elsewhere, were conducted through the 1960s.

In 1966, Lederman discovered that theorists had demonstrated that a high transverse momentum muon could *also* be created by an alternative reaction, involving the creation and decay of a virtual photon. He now realized that even if he saw large-angle muons it would not necessarily indicate W's. Reading up on the subject, he realized that virtual photons were extremely interesting, and worth investigating themselves. The cross-sections were reasonable, and the signal for a virtual photon decay was clean—a pair of leptons, from which one could calculate the mass of the parent particle, and they were good for looking at high energy and short distances. In 1967, he put together a scintillation counter experiment to study virtual photons by smashing protons on a uranium target. The result, published in 1970, was intriguing: the number of muon pairs decreased sharply from 1 to 6 GeV, except for a peculiar shoulder in the middle, around 3 GeV, where it nearly leveled out (Christienson et al. 1970). The next year Peter Limon, an assistant professor who had worked with Lederman, sent a proposal to the High Energy Advisory Committee to examine this shoulder at Brookhaven, modifying an existing detector able to detect narrow resonances. In justifying the experiment, Limon cited the "strange shape" of the continuum, which suggested the possibility of some peculiar physics: "one cannot rule out possible 'bumps,' particularly in the region around 3.5 GeV/c^2." [16] But neither Lederman nor Limon's Columbia colleagues were convinced that the experiment's potential was worth the time and cost, and they failed to back him. "Without the courage necessary to do really important science," Limon recalled, "I withdrew it, and went on to participate in a truly boring muon scattering experiment instead." [17] The bump remained a puzzle—curious enough not to fit, but not significant enough to qualify as an anomaly meriting immediate study. Lederman was far too preoccupied with the dilepton spectrum in his search for vector mesons, and was not about to get diverted. Years later, he would regard this experiment as "a big one that got away," for the explanation for the shoulder turned out to be an important particle of a new kind, discovered in 1974 simultaneously at Brookhaven and SLAC; this particle served as *another* process by which protons on a uranium target could give rise to dimuons, whence the increased cross-section. Lederman's shoulder was yet another example of a dim recognition, an enigma whose clarification would have to await another experiment. But the shoulder was significant enough in itself. About a hundred theoretical papers were written about the dimuon spectrum, including one that developed an important theoretical tool for understanding particle collisions, known as the Drell-Yan process, after the two theorists involved. While the measurement of the dimuon spectrum was important, it was not nearly as important as the new physics that would be signaled by the new particle.

Still another area of counter–spark chamber research at the AGS was leptonic (that is, weak) decays of hadrons. Experimenters had noticed that the rate at which hadrons decayed weakly into leptons did not fit the simple models of the universal weak interaction, but were off by a certain amount. In 1963, Nicola Cabibbo, an Italian physicist at CERN, produced a general formulation—known as the Cabibbo angle because it was commonly expressed in terms of an angle—for calculating the deviation between the observed rate of such decays and the expected rate if the simple universal model of the weak interaction held (Cabibbo 1963). Those interested in the structure of the weak interaction wanted to know whether Cabibbo's result was true of all hadronic weak decays or just some portion thereof, and several teams of researchers investigated various leptonic decays of K mesons and hyperons.

Thus the three major areas of weak-interaction research at the AGS in the late 1960s were: measurements of the Cabibbo angle, studies of CP violation, and W searches. But the decade would run out without further fundamental surprises. Cabibbo's result held up, no new insights were obtained regarding CP violation, and the W was not found. This last was especially galling, for theorists were sure it existed. Paraphrasing for his girlfriend a remark Joe Louis made of an opponent, Feinberg remarked apropos of W bosons, "You can run, but you can't hide" (Feinberg 1988, 13). The W hunters remaining at Brookhaven tried not to be discouraged by their lack of success. Goldhaber encouraged them with a piece of folk wisdom of his own, "Like many lion hunters, experimenters often return without a lion."

At the beginning of the 1970s, the reason why W's were still in hiding became clear when a sudden transformation in understanding the weak interaction took place. Events were set in motion late in 1967, when Steven Weinberg, who had been doggedly trying to apply the theoretical structure that Yang and Mills had developed in 1954 to hadrons, suddenly found that he had been "applying the right idea to the wrong problem" (Crease and Mann 1996, 245). The Yang-Mills theory would work brilliantly, Weinberg realized, if applied to leptons. That result, in combination with work by Abdus Salam and Sheldon Glashow, resulted in a theory of the weak and electromagnetic interactions in which they were treated as different manifestations of a single, "electroweak" force. For a while, the theory was regarded with skepticism due to certain mathematical inconsistencies and the failure to account for the masses of particles, but these, too, were resolved in the middle of 1971. Weak-interaction physicists suddenly had on their hands a comprehensive theory of the weak and electromagnetic interactions, called SU(2) × U(1). SU(2) × U(1) explained the Cabibbo angle, which was due to the interference or "mixing" of the weak and electromagnetic forces in the hadronic weak decays. According to SU(2) × U(1), too, the W boson had to be not 2 or even 5 GeV but much more massive. The W was ultimately discovered at CERN in 1983, in a spectacular example of a still bigger experiment with a single major quarry, at over 80 GeV—far beyond the capabilities of the AGS. The main remaining puzzle of the weak interaction was CP violation, which remains a puzzle today.

The Strong Interaction, Hadrons, and Bubble Chambers

If the 1950s, for hadron physicists, was dominated by the study of particles, and particularly of strange particles, the 1960s was dominated by spectroscopy, the study of particle resonances (and particularly strange-particle resonances). The 1960s began with the discovery of the first strange-particle resonances as well as a number of meson resonances. Hadron physicists would spend much of the rest of the decade seeking resonances and determining their masses, spins, parities, isospins, and other properties, and for the same reason that nuclear and atomic physicists had: in the patterns clues would presumably lie to the nature of such particles. Here, too, specific quarries would appear to focus the energies of experimenters, in the form of particles that would clinch one pattern over others.

Resonance hunting was largely bubble chamber work, and a particular interest of Alvarez's group at Berkeley and Shutt's at Brookhaven. (Because one sign of a resonance was a telltale bump in a cross-section, Alvarez liked to describe himself as a "professional bump-hunter" [Alvarez 1989, 304].) Shutt had staffed his group with people with a spectrum of interests, from construction to physics, and many of the more physics-oriented group members would propose experiments either on their own, or in collaboration with university groups. Two of the most active university groups were from Syracuse and Yale; both worked in collaboration with members of Shutt's group. Other important groups were from the University of Pennsylvania, MIT, Wisconsin, and Columbia.

A group would select an input particle, the two most popular of which were the antiproton (the favorite of the BNL-Yale group) and the K^- (favored by the BNL-Syracuse group). Antiprotons had the virtue that, in principle, they could produce any particle and its antiparticle, but the rates

of interesting events are very low and most of the time one simply produced pions. K⁻s could not produce antiparticles, but did tend to produce
a large number of negative strangeness particles. At the time, this was understood to be a consequence of associated production—strangeness conservation—and the fact that the hyperons have the same strangeness as the
negative K particles; thus these mesons could be absorbed by a nucleus of
strangeness $S = 0$, with production of a hyperon, while positive K mesons
could not. After the experiment, the hundreds of thousands of photographs
would be given to scanners.[18]

The BNL-Yale team turned up several new particles using antiprotons
in the 20″ in March 1962. The BNL-Syracuse team also turned up a number
of new resonances, including the phi (1020 MeV, a resonance of two K's)
and the xi star (1535 MeV, a lambda-pion resonance, called a cascade
particle because it decayed into strange particles.) In July 1962, several
members of the BNL-Yale and BNL-Syracuse teams left for the CERN
Rochester conference to announce these particles in one of the numerous
sessions devoted to hadron physics.

Murray Gell-Mann was an attentive listener to this news. For three years,
he had been seeking order in the apparent chaos of resonances with the aid
of a mathematical tool known as group theory. A *group* is a collection of
items related by mathematical rules, and is associated with *generators* that
give rise to the items in the group when certain operations are performed.
Gell-Mann was seeking, in effect, to find the rules and generators that
would place the known hadrons and resonances into groups. One, whose
mathematical name was SU(3), looked promising. SU(3) had eight generators, two of which corresponded to the properties of isotopic spin and
strangeness, the other six of which were rules for changing the values of
these first two. In joking homage to the teachings of the Buddha, Gell-
Mann called it the "eightfold way." [19]

While the SU(3) scheme was beautiful in principle, in practice it was a
mess. Many of Gell-Mann's particle families called for particles not known
to exist, and some properties conflicted with experimentally determined
values. Particularly embarrassing was a group of ten members: a decimet.
The decimet was shaped like a pyramid, with four tiers of particles of different amounts of strangeness: four particles of strangeness 0 at the ground
level, followed by three of strangeness −1, two of strangeness −2, and at
the apex, a single particle of strangeness −3. Gell-Mann could place particles only in the first two tiers: the four known delta resonances (strangeness 0), and three known sigma particles (strangeness −1). Moreover, only
the spin-parity (J^P) of the delta was known.

Gell-Mann had not planned to speak about his scheme at the conference,
but the news from Brookhaven emboldened him. The phi established one

of his meson families as a bona fide nonet (nine-member group), while the cascade star particle (whose discovery was also announced by a Berkeley team at CERN) fit neatly into the strangeness −2 row of the decimet, leaving only one particle in it—the apex particle, of strangeness −3— missing. At the end of one plenary session on strange-particle physics, he strode to the microphone. If the information he'd heard was "really right," Gell-Mann said, "then our speculation [about the decimet] might have some value and we should look for the last particle, called, say, Ω^- [omega minus]" (Prentki et al. 1962, 805).

Gell-Mann's scheme was still only one among many, and still contained an embarrassing number of holes. But the prediction focused the experimental possibilities sharply, and all the experimenters (and some theorists) who were listening recognized it. Only two accelerators were energetically capable of producing this mass (the AGS and the PS), with a third nearing completion (the 12 GeV ZGS at Argonne, the Cold War project that was supposed to be operational before the AGS). Experimental teams at all three facilities now had a quarry.

At Brookhaven, Samios had already submitted a proposal to the HEAC, AGS #198, to use a K^- beam to look for resonances in the 80″ bubble chamber then under construction.

> We had the highest energy accelerator in the world, the AGS, and were about to have the largest bubble chamber in the world, the 80″. I had noticed that K^-s made more negative strangeness particles, and thought that they were the way to go. My hunch was: "Get as many as possible, at as high an energy as possible, and run them into protons in the bubble chamber." That was the game I wanted to play. In our proposal we'd said basically, "K^-s at 2 GeV in the 20″ were great, so let's do K^-s at 5 GeV in the 80″." Why 5 GeV? Because that was the limit of the technology. Murray now came along and said, "Look for the omega minus, for which you need at least 3.2 GeV." That focused our efforts, and it gave us further justification for our project in competing for running time at the AGS.[20]

When the talk broke up a few minutes later for lunch, Samios cornered Gell-Mann for more information. Gell-Mann pointed out that his system made fairly good predictions of the properties of the particle: mass of 1685 MeV (because mass and energy are convertible by Einstein's famous equation $E = mc^2$, high-energy physicists generally find it most convenient to characterize a particle's mass in terms of energy), charge minus, strangeness −3. The particle might be produced in a collision between a K^- and a proton in a bubble chamber. The omega minus would be a "super-cascade" particle, for it would decay into a cascade particle, which in turn would decay into a strange particle. In each step of this cascading process, the strangeness would change by one unit until a proton or neutron, of strangeness 0, would be left. But the most interesting prediction was its lifetime. To decay strongly, the omega minus would have to conserve strangeness and decay into particles with a total strangeness of −3, but it was too light to do so. It would therefore have to decay weakly, with a lifetime, like that of other strange particles, of 10^{-10} seconds.

> That was the beauty of the prediction [Samios recalls]. All the other particles in the multiplet decayed strongly. Here was its brother, the omega minus, related to its siblings by this mass spacing and other properties, but with a lifetime *ten* orders of magnitude different![21]

Gell-Mann wrote down his predictions on a napkin, which Samios put into his pocket to take back to Goldhaber. Soon after returning, Samios and Leitner wrote a memo to Rau, the new HEAC secretary, providing additional justification for #198.[22]

But Berkeley, CERN, and Argonne also had teams preparing to look for it. Berkeley's 6 GeV Bevatron could not produce K particles readily at 3.2 GeV, but CERN's PS, where two chambers, a French 80 cm chamber (the scaled-up version of Brookhaven's 20″) and a British 150 cm chamber, were readied to look for the omega minus, the former in proton-antiproton (pp̄) collisions, the latter in a newly built K^- beam, by a British team led by Ian Butterworth. And Argonne's ZGS was nearing completion "on a crash schedule" with a high resolution 30″ bubble chamber at the end of a 4 GeV K^- beam (Day et al. 1980, 14). "Once the ZGS as up and running, the first order of business was to produce the $\Omega -$" (Holl 1997, 224). Samios's team also faced competition from Brookhaven colleagues. Jack Sandweiss submitted a proposal on behalf of the Yale group to use antiprotons, and another to use K^-s. The Yale group argued that antiprotons on protons might produce omegas and antiomegas; Samios argued that extrapolations from cross-sections of known particles indicated that the production rate would be higher with K^-s than with antiprotons.

Meanwhile, the BNL-Yale and BNL-Syracuse groups continued resonance hunting at the 20″, with the latter group contributing another piece with a measurement of the J^P of the Y star, or the excited sigma (Bertanza et al. 1963). A Berkeley group was not far behind.[23]

The new 80″ bubble chamber was the largest working bubble chamber in the world. It had cost $6 million, was housed in a separate building, and had taken four years to design and build. It was completed in early 1963 and took its first photograph on 2 June 1963, though months of calibration and other preparatory work followed. Meanwhile, the CERN team using the 80 cm chamber began looking for the omega minus in pp̄ collisions, while the British team readied the 150 cm chamber in the K^- beam. At Argonne, on 18 September, the ZGS reached 12.7 GeV, but at a low intensity and the machine was not yet able to support experiments. The Samios team kept abreast of developments at CERN thanks to Rau, who had arrived at CERN in September to serve as the Brookhaven liaison there, contributing expertise and helping out where he could, even helping to scan some of the thousands of pictures from the 80 cm for omega minuses. The British 150 cm chamber, however, was delayed. Rau wrote at the end of October:

> Now the high energy K⁻ competition here is as follows: The British chamber tried to take pictures during 3 weeks in October, but had problems with 1) expansion; 2) dirty optics and 3) stuff in the hydrogen which turned out to be H_2O ice plus other stuff. Also camera trouble. They are now warm. If they have to remove their large condensing lenses to clean them, they won't be able to try again before the middle of December. There is a slight chance that they might make it.[24]

Rau added, "I do hope that the K⁻ run at BNL will get off the ground as rapidly as possible."

The Brookhaven team was ready to run in November. But "God was very cruel to us," Samios says, for the team immediately encountered two big problems. The first was that, for some mysterious reason, the particles reaching the chamber from the accelerator were not Ks but pions. The team spent several weeks adjusting the magnets used to select Ks, but without managing to put a dent in the massive numbers of pions streaming down the beam pipe.

Early one morning, when Samios and Palmer were on shift together, they decided the protons had to be hitting something inside the machine, creating a spray of pions. They shut the machine down and set out inspecting the beam inch by inch. When they came to the place where the beam emerged from the AGS, they saw a machine part that stuck out suspiciously. They put a ruler to it; an inch or so smaller than it was supposed to be. Part of the beam was hitting the metal in a way that created a spray of pions that was able to flood the beam. Someone, it turned out, had substituted a metal piece of a different size during construction without noting the size change on the plans. "We could have killed the son of a bitch," Samios recalls. "It cost us two months."[25]

Another problem concerned the beam itself. The beam technology had been pushed to the limit, and certain instabilities in the separators were causing the beam to drift. Samios, Palmer, and company had to develop new techniques for managing the beam, eating up another month. These problems did not help in the competition for beam time. Samios recalls:

> We knew that if we kept flubbing around for another month the vultures would be on us. The AGS had a large research program, and there were many other experiments wanting beam time. Whatever time we took, other people didn't get.

Some people were already saying, "Why don't you turn off those guys for three months?" We had first priority for a certain window—but if we didn't perform, we might have been pulled. There was huge pressure on us.[26]

In December, the ZGS was dedicated at Argonne despite several machine difficulties, and the British chamber at CERN was still having trouble. Rau wrote Samios:

Now the latest is that the 80 cm chamber will be moved so as to be a standby for the 150 chamber & if the 150 is not ready by Feb. 1, then the 80 cm will most probably take 5.5 Bev/c K⁻ pictures. News as of today is that the British hope to start cool down in the middle of January.[27]

Early in January, the ZGS had a serious hardware failure when a section of the vacuum chamber collapsed; the machine had to be jury-rigged to work, but at reduced capability. Meanwhile, the Brookhaven team had began running in earnest. Shortly after the run began, potential disaster struck yet again, this time within the volatile chamber itself. A set of inch-wide black strips called retrodirectors that looked like a set of vertical venetian blinds had been hung against the wall of the chamber to serve as a kind of backdrop for the photography. The harsh jolting of the piston knocked ten of these hangers loose, which fell and came to rest against the window.

There we were, in the middle of the night [recalls Samios]—Shutt, myself, [William] Fowler, Palmer. And we looked at the window. The question is, Did we damage the glass? Because if we'd damaged the glass, and put pressure on it, then the glass breaks, we'd have a thousand liters of liquid hydrogen coming out—we'd have a real catastrophe. So there we were, with ten of the hangers down, and Shutt looking in. He asked all our opinions, but he had to make the decision. He said, "Expand." The piston expanded. Nothing happened. And he made the decision to continue. The other option was to dump the chamber, open it up, fix the slats, and lose a month. The logical thing would have been to stop, do it right. My feeling is, you get five of these things, then you've got to stop. But if you always stop at the first fix, then started again—you've wasted a month before you even know about the next ones. You go as far as you can without jeopardizing things.[28]

The experiment continued to run. Every few seconds, the AGS would fire protons into a target, creating a bunch of Ks that were sent down a beam pipe several hundred feet into the bubble chamber. Meanwhile, the heavy, 36″ diameter piston of the chamber would expand just after the moment a bunch of kaons sailed into the chamber, while four special cameras snapped a picture.

At the beginning of February, with the 150 cm chamber still misbehaving, the British team installed the 80 cm chamber in the K⁻ beam and began a check of operating conditions (bubble size, density, etc.), using a Polaroid camera to view the chamber liquid through a porthole. By chance, a Polaroid taken on 7 February caught an interesting event with the distinctive omega minus "double-break" signature, though what was recorded, according to one team member, "was not tightly enough constrained for this to be beyond all doubt." [29]

But by the time this was realized, the Brookhaven team had already found an omega minus. Samios himself discovered it, during the evening shift of 9 February. When Samios slid a negative across the white surface of his scanning table, the K's usually swept in lines from left to right across the frame, like the lines left by a group of skaters all moving in the same direction. But in the frame in front of Samios, one line suddenly darted off in a new direction. A K particle had hit a proton and created a new particle that veered violently downward. A foot away, a thin V appeared in midair, its two arms crossing lazily after a few inches—the classic signature of a strange or V particle. An old bubble chamber rule of thumb is that the line of flight of the invisible particle that created the V can be estimated by laying a ruler between the two intersections. If it misses the point where the K hit the hydrogen atom, that is a sign that some interesting physics had taken place. Samios set it aside to be measured. It was clearly something interesting.

What had happened was this: A K[−] had smacked into a proton inside the bubble chamber. The resulting collision had created a K[0], a K[+], and an omega minus. The omega minus had sailed a short distance and then burst apart, erupting into a pi minus at a wide angle and a xi zero. The latter had drifted invisibly for an inch or so, and then burst apart into a lambda and a pi zero. The lambda went into a proton and a pi minus; the pi zero went immediately into two gammas, each of which had turned into an electron-positron pair (fig. 11.4).

The next morning, 10 February, several physicists crowded around the photo. Some thirteen particles had been involved, and the scientists set about determining their identity. In their excitement, they failed to notice the two electron-positron pairs, which would be crucial for measuring the pi zero.

Someone—it wasn't me [recalls Samios]—said, "Look, there's an electron-positron pair!" And we said, "My God, that's right!" Someone else said, "Maybe there's another one from the pi zero!" and someone else said, "My God, there it is!" So there were two electron-positron pairs which I had missed. The significance of these pairs was that they were produced by pi zeros. We put the picture on the scanning table and measured the curvature of all these tracks and the angles. From the energies of the electron and positron and the angle between them, and I got the pi zero mass, and from that and the lambda, the cascade zero mass! Once I got *that* son of a bitch, I connected it to *this* [another pi] and I got a number for the omega that was within 10 percent of what Murray [Gell-Mann] said. Without the pi zero, we couldn't have done it—and without the electron-positron pairs, we couldn't have gotten the pi zero. The probability of these two appearing in liquid hydrogen is one in a thousand![30]

Meanwhile, in Paris, Rau gave a talk at the Collège de France, and stayed over afterward at Bernard Gregory's house outside the city. Gregory woke him at 3 A.M., saying he had a phone call from Brookhaven. Rau discovered an excited Samios and Shutt on the other end of the line, saying they'd just analyzed an event that they were ready to declare was the omega minus (fig. 11.5).

Samios took the paper, "Observation of a Hyperon with Strangeness Minus Three," to the *Physical Review Letters* office the next day, 11 February. "It has been pointed out," the authors began, that in Gell-Mann's scheme several of the new particles "can be arranged as a decuplet with

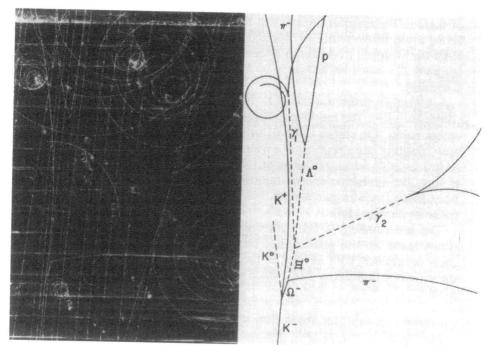

11.4 Bubble chamber picture of the first omega minus (ω^-) event, showing an incoming K^- interacting with a proton to produce a K^0, a K^-, and an ω^-, which decays in turn into other particles. The paths of neutral particles, which produce no tracks, are indicated by dashed lines; their presence and properties are determined by analysis of the tracks of their charged decay products, together with laws of conservation of mass and energy. The two pairs of more sharply curving lines are electron-positron pairs.

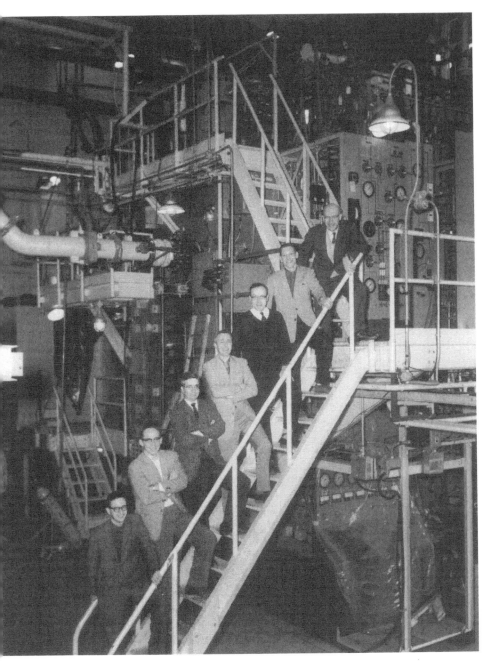

11.5 Team that discovered omega minus on the steps of the eighty-inch bubble chamber. From top right to bottom left: Ralph P. Shutt, Jack B. Jensen, Medford S. Webster, William A. Tuttle, William B. Fowler, Donald P. Brown, and Nicholas P. Samios.

one member still missing." In view of the new particle's charge, strange-ness, and mass, the authors wrote, "we feel justified in identifying it with the sought-for omega minus" (Barnes et al. 1964, 204, 206).

Gell-Mann, too, was alerted to the news by the scientific grapevine, which carries news at about the speed of light. Feigning ignorance, he called Samios for an ostensibly friendly chat, and when told gracefully pretended to be surprised. But the team was still not ready to make the discovery public. In accord with established scientific practice, the journal held to a rule that results submitted for publication could not be publicized until after they had appeared in print. The reason was the traditional one: to seek accuracy by allowing time for peer review. The team then scheduled a press conference for 24 February, the day the article was scheduled for publication.

But a premature disclosure angered the scientists and produced confu-sion over Brookhaven's role. P. T. Matthews, a professor at the Imperial College in London, had been working on an article on elementary particle physics, SU(3), and the predicted omega minus and its significance for the 20 February issue of *New Scientist*. Hearing of the omega minus news through the grapevine via CERN, Matthews could not resist intimating that a discovery announcement was about to take place, but slyly hinted his British colleagues at CERN might also be involved:

> February should go down in the history of physics as the time when a new fundamental law was established . . . so far, each new stride has seemed like a step backwards in that it has brought ever-increasing complexity and confu-sion . . . this month, however, the tide is turning and a coherent pattern is emerg-ing, to bring order and beauty into this sub-nuclear world . . . this is the type of situation about which physicists dream—a startling new theory at an absolutely fundamental level, which coordinates and clarifies a previously confused experi-mental situation . . . the technical requirements for the production of omega-minus are very difficult, and the necessary high energy beams of K-minus were only recently engineered at both the Brookhaven and CERN laboratories . . . an announcement may be expected any day . . . high energy physicists are walking around with a slightly hysterical look, as though they are actually witnessing the apple landing on Newton's head! (Matthews 1964, 458, 460)[31]

A London correspondent for *The New York Times,* armed with an ad-vance copy of Matthews's article, noticed that Brookhaven had scheduled a press conference, put two and two together, and promptly prepared an announcement of the discovery based on Matthews's account. Learning of the impending *Times* article, Samios pleaded with the newspaper to hold the article until 24 February. The *Times* refused and published the article on the front page. But the *Times,* failing to verify the source, erroneously attributed the announcement to Matthews and the discovery simultaneously to Brookhaven and CERN:

KEY PARTICLE FOUND IN THE ATOM, ENDING NUCLEAR PHYSICS 'CHAOS'

London, Feb. 19—The discovery of a new sub-atomic particle was disclosed today by a prominent British nuclear theorist. . . . The particle was found with the aid of high-energy apparatus at the Brookhaven National Laboratory, Upton, L. I. and the Center for European Nuclear Research (CERN) at Geneva, Switzerland. . . . (Hillaby 1964, 1)

Gnashing their collective teeth, the Brookhaven scientists agreed to talk to the *Times* on Saturday, 22 February, for an article to appear on Sunday, in order to straighten out the facts, and the Brookhaven public relations department bumped up its press release to Sunday morning. On Monday, 24 February, the BNL scientists held their originally scheduled press conference in New York City at the library of the American Institute of Physics. Goldhaber, Shutt, Samios, and Fowler showed up prepared to explain the discovery and its importance to the two dozen or so reporters present. The discovery, Goldhaber told them, "forms the capstone of a building that had previously been held together only by the bold imagination of Drs. Gell-Mann and Ne'eman. Physicists are so excited that they will even talk to journalists." Striving to find some way to express the significance of the discovery, one reporter asked how the discovery compared in importance to that of the electron. Goldhaber was gently reprimanding:

It is unscientific to make such comparisons. Permit me to invent a Chinese proverb: "In a journey of a thousand miles, you cannot omit a single step." If you look back afterward and say, "That was a giant step," or "That was a small step," then you are not making an objective statement (New Yorker 1964, 40).

Another reporter, overwhelmed by the forest of Greek-lettered particles that had accumulated on the portable blackboard, wondered aloud, "Where do you go from here? That is, when will all this end?" A more politically minded person would have seized the opportunity to stump for the value of federally sponsored science and the need for its continued support. Goldhaber's reply, utterly in character, was in keeping with his own view of the integrity of basic research. "It will end," he said, "only when it bores us" (New Yorker 1964, 40).

Boredom would not be a problem. The discovery of the omega minus was strong evidence for the validity of Gell-Mann's SU(3) scheme of classifying fundamental particles. SU(3) in turn came to have an impact on particle physics something like that of the periodic table on chemistry at the end of the last century. The eightfold way consolidated and coordinated what was known about hadrons, exhibiting the gaps and inconsistencies needing to be resolved. At the AGS and elsewhere, physicists set about trying to fill in the holes, finding new particles and determining their properties, with each new discovery providing additional support for SU(3).

Meanwhile, in an attempt to rescue morale, the British team circulated their Polaroid among themselves together with a memo entitled "A Possible $\Omega-$" that included diagrams and charts. The memo was carefully labeled "INTERNAL CIRCULATION, FOR INFORMATION ONLY." "Taking the evidence obtained at Brookhaven by Barnes et al. as establishing the omega-hyperon, this event is best described as another example of its production." [32] If the Brookhaven team had been less aggressive or gambled with poorer judgment, and if the British team had had fewer delays in commissioning their chamber, that group might well have been the first to see the particle. And when the ZGS at Argonne began to support experiments later in 1964, they turned up several omega minus particles.

A few months after the omega minus discovery, Brookhaven physicist Ed Salant retired, and his group, which had started out using emulsions and wound up analyzing bubble chamber film, needed a new leader. The time when high-energy experimental groups were small associations of a few like-minded individuals was long gone. They were now huge organizations with their own extensive equipment, laboratories, and infrastructure. It was much easier to appoint a new head than to dissolve the group and start over from scratch, and Samios, a young and obviously talented physicist, was chosen. He brought with him a few members of Shutt's group, including Robert Palmer, and turned the group's work toward spectroscopy and the hunt for new resonances in light of SU(3). In the next half-dozen years, the group made numerous discoveries validating the general thrust of the SU(3) scheme.

But as a mathematical structure, SU(3) had many potential families including some with twenty-seven members. Though searches for members of such families were carried out at the AGS and elsewhere, none were found. Gell-Mann had already proposed an explanation: if one assumed that the fundamental representation of SU(3) were a triplet of fractionally charged particles (which he called quarks), this would have the effect of limiting the number of SU(3) families and would explain much of the spectroscopy. The quark explanation was confused by a number of misleading developments, including the alleged discovery, eventually disproved, that one important resonance was saddle shaped.

At the beginning of the next decade, a series of major changes took place in the organization of the AGS. These included the replacement of Kenneth Green, accelerator department chairman for nearly all of the previous decade, who gave up the chairmanship on 29 July 1970, the tenth anniversary of beam day, by Frederick E. Mills. A few months later, in January 1971, Shutt became the deputy chairman of the accelerator department, and Samios was selected to replace him as leader of the bubble chamber group, which was then merged with his own much smaller team to create a huge new group. The era when experimental groups themselves comprised huge communities with their own subcultures had arrived.

Groups usually had a technical name and an informal name that was something of a point of pride among its members; Rod Cool's group, for instance, was known informally as the Cool group, and it had its own symbol, a tilted martini glass overflowing with bubbles. Precedent called for the new group to be named the Samios group. But Samios, even more than Shutt before him, was reluctant to identify the group with his own personality, refused to call it that, and for a while it, the largest group at the lab, was called, somewhat awkwardly, the new group. After a few months, a group of scanners and support personnel, feeling as though lack of a definite name gave them nebulous status, signed a petition asking that the group be given a real one. Samios called a meeting of the entire group, and in the ensuing discussion it was decided to call it the omega group, for a number of reasons: the fact that omega was the last letter of the Greek alphabet (implying some sort of historical finality to the group), the competition with Alvarez (an A group, as it were); and, of course, the eponymous particle.

Several other changes were also in store. Samios began to downsize the bubble chamber resonance search effort, preparing a major new program of work in neutrino physics in connection with the recently completed seven-foot chamber. Some downsizing had already begun to happen with the advent of automatic scanning equipment, but other factors were also at work, for it was clear that the bubble chamber physics of the 1970s was going to be much different from that of the 1960s. The spectroscopists of the 1960s had done their job, and picked out the symmetry that described hadrons. The purpose of the new big bubble chambers, which were made possible by new optics, was not hadrons anymore but weak-interaction physics, which required large volumes due to the low interaction rates. And large bubble chambers were superior to counters for studying the weak interaction because, while in counters neutrinos interacted with complicated material, in bubble chambers they interacted with the much simpler proton.

By 1972, as well, a remarkable theoretical development in high-energy physics that brought together the strong and weak interactions had occurred. SU(3) and SU(2) × U(1) were combined into one theoretical package, soon to be called the standard model, containing a comprehensive picture of all known particles and the forces that affect them. SU(3) had been shown to require *four* quarks, the fourth or "charmed" quark needed to suppress an effect that brought the theory into conflict with experiment. And there were probably *six* quarks, since two Japanese physicists would soon show that CP violation could be reconciled with the existing theory of the weak interaction only if at least three pairs of quarks existed.[33] These three families or "flavors" of quarks were matched by three families of leptons, described by SU(2) × U(1). The standard model also explained some puzzling aspects of quark dynamics: why they seem like free particles when close together in resonant states, and why they became more tightly bound the further they move apart.

Experiments at the AGS had played major roles in many of these developments. In retrospect, the two-neutrino discovery was the first clue as to flavor, or the generational structure of particles, with pairs of particles in each generation. CP violation may have been the first clue to the existence of three generations of quarks (though it remains so mysterious that the connection may be entirely fortuitous). And the omega minus was the key particle that had pointed the way to SU(3). But experimental physics had dramatically changed in the process. In 1952, when the Cosmotron was just completed, experimenters had led the first forays into uncharted territory, while theorists struggled to discern larger patterns in the chaotic thickets of particles and resonances. Two decades later, with the completion of the standard model, it seemed as though each new discovery was but a tiny piece of a much bigger, already existing structure that the theorists had firmly under control. "The experimentalists by now must feel like ants, or like pharoah's slaves building the pyramid," said one theorist at a conference held in 1972 at the opening of Fermilab (Polkinghorne 1989, 135).

"It's really remarkable to think about those years now," Samios recently recalled. "Once you can see the whole picture, yesterday's work seems trivial, but it wasn't; intellectually it was just as hard." The entire effort of all those years, he continued, has been replaced by the standard model, which graduate students can learn about without knowing anything of the furious and sometimes frustrating work that made it all possible.

> You had to have insight on the basis of fragmentary evidence. Like the K particles—remember the K particles? At first there were a dozen different kinds of them—the tau, the theta, the theta prime, the K mu 3, and so on—depending on how they decayed. You'd be proud of what you did, and people said, "So-and-so discovered the K star," and you felt important. And we studied those K particles and measured them and catalogued them and had huge debates about things like parities and spins and lifetimes. Huge debates! Then at the end of the 1950s that all got replaced. There was one particle in four versions—K^+, K^-, K^0, and K^0 bar—the K^0 and K^0 bar mix, and they all violate parity when they decay weakly. That's it! One particle! And then at the end of the 1960s it changed again. The

story was quarks. Now someone would ask, "What's the K star?" and the answer was, "It's an up-strange quark." You see? There were a lot of pieces to the puzzle to be gotten and everybody got a shot at it, and every little contribution helped. You got to play a certain amount of the game, and you did the best you could, and by the time you stop you had successors who took over. It took luck and ingenuity and skill and experience to do well at it.

Like stickball?

Well, in stickball, *real money* was involved. There was an immediate material payoff and the winners got to go out that evening. In science, the payoff is the high regard in which you are held by your peers. And there's a personal payoff, too. You do the experiment right, and it runs beautifully, and you and your colleagues see and understand what's happening—it's exhilarating. You do get replaced. And every time you solve something, you know it's not the final answer, either. But the fun's in playing the game. In science, the puzzle you solve becomes a tiny piece of a larger puzzle. Who knows if there's a final puzzle? I'm not even interested.[34]

Few outside the physics community appreciated how far it had come in twenty-five dizzying years. Few inside could foresee how little it would progress in the next twenty-five. The standard model would prove remarkably successful at solving existing puzzles, and the major discoveries in high-energy physics would all be predictions of the standard model, which would emerge from the next quarter-century essentially intact. Despite decades of effort, it is still extremely difficult today to see how it could be one piece of a much larger puzzle.

The year 1972 was transitional not only for high-energy physics, both theoretical and experimental, but for the AGS as well. It was no longer the most powerful accelerator; several with a more powerful physics capability had been switched on elsewhere, including Fermilab's 200 GeV machine, CERN's ISR, and Stanford's linear accelerator facility. In keeping with Goldhaber's caution, the lab had chosen to focus its energies on upgrading the AGS. The results disappointed. The AGS was down for a considerable amount of time, and the upgrade took longer, cost more, and resulted in less intensity than planned. Moreover, the standard model predicted the existence of a number of phenomena that redirected the interests of experimenters—weak-interaction physicists were interested in its prediction of neutral currents, while strong-interaction physicists were interested in charm spectroscopy, or resonances involving the charmed quark—which the newer accelerators seemed better able to reach. Some who had worked at Brookhaven moved elsewhere: Lederman left for CERN, Fitch for Fermilab. "We are not patient people," Lederman said.[35]

The High Flux
Beam Reactor

IN SEPTEMBER 1961, the same month that Brookhaven's second-generation large accelerator, the Alternating Gradient Synchrotron, was dedicated, ground was broken for the lab's second-generation reactor, the High Flux Beam Reactor (HFBR). Its project engineer was Joe Hendrie.

Hendrie was born in Detroit in 1925. As a youth he was good with his hands and had a strong physical intuition. Many of his male ancestors had been mining engineers; young Joe would build reactors. He entered college intending to become an electrical engineer, but soon discovered physics. In 1950, he began graduate studies at Columbia, where his teachers included some of the foremost figures in the field and several Nobel laureates or future laureates, including Polycarp Kusch (his thesis advisor), Willis Lamb, Rabi, James Rainwater, and Hideki Yukawa.

> If I'd gone to some place with a little lower intellectual aspirations [Hendrie recalled] I might have come away thinking I was a hotshot physicist. But one of the benefits of being at Columbia in those days was that you got a chance to measure yourself against the best. At Columbia, if you had any self-critical capability at all you could speedily determine where you fell in the scheme of things.[1]

Deciding that the scheme of things did not include a place for him as a physicist, Hendrie sought a job as an engineer instead. In spring 1955, with his thesis work done but not yet written up, he interviewed for a place in Kouts's group. Though the salary was well below offers he had received from private industries, Hendrie accepted it, liking the atmosphere, the sailing, and the people. Hendrie's design and procurement work impressed Kouts, while Hendrie's draftsmanship impressed the engineering staff, who were astonished to find a card-carrying Ph.D. physicist able to turn in quality engineering sketches.

When it came time to select a technical head for the HFBR project, Hendrie, all of thirty-three, was an obvious choice. He was charismatic, polymathic in interests and abilities, and a fanatic about detail. He was not only articulate, but liked to express himself with a folksy directness that often shocked: years later, as chairman of the Nuclear Regulatory Commission at the time of the Three Mile Island reactor accident, he brutally and impolitically described the ensuing investigation as "the blind leading the blind," landing him in trouble in many quarters (among them an organization representing the disabled). Nevertheless, he was the J. Robert Oppenheimer of the HFBR; he brought home a technically complex project through a combination of a deep appreciation for engineering, a thorough knowledge of physics, and an ability to motivate and even inspire.

In fall 1958, soon after agreeing to become HFBR's project engineer, Hendrie recalled a remark he'd once heard Lyle Borst make: "Every good engineer has *one* reactor in him." He wondered what that was supposed to mean. He was about to find out.

HFBR Construction

Haworth had been pushing the AEC to allow AUI to handle the contracting; poor communication between the AEC and its contractors had hampered several large construction projects at national labs (including the BGRR and BMRR), and Haworth was eager to avoid a repetition. At first it looked like he might succeed. But Haworth ultimately lost; the AEC decided that, as a matter of policy, its area offices would hold the contracts for significant construction projects though he resolved to increase the lab's diligence in inspection and technical supervision.[2]

Shortly after Labor Day 1958, Hendrie began to make sketches of what the reactor might look like, based on Chernick's design and General Nuclear's study, while others made calculations about flux and power distributions and neutron spectra. The modeling and calculating went on for several months, allowing Hendrie to work out a practical way to lay out the core, reflector, and general configuration. Meanwhile, Downes and Kouts constructed a critical facility to carry out benchmark measurements.

During the transformation of theoretical concept into workable real object, unpleasant surprises began to surface. The first was that General Nuclear had underestimated the power requirement by a factor of two; to get the planned neutron flux, the power of the reactor would have to be doubled, to 40 MW. Hendrie was thus forced to rethink the design: first to discover which parts had to be altered to double the flux, then how to redesign them for 40 MW. The change also meant, potentially, an unacceptably large increase in the reactor budget.

> The Congress had already appropriated $10 million. And it was in a bill which said that such projects can overrun up to 25 percent without the need to come

back to Congress for further authorization. That didn't mean that they would put up 25 percent more money—only that the AEC would have to find the extra money itself, if it could. The feeling was that under no circumstances did one want to go back to Congress. That would have been the kiss of death—both at AEC headquarters and at the joint committee. We had gone to the AEC and said, in effect, "We need a reactor and here's what it can look like and we need $10 million." If before we begin we run back and say, "We've reconsidered and we need $15 million," we would have been bounced right out of the room. OK. So our next question was, "Can we make it work?"[3]

Hendrie felt that the basic layout of the building, including the pit structure, floor loading, cranes, and air conditioning, could stay the same. But shielding and almost everything connected with the heat transport system, including heat exchangers, pumps, and flow rates, had to be significantly upgraded. To pull this off without cost overruns, the project group sought cheaper civil engineering contracts than intended, planning to zealously ride herd on the contractors to make sure of quality, and eliminated expensive items where they could. During the design stage, the group members did not think seriously about installing an extra liner around the spent fuel pool for several reasons: it would have been a costly addition to the budget at a time when they were worried about making the budget at all, was not required by contemporary safety regulations, and seemed unnecessary given that the BGRR's fuel pool of similar construction—reinforced and tile-lined concrete—apparently had not leaked.[4] At the end of the lab's second quarter-century, this decision would come back to haunt the reactor, and the laboratory.

Hendrie then had a second unpleasant surprise. The HFBR was to be moderated by heavy water. At the time, heavy water was considered a special nuclear material by the AEC, all stocks of which were owned by the government and loaned to government operations. After the HFBR project was approved, the AEC changed this policy, and suddenly Brookhaven was required to purchase the heavy water at $28 a pound. The unexpected additional expense, a whopping $1.25 million, further cut the overrun flexibility.

These surprises pushed the effective construction date ahead about two years, but construction proceeded quickly once it began in earnest in spring 1962. By the middle of 1963, one of the few key outstanding items was the safety report. Until then, such reports usually consisted of appendixes to the reactor design report, but Hendrie wanted something more substantial. He first tried dividing up the task and farming pieces out to various members of the safety committee, but was dissatisfied with the results: "I needed to write an integrated document." He took the project upon himself. "It was a bloody sweat. I got a staggering writer's block; it seemed to

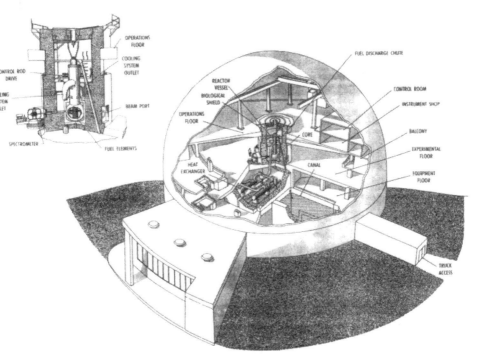

12.1 Cutaway of the HFBR, showing arrangement of building and reactor. The reactor, auxiliary equipment, and experimental facilities are housed in a steel hemisphere 176 feet in diameter. The building is kept at a slightly lower pressure than the outside, so that any air flows in the building rather than out.

me there was so much to be said that it was impossible to say it."[5] In September 1963, he began working on the report seven days a week. He'd work at the lab all day, return home for dinner and to see his family, and then be back to the lab to work until late at night, every day for nine months, including Christmas Day 1963. The result, a two-volume work of 612 pages issued in April of 1964, is a remarkable document, for it amounts to a complete design report of the reactor.[6] The report earned him fame in the AEC's Washington office, and became for them the model of what they were looking for in a safety report.

The reactor's construction was completed in September 1964 (at a cost of $12.5 million), when testing began on the various systems (fig. 12.1). Operation was scheduled for October 1965. Hendrie now felt he knew what Borst had meant.

It's something that occupies you totally. You pour energy into a project like that at a rate that a young, healthy man can maintain for four to five years, but you burn out. You juggle a thousand balls in your mind, all of them important, and keep an eye on how they all fit together. I was never able again to focus the energy and encompassing understanding of everything that went on that I did on the HFBR. It's not just the technical details; in some ways, these are not the most wearing. On a given technical problem you get good people together and they haggle and turn it and twist it until they get it. No, it's the administration: the reactor division, the steering committee, the users (who always want more flux),

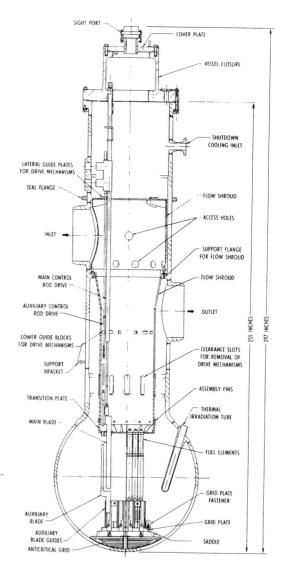

12.2 Cutaway of the reactor vessel and its internal components, including the core, control rods or blades, and control rod drive mechanism. The inlet and outlet are for the heavy water moderator/coolant.

the AEC (which always wants lower cost). And in this array of groups and "ethnicities" there are all kinds of personal quirks and egos, and a lot of them don't like each other, or have quite different views on things, and are not in a mood to be amiable about sitting down and compromising. You've got to keep that all in hand. I more than anybody else understood everything and worried about everything. I *possessed* the HFBR concept. This job is like a Swiss watch. Everything has to work, otherwise it won't go. And you've got to keep all these people patted into shape and working amiably together.[7]

The HFBR finally went critical on Halloween evening, 31 October 1965 (Oak Ridge's HFIR had gone critical that August). Several weeks later, the cover of the lab Christmas card sported a picture of the HFBR's silvery dome: snow-clad, silhouetted against a deep blue sky, framed by pine trees.

Like the AGS and any other large instrument, the HFBR required a testing period of a few months, during which a number of problems were ironed out. One of the most serious concerned the anticritical grid, a set of stainless steel bars that had been installed in the bottom of the core vessel. One of its functions was to ensure that, if some freak accident melted the core, the fuel would not coalesce in a critical mass and generate heat. Another function was to break up the jet of water flow and prevent erosion of the vessel; instead, the flow (thirty-five feet a second, about seventeen thousand gallons a minute) broke it up, jarring some bars enough to crack the welds and break free. This created the danger that some might strike the beam pipes. At first it looked like repair would have to involve removing the entire reactor vessel, but a way was discovered to reach inside the vessel and remove loose bars as well as inessential ones that might eventually break free.

The HFBR was water cooled and did not produce argon as a waste product, as had the BGRR. Defective fuel elements were quickly identified by the presence of fission products (usually iodine) in the heavy water; the machine would be shut down and the defective element found and replaced. The new, troublesome by-product of the HFBR was tritium, an isotope of hydrogen produced in the heavy water coolant/moderator when a neutron reacted with the deuterium. The reactor engineers established a limit of tritium concentration in the heavy water, and when this was reached they would send it to Savannah River and replace it with "clean" heavy water.

In early 1966 the HFBR had reached its design power and was ready for experimenters. The look of the HFBR experimental area was much different from that of the BGRR. Instead of a pegboard-like cube with its sea of regularly spaced holes, the HFBR was an eight-sided solid about twenty-five feet across, with one beam hole (occasionally subdivided) per face. The total number of neutrons produced by the HFBR, which would run typically at about 40 MW, was only a third more than that of the BGRR, which

12.3 Installing the HFBR reactor vessel, March 1964.

usually ran at about 22 MW—but since the HFBR's active core area is much smaller (90 liters) than the BGRR's (270,000 liters, in the original design), its flux (a measure of neutron density) was much higher. But only so many beam pipes could be installed around the undermoderated core before it would adversely affect the reactivity itself; thus the number of holes was limited to nine.[8] The HFBR was also much cleaner (i.e., having fewer emissions and waste products) than the BGRR, one reason being that the vertical irradiation holes were completely isolated from the experimental floor.

As long as a reactor operates reliably, operators and users are essentially decoupled; typically, many fewer conflicts between the two groups occur at a reactor than at an accelerator. But the vast shrinkage in the number of experimental holes, from sixty-one at the BGRR to nine at the HFBR, drastically changed the character of the experimental community and created a potentially serious management problem. Already in 1963, requests for experimental space seemed to exceed supply, and lab officials feared a fierce competition not unlike the mad scramble for beam time at accelerators. To head it off, Goldhaber decided to appoint what would amount to a program committee. Users resisted, arguing that the determination of who would use which hole should be done informally, in traditional Brookhaven style.[9] Goldhaber went ahead anyway to establish an HFBR Experimental Facilities Committee, chaired by Powell.[10]

Powell so loathed the idea that he resolved that it never meet. He succeeded. One reason was the way the hole charges were assigned. These fees were significant, due to the small number of holes: few, if any, outside users could undertake a financial burden of that magnitude (also, the fact that the HFBR had been optimized for beams rather than irradiations reduced its attractiveness for some potential users). The charges were assigned to the departments that paid for them, which made it relatively easy to put the decision of who worked at which hole in the laps of individual departments rather than the program committee.

Another factor involved instrumentation. Spectrometers and detectors were larger and more complicated than they had been at the BGRR, and had the effect (though to a smaller degree) that bubble chambers had on accelerator research: promoting greater group stability and long-range planning. Plus, the techniques of neutron scattering (unlike those of X-ray and electron scattering) were hard to learn and could be practiced only at a small number of holes with the right equipment. "It's like playing a cello," says Gen Shirane, later the head of the HFBR's experimental solid-state group, "they're rather difficult instruments to play, there are not many around, so not many people can play them well." As a consequence, the ability of an outside group to come in and start working from scratch was almost nil. For all these reasons, Powell succeeded in his effort to encourage prospective users into collaborations with existing groups. The HFBR was not a user facility, certainly not what the BGRR had been. A large visitor program was no longer possible, and no longer needed.

The HFBR is a classic example of a scientific instrument created for one purpose (nuclear physics) and used mainly for another (neutron scattering, especially inelastic scattering). The impetus for it was largely Hughes's quest for a facility with more intense flux for neutron physics research. But by the end of the 1950s, that field had changed. Mapping cross-sections for practical applications, in which so much of the exploration of the new world of atomic energy consisted, was largely completed. Many other traditional areas of nuclear research, including low-energy nuclear levels and states, had also been largely explored. And all of Hughes's gamesmanship could not obscure the fact that accelerator-based neutron sources were now rival facilities for most remaining cross-section measurements. Finally, Hughes's death in 1960 deprived Brookhaven of its most aggressive and influential voice for traditional nuclear physics. While Palevsky, Hughes's loyal lieutenant, assumed the mantle of the neutron physics group, Palevsky was no Hughes. Not only did he lack the commanding presence of his former boss, but his interests had changed to accelerator research and he was rarely seen on the HFBR floor.

Of the HFBR's nine holes (H-1 through H-9), three were for nuclear

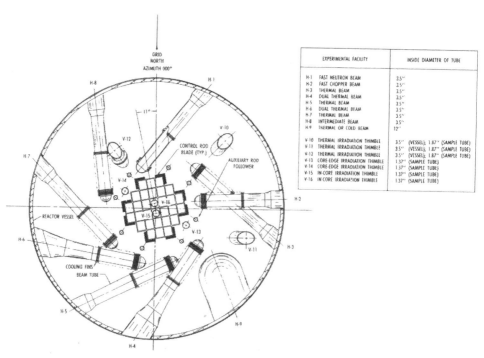

12.4 Diagram of the beam tube and irradiation thimble layout in the reactor vessel, showing the twenty-eight fuel element positions, and two irradiation facilities, of the core. The eight right-angle control rods or blades are alongside the core in black. H-9 is a large (twelve-inch-diameter) thimble-shaped hole for the cold neutron source.

physics and six for neutron scattering. (It also had seven vertical thimbles for irradiations, V-10 through V-16, with slightly different neutron energy spectra.) Each beam had a somewhat different neutron profile; the more directly its pipe pointed to the core, the harder (more energetic) its neutron spectrum. The instruments at each hole were tailored to specific projects.

H-1 to H-3

Nuclear physics was conducted at H-1, H-2, and H-3. But nuclear physics at the HFBR was a much different effort from what it had been at the BGRR. Gone were the days of cozy intimacy between basic and applied interests, and the Cold War "emergencies" of the early 1950s, when the AEC breathed down researchers' necks to churn out quickly new cross-section data and was willing to contribute endless resources for the cause. The cross-section work of the 1950s at many of the national laboratories, and completion of the Bohr-Mottelson model, had stabilized the picture of the nucleus for basic researchers and provided reactor and weapons engineers with much of the essential information they needed. Accelerator-based neutron sources had substantially reduced the viability of reactor-based cross-section work. Nuclear physicists at the HFBR therefore applied the new flux in expanded research programs, usually in nuclear excitation spectroscopy; the study of high excited state levels and their properties.

Vance Sailor's nuclear cryogenics group took up residence at H-1, where it used the HFBR's more intense neutron flux to study spins of resonance states of nuclei. Because he needed neutrons of somewhat higher energies than neutron scatterers desired, he used a beam tube that pointed more directly at the core than theirs, but still tangent. That it was tangent meant neutrons in the beam had to have at least one collision with the moderator (eliminating gammas and very high-energy neutrons from the reactor core), while the fact that it was pointed more directly at the core meant the neutrons would not have had a large number of collisions. To polarize the targets, his group built another refrigerator, much larger than his previous one, with a superconducting coil able to handle bigger samples and cool them to lower temperatures. It was massive, a two-story piece of equipment that took more time and money to build than expected, far and away the largest piece of equipment built during the first years of the HFBR.[11]

But the group members ran into unanticipated difficulties once they began to tackle higher excited states of uranium. One was that the uranium resonances were extremely close together or even overlapped, requiring exceedingly high resolution to disentangle. And after all the labor involved in building the refrigerator, studying such states seemed anticlimactic and to have run its course, even to Sailor himself. Lab administrators took note.

When, in the late 1960s, they were forced to cut back on research areas deemed unlikely to move forward, Sailor's was one. With the wind taken out of his sails, Sailor grew unhappy, a little resentful, and seemed to lose his sense of purpose. "I'd been at it for almost twenty years," he recalled, "and we were repeating the same kinds of things we'd done before but with different targets. Basically, we were contributing more data to the "Barn Books." After a certain point you lose enthusiasm." [12]

In 1970, he saw the opportunity for a change. Leland Haworth had returned to the lab as a consultant to the director and established a small group to carry out a series of studies on U.S. energy consumption and resources. With the encouragement of the administration, Sailor left H-1 and the physics department to join Haworth's group in the department of applied sciences.[13] David Rorer, a member of Sailor's group, ran the program for a while, but the huge refrigerator was soon dismantled. Sailor also became interested in nuclear power. The Long Island Lighting Company (LILCO) was planning to build a nuclear reactor in the nearby town of Shoreham and hearings on the construction permit began in late 1970. Some conservation groups opposed to the plant became involved in the hearings; Sailor, an advocate, founded a private organization called Suffolk Scientists for Cleaner Power and Safer Environment that became involved in support of the reactor. He began to invest considerable time and energy in the proposed Shoreham plant—but he would be bitterly disappointed, too, in the fate of that project.

The neutron physics group was stationed at H-2. Its head, Harry Palevsky, worked at the Cosmotron almost full-time, and Robert Chrien was in charge of the group's HFBR effort. Chrien had arrived at Brookhaven in 1957 to work with Hughes, only to be sent to Chalk River in 1959, where the flux was higher and resolution better. There, Chrien measured cross-sections using a copy of the fast chopper, built by Hughes and shipped off to Chalk River a few years previously with group member Robert Zimmerman, in the days when the resolution of accelerator-based neutron sources was just beginning to overtake that at the BGRR.

> In those days [recalls Chrien], we had hunks of uranium all over the place and carried it around and nobody paid any attention, except once, when Zimmerman got into trouble crossing the border into Canada with some uranium in his car. We had all these rare earths, *pounds* of them, and stored them in the cabinets and desk drawers, wherever, along with a lot of gold and platinum.[14]

In 1962, Chrien returned to Brookhaven to begin building a fast chopper for use at the HFBR. Because accelerator-based systems were now superior for traditional cross-section research, they modified the chopper's design to study what were called capture gamma reactions, which were one way of examining high-energy nuclear levels. While at low excitation energies, nuclei occupy a relatively small number of clearly defined fixed states, at high energies the states are numerous and apparently chaotic. A highly excited nucleus typically sheds its excess energy by dropping from state to state to state in a cascade process, via emission of a series of gamma rays and often by different routes, until it reaches the ground state. By using a gamma counter and coincidence electronics, nuclear physicists could examine these cascades to see what patterns (selection rules, for instance) might lie in them. Hughes's group had done such work at the BGRR, and Chrien carried it forward with the new high flux. A new type of "lithium drifted" gamma detector, made from germanium crystals doped with lithium, had recently been developed that vastly upgraded the ability to detect gammas over the old sodium iodide detectors.[15] Chrien's was the only group principally interested in high-energy neutrons, and the H-2 beam tube was pointed directly at the core to give it a beam rich in high-energy neutrons.[16]

Because Chrien had sacrificed much in the way of resolution for intensity, in 1968 he improved the neutron resolution by building a longer flight path (fig. 12.5). A hole was cut in the containment vessel through which neutrons would coast along a forty-eight-meter flight path to a special station (Chrien 1980). Eventually, accelerator-based sources (notably ORELA at Oak Ridge) became intense enough to overtake the resolution of reactor sources for capture gamma studies, too. In 1979, the chopper was removed and (at the request of the Department of Energy) replaced by an isotope separator called Tristan, whose purpose was to produce various fission products for collection and study in a mass separator. The chopper, its cobalt-loaded steel core slightly radioactive from years of assault by neutrons, now sits in a storage field.

The nuclear structure group, founded by the Goldhabers, was the first occupant of H-3. The group consisted of Gertrude Scharff-Goldhaber (the senior member), Ed der Mateosian, Walter Kane, and Andrew Sunyar. This group, too, was doing nuclear spectroscopy, but its effort at the reactor was small. The group had only enough resources to maintain one of the two ports at H-3, at which they built a crystal spectrometer. Thus nuclear spectroscopy at the HFBR, as at the BGRR, was carried out by two complementary techniques: time of flight (Chrien's group at H-2) and crystal spectrometer (the nuclear structure group at H-3). Otherwise the nuclear structure group employed the same basic method as had Chrien's, using the HFBR's neutrons to pump energy into nuclei, creating excited states that returned to the ground state via a cascade of gammas that were used as a clue to the resonance structure. When Sailor left H-1, Kane became a co-group leader with Chrien at one of the ports there. The nuclear structure

12.5 Cabin at the end of a long time-of-flight path to improve resolution for high-energy nuclear level studies.

group's H-3 port and crystal spectrometer were given to Swiss biologist Benno Schoenborn, who had been making substantial progress in neutron crystallography and needed a low-energy hole for his studies.[17]

In 1964, Schoenborn received a research fellowship to study at the Cavendish, a center of X-ray crystallography of proteins (now called structural biology). Two years previously, Cavendish scientist John C. Kendrew had received the Nobel Prize in chemistry for his X-ray crystallography of sperm whale myoglobin, a protein consisting of a single chain of some 150 amino acid units with a combined total of about twenty-six hundred atoms. Kendrew chose sperm whale myoglobin partly because it was a small and stable, easily crystallizable molecule readily available in large quantities, and partly because myoglobin was structurally related to hemoglobin, a protein that was not only a vitally important biological compound, but also had been closely studied by Kendrew's colleague Max Perutz, making it easy for Kendrew to tap a wealth of information about it. "[S]perm-whale myoglobin," Kendrew and company wrote, "possesses a structure the significance of which extends beyond a particular species and even beyond a particular protein" (Kendrew et al. 1960, 422). Kendrew had determined fairly well the location of everything but the hydrogen atoms, which X rays could not detect (Kendrew et al. 1960). As Donald Hughes had noted already in his *Pile Neutron Research* (Hughes 1953b), neutrons hold a great advantage over X rays in this kind of work given an intense enough beam. Schoenborn was one of those impressed by the significance of locating the missing hydrogen atoms, and aware of the possibilities of neutrons.

> About half the atoms in a protein are hydrogens, and virtually all functional activity in a protein is mediated by hydrogens. So every time you have an enzymatic reaction in a protein, a hydrogen is involved. In a lot of cases, one could speculate what's going on, but you didn't really know if its there or not. The only way to do it properly was to know where the hydrogens are. And the only way to find the hydrogen atoms, I began to realize, was neutron scattering.[18]

When his fellowship expired, he returned to San Francisco to apply neutron diffraction to proteins (diffraction being the part of scattering concerned with structural studies by elastic scattering), and figured that the required flux was about that of Brookhaven's new HFBR. Biology department chairman C. H. W. Hirs was encouraging and invited Schoenborn to visit. Schoenborn's project amounted to an entirely new direction both for biologists and for the HFBR. Previously, all crystallography of large biologically interesting compounds such as proteins had been done with X rays; when biologists used reactors it was either to irradiate samples or create tracers. All the compounds that had been studied with neutrons had

12.6 Neutron density map used by Benno Schoenborn.

at most a few dozen atoms; Schoenborn's project involved a structure with twenty-six hundred atoms.

Schoenborn's first major problem was poor instrumentation. While the technology of high-energy physics detectors was constantly being pushed to the envelope, that of neutron detectors suitable for crystallography lagged by comparison. Existing neutron detectors were mounted on an arm that was slowly moved, point by point, through a series of angles, stopping at each to scan a reflection or peak; this would be a hopelessly slow way of measuring the tens of thousands of reflections that Schoenborn needed. Members of the department of instrumentation and health physics developed an electronic refinement that dropped the resolution from several millimeters to 1.1, a vast improvement; they also applied their expertise in high-energy charged particle detectors to help him develop a detector that could collect data in two dimensions, reading out the coordinates on both the X and Y axis (fig. 12.6).

As his first major target protein, Schoenborn like Kendrew selected sperm whale myoglobin, which plays for protein crystallographers somewhat the role sodium chloride has for chemists studying bonds; for various historical reasons it is *the* compound on which to test a new technique, about which the most on the subject has been worked out. By 1972, he had managed to use neutron diffraction methods to determine the positions of all its twenty-six hundred atoms, including those of the all-important hydrogen atoms—a breakthrough in protein crystallography. That year, a detail of Schoenborn's picture of the molecule graced the cover of the lab's Christmas card.

Neutron Scattering: Holes H-4 to H-8

By the time the HFBR was completed, the field of inelastic scattering was on the rise. In inelastic scattering, neutrons gain or lose energy as they rebound from the lattice excitations, meaning that energy transfer has occurred. If it is an energy gain, they have picked up energy from an excitation of the solid; if a loss, they have given it up in creating some excitation. This reveals information about lattice vibrations, which may be of several kinds, including phonons (density fluctuations of the lattice, analogous to the fluctuations caused by a sound wave), and spin waves (a similar kind of wave effect, but involving the spin orientation of the atoms rather than their locations). Phonons and spin waves, in turn, are crucial for understanding such phenomena as phase transformations, superconductivity, and magnetic properties.

The field was booming for several reasons. First, the instrumentation had considerably improved, thanks largely to the development, by Bertram Brockhouse at Oak Ridge in work that would later earn him a Nobel Prize, of the triple axis spectrometer, soon to become the basic tool of inelastic scattering (described in Brockhouse 1986). In addition, the theory of neutron scattering had advanced to a new level of sophistication, thanks to the cumulative work of several people, among them Fermi, J. Schwinger, O. Halpern and M. H. Johnson, G. Placzek, and Van Hove. Walter Marshall, a theorist from Harwell (then a world capital of neutron scattering), synthesized and extended this work in an influential set of lectures at Berkeley and Harvard in the course of a year spent in U.S. in 1958–59 (Marshall and Lovesy 1971). Finally, in the early 1960s, a number of breakthroughs had taken place in the understanding of some of the processes examined by inelastic scattering.

Inelastic scattering had been possible, just barely, at the BGRR, but the low flux meant that research was effectively restricted to detecting effects and the most important research took place at reactors elsewhere. By the mid-1960s, it was clear that it would be the field of promise of the HFBR. One sign of the HFBR's newly reconceived purpose was a conference held in September 1965, a month before the reactor was scheduled to go critical. The conference, Symposium on Inelastic Scattering of Neutrons by Condensed Systems, was attended by over 160 of the most renowned experimenters and theorists in the field.[19] The opening talk, "Comparison of Electromagnetic and Neutron Studies of Solids," was delivered by a young neutron scattering theorist, Martin Blume.

As a graduate student, Blume had shared an office with Marshall, from whom he picked up an interest in neutron scattering, during a visit by the latter to Harvard in spring 1959. When Blume arrived at the lab, in June 1962, he joined Dienes's solid-state group at a time it was undergoing both an expansion and change of direction. While initially it had been preoccupied with radiation damage in solids, a subject closely linked to applied engineering questions related to the reactor, it was now moving more toward the mainstream of solid-state physics, which involved many-body problems. These involve properties that are a consequence of the interactions of large numbers of particles: superconductivity is a many-body problem (there is no such thing as a single superconducting electron), as are phase transitions of all kinds (magnetic and structural), crystal growth, and applications of statistical mechanics. As it happens, a large number of these problems can be addressed through neutron scattering. Blume's arrival in 1962 both reflected and reinforced a reorientation of Brookhaven's solid-state program in this more mainstream direction.[20]

The five holes H-4 through H-8 were to be used for various solid-state research projects by members of both the chemistry and physics departments, solid-state physics still having somewhat ill-defined interdisciplinary borders ("squalid-state physics," in Gell-Mann's phrase). Similar equipment was needed at each. To economize on design and construction, a basic spectrometer was built, known as US-1, which (though sounding like the name of a highway) stands for Universal Spectrometer 1 (fig. 12.7). US-1 was designed to be readily convertible to a number of applications: single-crystal spectrometer, double-axis spectrometer for elastic scattering, or triple-axis spectrometer for inelastic scattering. Its physics conception was by Harvey Alperin, its engineer was Andrew Kevey.

Kevey was born in Hungary in 1923, and his route to Brookhaven was, literally, material for a novel (Kevey 1991). He had a childhood ambition to enter the Ludovika Academy (the Hungarian West Point); was refused until the war against the Soviets depleted Hungary's military academies and created a scramble for cadets; was accepted into the Ludovika and sped through in barely a year and a half; trained briefly in Germany be-

12.7 US-1, the Universal Spectrometer developed for the HFBR.

fore being sent against the British and captured without ever having fired a shot against the enemy; spent a year and a half in Allied POW camps; languished another year and a half in rat-infested displaced-person camps; found work in England as a domestic servant; took the boat to New York City with no marketable skills and poor English; and went to night school and eventually transferred to Brooklyn Polytechnic Institute, from which he graduated in 1959 with a degree in mechanical engineering.

That year, he was hired by Brookhaven, was responsible for assembling the AGS injector, and then was put to work on a neutron spectrometer under Chalmers Frazier and Robert Nathans. Kevey had to learn the need to anticipate the rapidly changing demands of experimental physicists. One day, Nathans told Kevey to design a small calibrated turntable for mounting a crystal, telling him that it would have to bear a weight of a few ounces—certainly never more than a pound. Three months later, Nathans asked Kevey to install a fifty-pound magnet on the table. Kevey protested. "But Andy," Nathans replied, "don't you know that, in physics, *never means three months!*" [21]

After construction began at the HFBR, the former domestic servant was put in charge of engineering US-1, the huge new spectrometer. One of Kevey's biggest engineering challenges was to provide a suitable bearing to support the weight of the drum (about five tons), the overturning moments from the second (and possibly third) axis, and the detector itself, all the while with gearing of high accuracy. Kevey also worried about the Nathans syndrome; future loads placed by cryostats, magnets, and detectors much heavier than currently conceivable. In addition, he knew that building a bearing and drive from scratch would be prohibitively expensive given the budget.

One day Kevey gave a lift to an old retired navy captain named Mead who worked at the lab in procurement. Kevey happened to mention his bearing problem, and Mead said he would inquire through his old navy channels. A few days later, Mead called back with good news. The navy had nine 40 mm World War II antiaircraft gun mounts in its warehouse near San Francisco, and it might be possible for Brookhaven, as a government laboratory, to obtain them. Kevey flew to San Francisco, inspected one of the mounts, obtained the drawings, and saw that these would make perfect bases for the new spectrometers. The bearings were rigged to withstand the kick of the guns when fired, and the gears were cut with high precision to orient the guns properly. He flew back and put in paperwork for all nine. The gun mounts were free; the lab paid about a thousand dollars for cleaning, regreasing, and shipping. Although many of the HFBR spectrometers have since been modified to accommodate new demands, the gun mounts have never been replaced.

Nine spectrometers were under construction in the Brookhaven shops: five for the HFBR, two for the naval ordnance laboratory, and one each for labs in Puerto Rico and Israel. The HFBR spectrometers were hooked into a large-scale time-shared computer data acquisition system, in which the spectrometers were serviced by SDS 910 (neutron physics) and SDS 920 (neutron diffraction) computers, and in 1970 by a PDP-11. The computer would not only take data on line, but instruct each spectrometer how long to stay at one stop, how many degrees to move to get to the next stop, and so forth, a far cry from the manual settings and pencil data taking of just fifteen years previously. The prototype for this system, at Brookhaven, had been the time-shared slow and fast chopper data acquisition facilities at the BGRR.[22] By 1966, the spectrometers were ready on schedule and within budget, and worked well (fig. 12.8). It was the first major project Kevey led himself.

> It meant much more to me than my degree. That's only a dead piece of paper with a stamp and several signatures now hanging in my den. But those spectrometers are alive. They hum day and night, spewing out data for a dozen scientists. And they were my creations. I had an inner glow within me. (Kevey [n.d.])

12.8 HFBR experimental floor, showing three neutron spectrometers in position.

H-4, H-7, H-8

In 1965, the solid-state group had four senior staff members (Dienes, Frazier, Nathans, and Shirane), as well as several postdocs. Dienes was head of the group. But, as was often the case at Brookhaven, the actual functioning of the group bore little resemblance to its official structure. An informal, anarchic situation prevailed on the HFBR experimental floor, with power in the hands of the most ambitious, notably Nathans and Shirane. For a while these two worked relatively independently of each other, with Frank Langdon, the chief technician in charge of the experimental floor area, serving as a key intermediary in deciding the chief priorities. Shirane became head of the solid-state neutron group after Nathans's departure in 1968.

Shirane was a leader in the Donald Hughes mold. He liked nothing better than to focus deeply on a technically demanding experiment, and was notorious for driving subordinates hard. Economical with both words and time, he was known for scheduling not only the start of a meeting but also its end. If at the end of twenty-five minutes or however long the meeting was scheduled to last the issue in question was unsolved, Shirane would dismiss the group, sending them back to their offices to work on it.

Shirane was another who followed a circuitous route to Brookhaven involving an extraordinary amount of luck. Born in Japan in 1924 in a small town between Osaka and Kobe, he began to study science after passing an exam that entitled him to enter a fast-track high school. "Therefore," says Shirane, "I didn't die. All my friends in the soft sciences, liberal arts and literature, were drafted. Most went into the navy, became pilots, and died. But students in science and technology were not drafted." [23] Even so, he came perilously close to not surviving the war. When he entered the University of Tokyo as an engineering student, he and his classmates were randomly assigned to different campuses. Shirane's happened to be outside Tokyo proper, which was not subject to the massive bombings that devastated most of that city. He studied aeronautics, intending to become an airplane designer, and among his instructors was Itokawa, designer of the famous Japanese Zero fighter. The toughest year of the war was the last, when Shirane and his colleagues had little to eat except the pumpkins that grew even in ravaged soil. But the war ended abruptly. The occupation force disbanded all activities it decided were war related, among which was the University of Tokyo aeronautics department. What was left of the department was converted to applied physics, and Shirane suddenly found himself in a new field. He graduated in 1947.

> For me, it was lucky. Only about one out of hundred aeronautics graduates design airplanes; the other ninety-nine calculate things like the strengths of materials. Therefore, you might say I became a physicist by chance. I didn't want to become a physicist. But I was not disappointed.

When Ray Pepinsky, a scientist at Pennsylvania State University, asked a senior Japanese scientist visiting Penn State to find a young scientist from his homeland who might be recruited, a friend of a friend recommended Shirane. In 1956, Pepinsky sent Shirane to Brookhaven for a year, the same year that Clifford Shull was at Brookhaven. Shull was using a spectrometer at the BGRR to study the magnetic properties of single crystals. Shirane was impressed both by Shull and by the work he was doing, and decided that he could figure out no better way to advance his scientific career than to hang around Shull and pick up whatever he could. Besides Shirane, two other young associates were smart enough to have reached a similar decision: the already-mentioned Robert Nathans (another Pepinsky recruit from Penn State) and Tormod Riste from Norway.

> It was wonderful [Shirane recalled]. It happens only once in your lifetime, if you are lucky; you are young and fresh and just happen to meet by chance somebody who is really great. The only trouble was, Shull would rather work alone, though doesn't mind teaching. And there were three youngsters wanting to learn from him—me, Nathans, and Riste. That was hard. Only one or two people could really work at once with him, and it was his experiment and he had all the ideas

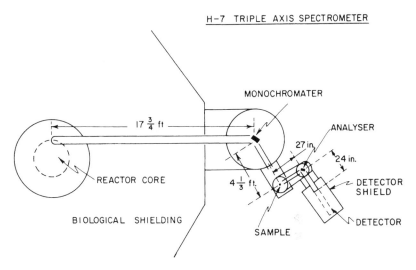

12.9 Drawing of triple-axis spectrometer at H-7.

and you couldn't learn anything until he showed up. I had a strategy. I tried to show up just as Cliff showed up at the reactor, and if he was alone I would go work with him. If one other guy was there, I would join them, so that's Cliff and two youngsters. But if the two other guys were already there, I wouldn't even try, I'd just go home. That strategy worked extremely well.

At the time, solid-state physics was regarded as an inferior cousin to high-energy physics at Brookhaven, and there were no opportunities for Shirane to remain at the lab. He left for the Westinghouse research lab outside Pittsburgh, one of the best industrial laboratories, which had its own materials-testing reactor where Shirane could conduct neutron-scattering experiments. Westinghouse had planned a research reactor that would vastly improve the neutron-scattering facilities, but abandoned the project early in 1962, in a widely publicized decision that made the front page of the *New York Times*. Physics department chair George Vineyard saw the article and called Shirane, telling him of Brookhaven's commitment to a new reactor, and successfully recruited him.

He arrived in 1963 and was assigned to Dienes's solid-state group, now much expanded and with a bright future. At the soon-to-be-completed HFBR, the group shared three facilities, H-4, H-7, and H-8. The facilities at each hole were slightly different: H-4 was optimized to support very heavy accessories, H-7 had a lower background, and H-8 had a slightly higher energy. While at the beginning H-7 had the only triple-axis spectrometer, a number of others were soon converted (fig. 12.9).

No one guessed how big a field inelastic scattering was going to be [recalls Shirane]. It became a major field at Brookhaven for two reasons. The first was the HFBR, which gave a factor of ten increase in the neutron flux; the second was that a new type of crystal was invented around that time, a pyrolytic graphite monochrometer which can be shaped so that it not only reflects but also focuses the beam, which gives you another factor of ten. A factor of ten is a big factor. So all of a sudden you get a factor of a hundred.

Shirane keeps several of these crystals in his office. Each is a black and glossy lens-like surface an inch or so square. "Fantastic stuff," he said, lifting one carefully out of its box.

This one costs $10,000. For what it does, $10,000 is *cheap*. The old way, we made three phonon measurements in a given year. With these, we can do thirty-five in that time in much more detail. It made a very difficult type of experiment easier, and by doing so opened up a new field.

Shirane achieved a level of skill at triple-axis spectrometry that has rarely been matched, and applied the skill to a number of issues involving inelastic scattering. He liked working in collaboration with one or two others, and his early collaborations with Robert Birgeneau proved especially fruitful. While Shirane was deeply focused on technique, Birgeneau was proficient in both theory and experiment and had an ability to identify model systems whose exploration would shed light not only on themselves but on related systems as well. One involved one-dimensional magnetic materials, whose magnetic ions are arranged in one long chain so that each ion interacts mainly with its fore and aft neighbors, and each chain is relatively insulated from the next. Theorists were extremely interested in such systems because a number of many-body problems that were hopelessly intractable in three dimensions were solvable in one, allowing the building and testing of models. In 1971 Shirane, Birgeneau, and others began exploring one example of such a system, tetramethyl ammonium manganese chloride or TMMC, using neutron scattering, taking one of the first steps in what would soon become a large field.[24] Shirane, Birgeneau, and Roger Cowley also studied two-dimensional systems, or planes whose atomic moments were separated so much that they acted like two-dimensional entities, which again interested theorists because of the prospect of model building.

Another important effort of Shirane's involved understanding certain kinds of phase transitions. Phase transitions, which include the melting of ice and the boiling of water, are familiar phenomena in nature. What happens when a phase change takes place—substances metamorphosing into different forms—is the sort of mysterious question that can inspire child-like wonder even in seasoned scientists. One important phase transition occurs in magnetic materials in the transition from paramagnetic (i.e., disordered) to ordered (ferromagnetic or antiferromagnetic) states. How this process takes place was vital to understanding what makes one material ferromagnetic and another antiferromagnetic. Another kind of phase transition, structural phase transition, involves soft phonons; when a structural phase transition occurs, the phonons or lattice motions get slower and slower and finally "freeze out," with the atoms locking into a fixed position. The study of this process was a major focus of Shirane's work.

H-5 and H-6

Two solid-state holes, H-5 and H-6, were assigned to the chemistry department. Corliss and Hastings moved their collaboration to H-5, where they continued their work on the magnetic properties of solids, especially the behavior around the critical point. In their first years at the HFBR they sought out and explored magnetic systems that were experimentally challenging and of theoretical interest.[25] These studies led to the determination of the critical exponents associated with second-order phase transitions. H-6 was the domain of Walter Hamilton, whose introduction to Brookhaven was via a postdoctoral appointment with Corliss and Hastings and who within a short time became a legendary figure at the HFBR.

Hamilton, a crystallographer, was a commanding presence. Texas born and Oklahoma raised, he was over six feet tall and deaf in one ear with a booming voice. He would often crack up his audience at the beginning of talks by setting the microphone aside with the friendly words "Don't think I'll be needin' this!" He had endless energy, was an avid hiker, and whenever he attended a conference would seek out a nearby mountain to scale. He seemed in the thick of every discussion, every subject he touched seemed exciting, and he had an intuitive grasp of even remote aspects of his field. He attracted many younger assistants, whom he cared for well and looked after. One collaborator, in a remark typical of others, would later say that working with Hamilton was "the high point of my crystallographic life" (Kamb 1983, 337). When describing Hamilton, those who worked with him rarely fail to use the word "genius."

Hamilton grew up in Stillwater, Oklahoma, and in 1950, at the age of nineteen, graduated from the Oklahoma Agricultural and Mechanical College, where his father was a professor of mathematics. He did graduate studies at Caltech, a center for the structure determination of chemical compounds, where Linus Pauling worked. Hamilton earned a Ph.D. from Caltech in 1954 with a thesis concerned with electron diffraction, and spent the next year on an NSF postdoctoral fellowship working at Oxford with

Charles Coulson, an important chemical theorist. When Hamilton arrived at Brookhaven as a postdoc in 1955, he joined Corliss and Hastings at the BGRR, and quickly added neutron diffraction to his skills.

The field of crystallography involves handling massive amounts of numbers (it has provided a significant impetus for computer development), and its practitioners often rely heavily on tables, preferring techniques where one value can be inserted and another extracted without having to worry about what happens in between. Hamilton was different. Steeped in mathematics from his youth onward, he had a profound understanding of the mathematics and the physical properties on which the routine work in the field was based. He spent much of his career developing and improving methods for the computation and refinement of structures and the analysis of diffraction data. One of his papers on methods of crystallographic interpretation, written in 1964 and entitled "Significance Tests on the Crystallographic R Factor," was once at the very top of the Science Citation Index and is still cited today (Hamilton 1965). He was also versatile at computing. During construction of the chemistry building in 1965, the architects ran into difficulty figuring out how to arrange the bricks in the walls of the lecture hall, constructed in the form of a truncated cone; Hamilton promptly developed a computer program to solve the problem. He was also the driving force in the development of the single on-line computer that controlled the HFBR spectrometers. By then he was one of the most prominent scientists at the lab; in 1968, when Friedlander became the chemistry department chairman, he made Hamilton deputy chairman.

At the HFBR, Hamilton was assigned H-6, which had two ports. At one was a double-axis spectrometer used for elastic scattering; at the other was a triple-axis spectrometer, which would also be used for inelastic scattering. Both spectrometers had a chi circle, or device that could orient a crystal sample in any particular direction in space. If a crystallographic problem could be addressed using neutrons, Hamilton used one of these two instruments to tackle it. When a Caltech crystallographer approached Hamilton with a puzzle concerning the structure of high-pressure ice, Hamilton promptly joined him in a collaboration. For the first time, thanks to the HFBR's neutrons, they were able to detect the proton position in high-pressure ice (Kamb et al. 1971).[26] When a controversy broke out over whether a hydrogen atom that had a chemical bond to a metal had a normal bond distance or was somehow buried in the metal electrons, Hamilton and his coworkers decided the issue (La Placa et al. 1969).

One of his most important programs concerned amino acid structures. Hamilton foresaw the long-range need to get the best possible structures of the amino acids, including all the hydrogen positions, to construct a kind of directory to help determine the structure of proteins. Together with

Thomas Koetzle, a postdoc from Harvard, Hamilton mounted a major effort in determining amino acid structures at Brookhaven.

By the end of the 1960s, Hamilton was not yet forty but at the peak of his profession. He thrived amid all his projects, even seeming more youthful than ever. His appearance changed; while until 1968 he had short hair and looked like a marine, thereafter he let his hair grow long and unruly, adding a broad, friendly beard. He was one of the central figures in the profession: coeditor of *Acta Crystallographica,* a key member of many committees, and in 1969 the youngest president of the American Crystallographic Association. For many, he personified crystallography.

Early in 1971, one of a series of meetings called the Cold Spring Harbor Symposia on Quantitative Biology was held at Cold Spring Harbor, organized by James Watson around the subject of protein crystallography. Those who were on hand recall that something of a sea change took place among those who attended. It was not just that the great Nobel laureate James Watson (who, with Francis Crick, discovered the structure of DNA) was sanctioning protein crystallography as part of biology; there was a sense that it was a field about to explode. At the meeting, Hamilton was approached by Helen Berman, a postdoc from the Institute for Cancer Research in Philadelphia. Berman was one of a small group of crystallographers, which also included Edgar Meyer at Texas A&M, who had been agitating for a special database for proteins. Crystallography was not being deluged by protein structures; exactly seven were known. Also, crystallography, historically a close and well-organized discipline, already had an important database, the Cambridge Crystallographic Data Centre (CCDC), which had been founded in 1965 and was ably run by Olga Kennard, one of the movers in the field. But Berman and company were convinced that the number of known protein structures would soon skyrocket. They also felt that while the CCDC was suited for small molecules, databasing much larger structures like proteins would pose a special set of problems requiring larger data handling, larger amounts of storage, and a state-of-the-art computing facility. Berman knew the project required some politicking to get off the ground.

Hamilton was the right person to approach. He not only had the contacts and interest, but also the authority to compel people to pay attention to the subject and trust that it was important. He was already working on a grant proposal to establish a computer network for crystallography, the Crystallographic Computing Network (CRYSNET). Hamilton incorporated the Protein Data Bank (PDB) concept into the grant, which involved one central site, Brookhaven, and two remote sites, College Station, Texas (Meyer), and Philadelphia (Berman), and the grant was submitted in the fall of 1971. Berman began making the three-hour trip from Philadelphia to Brookhaven about once a week to help set up the database. By then, the basic organization was effectively in place, with the seven known protein structures (Schoenborn's sperm whale myoglobin would become the first new protein added, the eighth overall, in May 1973). Hamilton and company began organizing to start public distribution in the summer of 1972, produced a standardized format, and developed a computer graphic system to store and disseminate the structures.[27] "The reason the Protein Data Bank ended up at Brookhaven had little to do with the obvious fact that neutrons were good for studying proteins," says Koetzle. "It had *something* to do with the fact that we had a state-of-the-art computing center, as well as a good computing infrastructure for the time. But mostly it was because we had Walter."[28]

In the fall of 1972, Hamilton went on an around-the-world trip that included talks and conferences in Denmark, England, and Japan, where he climbed a mountain in Hokkaido, but a last-minute meeting kept him from climbing Mount Fuji. When he returned, he admitted to not feeling well, though the last thing he had time for was a doctor. In November he went to Hawaii for a meeting, where he went snorkeling and did more hiking. On his return, his unprecedented complaints about his health were even worse. Koetzle and Berman nagged him to see a doctor, which he finally did in December. Tests revealed colon cancer, far advanced.

Hamilton was admitted to New York University Hospital a few days later, just before Christmas, and was operated on the following day. From then on, every piece of news about his condition was worse than the last. Still, Hamilton tried to disguise his condition from associates; phone callers could pick up little from the familiar booming voice he now had to affect. Only those who visited the hospital regularly, like Berman and Koetzle, knew how ill he really was. For them, it felt like a bad dream; neither could understand how anybody so full of volcanic energy just a few weeks before could possibly be so catastrophically sick. The sight was almost too much to bear, or even believe.

In mid-January 1973, the American Crystallographic Association held a meeting at the University of Florida in Gainesville. A few days before, from his hospital bed, Hamilton talked to Berman and Koetzle of going, and even made a poster that he said he wanted to present, while they spoke of seeing him there. By then he was too physically incapacitated to get out of bed. Berman stayed behind at the hospital, while Koetzle went to Gainesville. By the time Koetzle returned, Hamilton had lost his lucidity, and within a few days slipped into a coma. He died a few days later, on 23 January.

Meanwhile, the CRYSNET grant containing the Protein Data Bank had come through. Today, it is still at Brookhaven, a world-wide resource containing some five thousand entries. But in the second quarter-century of the lab's history, Hamilton's death would continue to exert an impact on the PDB.

H-9

H-9 was a fiasco.

When the HFBR's beam tubes were laid out in 1958, the designers envisioned a source for "cold" neutrons at H-9, those with lower energies and longer wavelengths than thermal neutrons. Cold neutrons are superior to thermal neutrons for studying low-energy collective motions, and relatively large-sized objects like biological tissues. Today, cold neutron sources are standard; it would be unthinkable to design a new research reactor without one. At the time, it just seemed like a good idea.[29]

In the early 1950s, Hughes and Palevsky had generated their beam of cold neutrons at the BGRR by simply filtering out the more energetic ones, but this produced an extremely weak beam. By the end of the decade, a far more effective technique was being developed that involved using a vat of liquid hydrogen to moderate the beam, which shifted the entire spectrum of energies downward, producing a much greater intensity. A neutron banging into a very cold (low-energy) proton tends to transfer more of its energy to that proton than a neutron hitting a room temperature proton vibrating with customary thermal energies.

Palevsky took nominal charge of the project but did nothing about it for a long time; his interests lay elsewhere and the engineering challenges involved were daunting. Liquid hydrogen, highly explosive in contact with air to begin with, becomes still more volatile in the presence of ozone, which is created when neutrons interact with air. Placing it near a reactor's hot core raised serious safety concerns. Sailor found the idea such "an appalling engineering task" that he suggested building an entirely separate reactor just as a cold neutron source.[30]

When Kevey had finished the US-1 spectrometer project, Palevsky asked him to design a liquid hydrogen refrigerator to go in the H-9 hole, telling him that the French had worked out all the details in connection with a cold neutron source at their Saclay reactor. Costs, Palevsky said, would be modest; about $60,000 to $80,000. Palevsky neglected to tell Kevey that the Saclay reactor had a much lower power than the HFBR, that the amount of liquid hydrogen they were using was tiny, and that the French had a notoriously cavalier attitude toward reactor safety. But Kevey gamely began to tackle the engineering, scaling up the Saclay design. It did not take him long to realize the project was almost hopeless, due to pump reliability and the huge inventory of liquid hydrogen that needed to be pumped all the way into the beam hole and then out again.

Meanwhile, Palevsky's group began building a slow chopper to station at the hole. The slow chopper was a one-of-a-kind device designed with a tricky arrangement of two rotors to increase accuracy. The device had problems almost from the beginning. Some were mechanical; in 1963, during a test, the rotor suddenly failed at 9,500 rpm, destroying the housing and much test equipment. Others were design flaws: when the rotor was rebuilt to handle the stresses, the machine never worked right. A defect in the motor-generator system adversely affected the speed control unit, and the two rotors were rarely able to stay in phase.[31]

Robert Nathans then entered the picture. Nathans was an experimental impresario; he was ambitious, dreamed big, and was never intimidated by the magnitude of a task, for he felt that if somehow he got into something over his head he could recruit enough talent to pull him out. Nathans decided that the scientific rewards of a cold neutron source were worth the effort and that he wanted control of the project. "I don't ask; that's not my style," says Nathans. "I just take, and see if anybody protests."

Palevsky was only too happy to be rid of it, and Nathans would soon feel like an advertisement for the old adage about having to be extremely careful what one wishes for lest the wish be granted. He made two assumptions that would come back to haunt him. He assumed that he had a commitment from Shutt's cryogenic group in the physics department, the only group on-site with experience in handling liquid hydrogen in large quantities, to help him out. And he assumed that since he was using the technology of a major cold neutron project at Harwell, where an extensive safety evaluation program had already been conducted, he did not need to arrange for safety evaluation studies.

Both assumptions fell through. The cryogenics group said it was too overloaded with its own work to spare any help. After major remonstrations, it finally agreed to lend one of its members to supervise construction of the cold neutron source's cryogenic system—but the person they sent was not up to the huge task now thrust upon him. And with reactor and liquid hydrogen safety concerns on the rise, the AEC declared that the Harwell studies would not suffice and that Brookhaven would have to carry out an entirely new set of safety studies. (An important contributing factor was an explosion, on 5 July 1965, at the forty-inch liquid bubble chamber at the Cambridge Electron Accelerator, killing one person and causing over million dollars in damage [Galison 1997, 353–62].) Few guidelines or precedents existed for cryogenic safety at reactors, and every time the cold neutron group made a move, the AEC officials, growing ever more safety conscious, found fault with it. The unexpected safety studies alone ate up most of the initial budget.

Then, in 1968, Nathans abruptly walked away.

I'm an opportunist, and there's only so much time in your creative career. The cold neutron source was a tougher technical problem than we thought, especially concerning the cooling system. And then the AEC started hounding us about safety, and asking us to file report after report. Remember, we were antibureaucracy and not used to that. We did what we thought was right, and Washington was Mars! I said, I ain't writing a goddamned report to anybody, I wanted to get on with the physics, and if I can't do that I'll go on to somebody else. I'm not going to spend the rest of my life worrying about a goddamned piece of apparatus. So I dumped the project. I don't piss into the wind; that's one of my talents. I got better things to do. And I know that what I did was bitterly resented by the others at the time.[32]

Nathans left the lab and accepted a position with the State University of New York at Stony Brook. The overwhelming, undersupported, and now leaderless project dragged on without him, its entire budget exhausted before construction had even begun. Meanwhile, the group working on the slow chopper discovered that its speed control problem could not be fixed, though they managed to nurse it along for about two years. But the effort was proving more trouble than it was worth; to operate the slow chopper required more technician time than was used up by all the other spectrometers combined. It was finally junked and a triple-axis spectrometer installed in its stead.

In 1972, physics department chairman Joseph Weneser was able to turn an unfortunate situation in another department into a stroke of good luck. At the time, Ralph Shutt was clashing with accelerator department chairman Fred Mills. The conflict was so unpleasant and destructive that lab officials transferred Shutt out of the accelerator department into the physics department. Weneser then coaxed Shutt and his group into working on the cold neutron source. In 1972, two members of the cryogenic group undertook a thorough review of the cryogenic system, and came up with a neat, well-calculated conceptual solution to the problem.[33] They suggested not circulating the liquid hydrogen in and out the reactor, but liquefying it inside the reactor—which at a stroke cut the amount of liquid hydrogen involved from 40 to 50 liters to 1½. Once this idea was accepted, the design and engineering of the cold neutron source could begin in earnest. Over the rest of the 1970s, Shutt applied his meticulous engineering skill to bring the project to completion, though on money quietly borrowed from other projects (mostly by paying those involved out of the operations budget), a practice sometimes, and not unsensibly, resorted to on occasion but rarely to this degree. The HFBR cold neutron source finally became operational in 1980.

AFTER THE HFBR was completed, Brookhaven's nuclear engineers talked about the possibility of another reactor. The next research reactor should have a significant increase in the available neutron flux. But Hendrie and others in the nuclear engineering department did not see any realizable way to improve the power density capability of an HFBR/HFIR-type reactor by a significant factor. One possibility was to build a "pulsed" device, which delivers its neutrons in extremely intense bursts, but with a small duty cycle.[34] If one generates a 100-millisecond-long pulse of neutrons every 10 seconds, it can be run at 10,000 times the power density of a steady-state machine and achieve a flux of about 10^{16}. (Arguments about the relative merits of pulsed versus continuous neutron facilities continue today.)

Hendrie became nominal head of the project, which dragged along for a few years on reactor physics money. At the time (the mid-1960s), about a dozen reactor prototypes were being worked on at various places in the U.S. Every national laboratory, and several major private laboratories with AEC contracts, had a power reactor concept, and some two or three: Oak Ridge was working on molten salt, Argonne on advanced boiling water, Atomics International in Los Angeles on sodium, while several laboratories toyed with schemes to use organic fluids with good heat transport properties. But at the AEC, Milt Shaw, the head of the division that developed new reactors, had become "not only the hardest working but also the most despised administrator in the history of the AEC" (Weart 1988, 306). Shaw loathed the loose research atmosphere in the national labs and promoted, with the accord of AEC administrators, the idea that all these more exotic types of reactor designs were draining money away from the AEC's more important long-range goal: fast breeder reactor development. Beginning around 1964, several other reactor development programs were terminated to consolidate resources to that goal, including Argonne's A^2R^2 (which was far along to the tune of $80 million; Holl 1997, 257–59, 268) and Brookhaven's much more preliminary work on plans for a pulsed reactor.[35] "It was a decision," Hendrie says, "that brought to an end a grand time for national laboratory reactor engineers." Hendrie had to fire about half of the scientific and professional staff of the engineering division—and had to be clever about how to support who was left, taking advantage of revisions in the Atomic Energy Act allowing outside funding:

> We had all kinds of schemes. We got into rock bolts as an aid to mine stabilization because it looked like the Bureau of Mines had a little money. We got into polymer-strengthened concrete as a structural material for big desalinization vessels because it looked like the Bureau of Reclamation had a little money. Jim Powell had a scheme for using a technique that had been developed for reading cloud chamber pictures for putting targets on the walls of mines and then you'd scan them with this device and if they moved it would tell you things were about

to happen. Things like that kept us alive for about three years while we waited for something to happen to the pulsed machine idea.[36]

The pulsed reactor project did not crystallize. Most reactor users were not that enthusiastic, because the pulsed operation limited the types of experiments that could be done. "There was hardly a tumult of enthusiasm for it," says Hendrie. "And even if there had been, whether one could have gotten development work funding, let alone construction funding, in the climate of the times was another question." [37] Tired of the fight to keep the department alive, Hendrie accepted a job in Washington as the first chief engineer for the AEC's nuclear regulatory staff. He left in May 1972.

Meanwhile, the RILL (Reactor Institut Laue-Langevin) in Grenoble came on that year. Based on Chernick's design for the HFBR, with an undermoderated core and tangential beam tubes, its flux was about 30 percent higher than that of the HFBR. More significant, its accessories were also well designed: it had both cold and hot sources, as well as a new kind of high-speed chopper, developed in Germany, that relied on a magnetic bearing suspension system in which no mechanical contact took place between rotor and housing. And the Grenoble facility clearly had more resources than Brookhaven for developing and upgrading its instrumentation. Though the RILL experimental program was managed more formally and inflexibly than the HFBR, this was outweighed for many experimenters by the advantages. For Brookhaven's reactor experimenters, as for its accelerator experimenters, 1972 was a bittersweet year: both groups saw their utilities surpassed by a superior version with no significant new one on the horizon.

The HFBR would also exert a major influence on the lab's history during its second quarter-century. In 1994, a small electrical fire at the Tristan experiment stationed at H-2 resulted in minor contamination to the building and to several emergency personnel, but raised major concerns about the lab's experimental review process and about safety at the laboratory in general. And in 1997, at the beginning of Brookhaven's fiftieth anniversary year, the announcement of the discovery of a small leak of tritium-containing water from the spent fuel pool of the reactor helped trigger a dramatic chain of events, including the cancellation by the Department of Energy of AUI's contract to manage the lab, the introduction of legislation in Congress to prevent the reactor's restart, and the search for a new contractor—in the process signaling what is no doubt a turning point in U.S. science policy.

CHAPTER 13

Crossroad

A LUCKY leader is one whose strengths coincide with the abilities needed to address the problems one faces; in an unlucky leader, these abilities coincide with weaknesses. Maurice Goldhaber was a lucky leader for the first half (1961–1967) of his directorship, and an unlucky one the second (1967–1972). In the immediate wake of Sputnik, the science-government-industry partnership forged during World War II was revitalized. But by the end of the decade, the enthusiasm for basic research had crumbled, funds for scientific research and development had fallen for the first time since World War II, and the ties that bound the partnership had begun to unravel. Successes of the space program made NASA, not the beleaguered AEC, the agency symbolic of science. More and more, scientific leadership required exercising the skills of a promoter and a politician, skills in which Goldhaber was untutored and uninterested.

Goldhaber's last years as director were the bleakest yet for postwar U.S. science. Policy analyst Bruce L. R. Smith has called 1966 the year in which the threshold was crossed into a new, "more troubled" era in the relationship between government and science:

A darker vision replaced the innocence and optimism. Regulating dangerous side effects of technology became an urgent challenge. Such social priorities as preserving the environment and protecting consumers became important objectives. . . . The assumptions of science policy became subject to intense scrutiny and doubt. Although cushioned somewhat from the shocks affecting the political system by virtue of the esoteric nature of their work, scientists were forced to confront an insistent and clamorous attack on premises that had once appeared self-evident. (Smith 1990, 71–72)

The role of science in society was hotly debated on many fronts. Many liberals criticized science for its alleged remoteness from human needs (e.g., for being dehumanizing and failing to contribute to human self-understanding), for its political role (e.g., for its ties to the defense industry in projects related to the war in Vietnam, to the segregationist government of South Africa, and to Defense Department advisory groups such as Project Jason), and for introducing new hazards into the environment (e.g., radioactive substances and toxic chemicals). Such critics no longer viewed science as the means by which society identifies and overcomes dangers to humanity, but rather as the powerful agent of the "establishment."

Conservatives, too, objected to science's alleged remoteness from human needs (e.g., for failing to deliver on promised spinoffs). They, too, distrusted its political role (e.g., for the perceived liberal or leftist tendencies of many scientists). To this set of critics, many scientists were unpatriotic for their outspoken opposition to the Vietnam War, as well as to such military projects as the antiballistic missile system.

Under the force of these attacks, science was being inadvertently reshaped. The rationale for the postwar science-government-industry partnership had been articulated in Vannevar Bush's report, *Science: The Endless Frontier*, which described the relation between basic research and applications by analogy with a bank or common fund—a necessary investment that grows over time, from which withdrawals may be made to support specific projects, and without which the money for withdrawals eventually dries up. Money invested in science was thus sure to bring welcome returns.

> Basic research leads to new knowledge. It provides scientific capital. It creates the fund from which the practical applications of knowledge must be drawn. New products and new processes do not appear full-blown. They are founded on new principles and new conceptions, which in turn are painstakingly developed by research in the purest realms of science. (quoted in Smith 1990, 43)

By 1966, many no longer found this analogy convincing. Funds allocated for science now needed to be scrutinized and defended, with guarantees that the money be well spent. That June, President Lyndon B. Johnson, speaking at the dedication of the National Institute of Medicine, announced that the federal government needed to refocus medical research on obtaining quick results: "The time has now come," he said, "to zero in on the targets to get our knowledge fully applied" (Smith 1990, 75). This view was echoed in Congress by Representative Emilio Q. Daddario (D.-Conn.), a member of the House Science and Astronautics Committee and chairman of the Subcommittee on Science, Research and Development—an influential voice among those questioning traditional arguments for basic research.

> When scientists insist that their work is pure and devoid of application motives, they are naive to expect substantial portions of our tax revenues to be devoted to their projects. These same scientists sometimes represent science as a peculiarly noble human endeavor, ranking with the fine arts in challenging the intellect and talents of man. Within the framework of our political system it *is* difficult to justify expenditure of large amounts of public funds for the purely personal satisfaction of curiosity—merely for the sake of knowing.[1]

In March 1966, Daddario introduced the first version of a bill to alter the charter of the National Science Foundation to allow it to support applied as well as basic research. The corresponding Senate bill, introduced by Edward Kennedy (D.-Mass) easily passed. The Daddario-Kennedy bill was signed into law by Johnson in 1968.

Basic researchers have sometimes found it convenient to justify themselves by citing not curiosity or the sheer desire for knowledge but the (often unexpected) applications that have arisen from their work—a justification that seemed readily defensible. During the 1960s, this strategy became increasingly common in congressional testimony, leading one Brookhaven administrator (Martin Blume) to nickname it the Yalu River defense, alluding to the border between North Korea and China, behind which Chinese troops would scurry for safety when pursued during the Korean War, much to General MacArthur's chagrin. The national labs even began to reshape their research programs to emphasize applied research. Brookhaven's nuclear engineering department committed itself to applied programs, and at the end of 1968 a decision was made to rename it the department of applied science. In 1971, President Richard M. Nixon's science advisor, Edward David, asked Brookhaven's deputy director, Ronald Rau, for a list of practical applications resulting from basic research at the lab to help him support the case of the national labs in Congress. Rau provided a thirty-page description that ranged from new types of physics instruments, to development of tritiated thymidine for biology research, technetium 99m (then, as now, the most widely used radioisotope for medical scans), and the development of the first therapy for Parkinson's disease. Laboratories, especially those engaged in basic research, are often approached by skeptical outsiders seeking proof of the practical value of their product; what was new was that the person needing to be informed was the presidential science advisor.[2]

By 1968, conflicts about the role of science were reaching the scientific community. At the American Physical Society's winter meeting in Chicago, protesters from the scientific community disrupted talks by invited speakers from weapons laboratories, and argued that the society should take a stand against weapons work and against the administration's policy in Vietnam. A new insecurity arose inside the national labs. Not only was federal support openly questioned and the relevance of the work challenged, but, most disturbingly, the work itself was regarded with suspicion. The question, "So do you make bombs?" and the fears about radiation were no longer funny—no longer likely to reflect innocence or ignorance, but rather to indicate of deep suspicion of the role of science in society. It was growing more difficult than ever to maintain a dialogue between the scientific community and the public in which consensus on issues such as safety and relevance could be established.

Though Brookhaven had had trouble creating that dialogue since its inception, its difficulties deepened in the 1960s. One reason was the growing change among the public from what sociologists have called the human exemptionalist paradigm, in which human beings are seen as exempted from nature and its constraints, to the ecological social paradigm, in which human beings are viewed as part of nature, conceived as having limited room and resources, and with modern industrial-technological practices threatening to destabilize what should be a harmony in nature (Catton and Dunlap 1980). In particular, the public was becoming familiar, through nonfiction (e.g., Carson 1962) as well as fiction (Vonnegut 1963), with dramatic descriptions of how small and apparently innocuous effects introduced by human beings (especially scientists) into the environment might connect with other processes to give rise to major and even disastrous consequences. Environmentalists had a new political clout, as revealed in such actions as cancellation of the supersonic transport (1971). Aware of this, the AEC tightened the monitoring and evaluation practices at national laboratories.[3] But convincing the public of the safety of the small amounts of certain contaminants still released into the environment was often complicated: a knowledge of statistics, geology, and background effects was often required, which could be hard for scientists to do in the sound bite format demanded by politicians and the media. The day was gone when a Willy Higinbotham could calm the fears of a farmer by pulling out a Geiger counter to prove that ducks or cows were not radioactive.

Another factor was a growing demand for dissemination of information about, and public participation in, risk-related processes (which received an impetus from the National Environmental Policy Act of 1969), coupled with a growing skepticism, in the Vietnam War era, of government and authority. One result of the latter was the media visibility of people like Ernest Sternglass, a professor of radiation physics at the University of Pittsburgh. Sternglass made something of a career of alleging correlations between low levels of radioactivity and everything from high infant mortality rates to low Scholastic Achievement Test (SAT) scores. Sternglass, according to a recent review—though admittedly one carried out by Brookhaven scientists—did not "follow accepted scientific methods, never calculated the dose equivalent to the population studied, misinterpreted the raw data, and did not evaluate any possible confounding factors" (Musolino et al. 1995, 475). A scientist whose work was as repeatedly discredited as Sternglass's would be annihilated—but that was not the game Sternglass was playing, and he flourished as an expert with an alarming message on TV and radio. A science writer, Phillip Boffey, attributed this to the fact that "Sternglass makes good press copy—he has a startling theory that relates to important public issues," and was attuned to a deeply felt public mood: "the revulsion against the military, the desire to end contamination of the environment, and the tendency to disbelieve the rosy reports emanating from government agencies" (Boffey 1969, 199). That mood was strongly held on eastern Long Island, once farm country, where Brookhaven National Laboratory was not merely an interloper, but a government interloper. For some, as if in the grip of some conceptual hysteresis fixed in 1945, it was as if all the social, political, and scientific transformations of the previous quarter-century had not occurred, and they viewed the lab's reactor as essentially a vestigial remnant of the Manhattan Project, a part of the military-industrial complex against which, in the sensibility of the 1960s, it was morally right to object.

In his article, Boffey posed to himself the question of what harm there is an a scientist's yelling "Fire!" when there may be none. Aside from stirring up unnecessary fears, it creates the danger of society's losing its ability to determine what the real threats to it are, and in squandering limited resources fighting the least dangerous threats while continuing to expose itself to more pernicious ones. If it becomes impossible to distinguish real from unreal threats, reach solutions to the real threats, and choose real solutions over only seemingly safe ones, then problems that could be resolved are guaranteed to continue. Inciting society to target significant resources on extremely minor environmental hazards can be politically popular but lead to what I would call an *environmental Maginot line:* harmful because it squanders resources on the wrong places, lulls people into a false sense of security, is vulnerable to political manipulation, and contributes only negligibly to the wider social purpose—public safety—that is the ostensible goal.

Still, when Sternglass was paired with a professional expert on radiation safety, Boffey pointed out, "it was impossible to tell who was right," because both cloaked themselves in plausible-sounding jargon (Weart 1988, 314). And why shouldn't an unsuspecting person trust Sternglass? Assurances by experts were even less convincing now than they had been in the previous decade because the public was all too familiar with the fact that some dangers escape detection by scientists for years. If, at the end of the 1950s, the public and its representatives began to fear that the authorities in charge of certain kinds of scientific projects were out of control, at the end of the 1960s a deeper worry began to dawn about whether technological society itself was out of control. The worry was no longer about "their" control over "it," but "its" control over "us."

The changing public attitude toward science was sharply felt inside

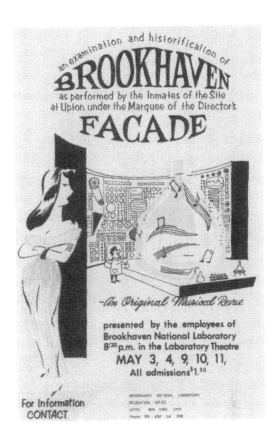

13.1 Advertisement for
Brookhaven/Facade, 1968.

Brookhaven. In 1968, the lab's Theatre Group staged an original musical
revue, *Brookhaven/Facade,* to mark the twenty-first year of the lab and the
fiftieth of Camp Upton (fig. 13.1). An introductory note to the script by the
revue's director called it "expressive of the spirit of Brookhaven," and
"primarily a humorous view of the prevailing attitudes here." [4] The revue
was a pastiche of silly songs, topical references and puns (the lab's master
plumber was the "drain brain"), that went out of its way to avoid political
statements. Still, for that very reason, it was a trustworthy gauge of the lab's
self-esteem, and of a new feeling of beleaguerment, isolation, and being
misunderstood.

In the first skit, a group of Druidic priests gather in a remote place called
Stonehank, a barren land of sand and swamps that supports only "scrubby
little oak and pine trees," chosen because it is equally inaccessible from the
priests' nine respective Temples of Wizardry. The priests have come to construct a new Pile of Stones, represented onstage by a pyramid-cum-tower, an

immediately recognizable send-up of the reactor. The priests immediately encounter hostility from the neighboring inhabitants of Patchenge, who do not believe assurances that, at Stonehank, the best brains in the country are working on peaceful research. The Patchengians fear not only that the priests are really chipping flint for weapons of mass destruction, but also that magic rays from the Pile of Stones will damage their crops, poison their waters, and even may blow up the countryside. They have also heard rumors that the priests are actually subversives. To the tune of Cole Porter's "Friendship," the neighbors sing:

> Radiation in our soil and in our crick
> Makes us sick
> We're selfish, selfish
> Don't kill all our shellfish
> And we don't want any scientists—
> They're all communists!

Later, during Roman times, successors of the Druids are forced to host a set of governmental officials who descend on the site for an inspection. What practical applications, the government officials ask loudly, do these devices have for the Roman Empire? When alone, however, the officials fawn over items of personal interest; Caesar is particularly enthusiastic about the Tandem Vanity Bath and the fourteen-foot Bubble Bath, which make "a really big splash." [5]

As at other academic institutions, a certain amount of mild political activism arose within the lab. This was one of the few times when the lab was roughly divided along scientist-nonscientist lines, with the activists (or doves, in the terminology of the day) tending to be scientists and their opponents (hawks) tending to be support staff. (A conspicuous exception was the medical department, several of whose members hailed from the navy, including the chairman, Eugene Cronkite, who was promoted to rear admiral in 1969.) On 15 October 1969, a national "moratorium" was held in protest against the Vietnam War. The policy of most national laboratories, including Argonne and Oak Ridge, forbade demonstrations, and a smattering of black armbands was the limit of protest. Fermilab's moratorium day activities consisted of a lunch-hour meeting of about twenty people in the auditorium of the director's complex, mixed with a number of counter-demonstrators demanding to know who paid for the memo advertising the event, and on whose time it had been circulated. Brookhaven had more activity than elsewhere, for Goldhaber adopted a lenient stance and allowed a group of lab scientists, who had formed a moratorium committee, to use Berkner Hall for a lunchtime meeting (fig. 13.2). The decision involved a deeper and more complicated management decision than it appeared to on

13.2 Moratorium day activity at Brookhaven, October 1969.

the surface, given that suppression often wound up exacerbating such conflicts. But Goldhaber was characteristically genial, open-minded, a little idealistic, and interested in running the lab in a more or less democratic way. He prepared a memorandum to deliver at the moratorium meeting: "I have allowed an orderly discussion in Berkner Hall because I believe that Brookhaven cannot be a great laboratory if its members remain forever isolated from issues which concern us all."[6] As a result, the lab had its own moratorium demonstration, with a large group of protesters in front of Berkner Hall, heckled by a large group of counterdemonstrators. The next few months witnessed a number of antiwar programs, and in April, the appearance of a four-page radical newsletter, the *Brookhaven Free Press* (fig. 13.3).

At the safe distance of a quarter century, the *Free Press* is a true artifact of the 1960s: it contained a mixture of theatrics, a large dose of self-importance and moral righteousness, and a certain amount of self-mockery and even fun. The first issue prominently displayed a cartoon on page one, captioned "Proud Parents," depicting two bald eagles, one labeled "Bill of Rights," the other "U.S. Constitution," gazing down at their tiny offspring, labeled "Brookhaven Free Press." Articles took stands against racism, sexism, antiabortion legislation, weapons research, and the draft. But the *Free*

THE BROOKHAVEN
free press

Statement of Purpose

The function of the FREE PRESS is to provide a free press; to publish information concerning the Brookhaven community such as union activities, civil rights actions, etc; to provide a forum for any discussion concerning the community such as the pension plan, local pollution issues, etc; to provide an outlet for grievances of general interest in an open and non-paternalistic fashion.

The aim of the FREE PRESS is to have a free and open forum. All written contributions (articles, jokes, excerpts) and cartoons will be published on a first come basis provided they are signed, typewritten and within the size limit. Any article by a non-lab member must be sponsored by a lab member. All names will be printed. The newspaper will expand to fill the demand. Any articles not published due to lack of space will be included in the subsequent issue. There will be no more than one free article per person per issue as long as there are any other available.

There will be no editorial board and no editing. The staff is completely open. All are welcome and encouraged to help. Defamation, personal abuse and general lack of decorum are strongly discouraged.

The FREE PRESS will be financed by contributions, subscriptions mailed to the home, and paid advertisements. There will be no charge for any written contribution. The distribution will be lab-wide.

All contributions must be typewritten in columns 4 inches wide and no more than 10 inches long (a maximum of 375 words, single spaced). Since we use photo-offset, the copy must be immediately photographable. Each copy must be signed and a telephone extension given for verification.

The written and monetary contributions are to be sent to the Brookhaven FREE PRESS, P.O. Box 395, Upton, L.I. Subscriptions costing $1 per three issues are obtainable at the same address. Paid advertisements are charged at four times the printing cost, or $2 per column-inch. Checks are payable to the Brookhaven FREE PRESS.

Volunteers are encouraged and solicited for the distribution and the work in general. They should write to the above address.

Support your local FREE PRESS through written and financial contributions.

We need each other.

PROUD PARENTS

Union News

During late 1969 and early 1970, Local 8-652, O.C.A.W. attempted to impress the management of Brookhaven National Laboratory with the necessity for having a forum to discuss the effectiveness of its current retirement plan. We solicited the assistance; by open letter, of all interested employees, in order to indicate to management that this was not a case of the 'tail wagging the dog', but a matter of deep concern with a majority of Brookhaven employees. You responded by giving us 1200 affirmative replies suggesting many proposals for changes in the system.

The Laboratory responded by refusing to create or assist in the creation, or joining in any discussion on any level, other than; the small group captive audience establishment-type meeting. They have said that they do not wish to create a rule by referendum at the Laboratory. They further indicated that they were satisfied with this paternalistic fiduciary method and had no immediate intention of changing it, regardless of how many people objected to it.

After four months of negotiations with the Laboratory, Local 8-652 broke off negotiations with management without ratifying a new contract. After having been threatened with lockout and the possibility of lay off, at their discretion and not by seniority, the Union requested,and received a stay of execution and now functions under the 1967 contract; temporarily.

Cont d

13.3 First issue of the *Brookhaven Free Press,* April 1970.

Press was also a forum in which apparently unrelated issues, problems, and gripes about the lab were thrown together, intertwined, and discussed in conjunction. The newsletter thus came to play what Michael D. Cohen and James G. March, in a book about the American college presidency, somewhat unfortunately call a "garbage can" role. A garbage can is a forum "into which various problems and solutions are dumped by participants" reflecting "the tendency for any particular project to become intertwined with a variety of other issues simply because those issues exist at the time the project is before the organization." Garbage cans perform an important function in an institution, Cohen and March argue, and "Provide garbage cans" is one of their rules for leadership. If garbage cans do not exist, Cohen and March write, they create themselves (Cohen and March 1974, 81–91, 211–12). The *Free Press* became a much-needed garbage can for the lab. Articles attacked the absence of day-care facilities, on-site meetings of the Naval Officers Reserve Group, the retirement plan, and the cafeteria's architecture. The editors seemed motivated by a spirit of fun as much as of politics, with a self-consciously antielitist policy proclaiming that the *Free Press* would accept "all written contributions." Its anarchic voice varied widely in content and coherence, including sometimes downright silly stories. One editorial urged coping with gypsy moth infestation by shunning pesticides and enlisting "boy scouts, girl scouts, students, and senior citizens" to remove the insects from public property by hand. A masthead? God forbid! The staff squeezed their names together in as non-hierarchical a way as Western linear writing would allow:

```
thestaffforthisissueiscompos
edofthefollowing:audreybiitt
nermaryhallmargarethindclair
ehullwilliamkeatingtomkitche
nssandylacksgeorgeslondonmar
ksakitternieurvatertedwerntz
```

The *Free Press* was designed in large measure to provoke. It succeeded. "Tune In!" the lab's anonymous suggestion box, was deluged by outraged letters demanding to know whether any lab resources had been sapped by the newsletter's production; some asked it be banned from the site. Allowing it to continue, complained one individual, "is putting a fuse in a powder keg"; another, noting that *Newsday* had published an article about the *Free Press* ("Press Freedom Comes to Atom Lab"), wrote that it was bound to turn public opinion against the lab, and that, as a result, "all of us may lose our jobs, through diminished public support for BNL." Cronkite refused to allow its distribution in the medical department. At the first meeting of department chairmen following the appearance of the newsletter, Goldhaber again took a relaxed stance: as long as the newsletter was not produced with any of the lab's resources, it could be left in bulk at standard mail drops.[7] Still unhappy, Cronkite took the day off the day the next issue was due to appear. After a dozen issues, the newsletter ceased publication. "It ceased being fun," said Mark Sakitt, one of those involved.

In truth, the nonconformist political activity on-site was mild, though it did not seem that way in conservative Suffolk County. And with lab employees tending to be more liberal and politically active than their neighbors, the Vietnam War was not the only issue to generate friction in local communities. Brookhaven staff members were a driving force in the foundation of the Environmental Defense Fund, and its certificate of incorporation was signed 6 October 1967 in a conference room at the lab, a key moment in the history of the U.S. environmental movement (Rogers 1990, 44). The Suffolk County Planned Parenthood League consisted mainly of Brookhaven employees. An incident involving racial tension among high school students was blamed on lab employees encouraging black activists.[8] Occasionally, the friction between conservative community members and "lab people" grew heated. The *Main Street Press,* a local newspaper, facetiously compared Brookhaven to Sweden as a favorite destination for draft dodgers.[9] This provoked one ultra-right-wing reader to write President Nixon that Brookhaven, "dominated by pinks, finks, and other democrat [*sic*] types [was] infecting Republican Long Island. . . . Correct this situation please." [10]

Nixon was already on the case, in the form of sharply reducing the number of federally supported U.S. scientists. The federal budget was tight, thanks to the huge cost of the Vietnam War. Federal research and development money declined in current dollars in 1969 and 1970, when, for the first time since World War II, steep cuts in the AEC budget forced staff reductions at the national laboratories. Goldhaber was forced into the uncomfortable position of having to announce that the lab would lay off employees; at the beginning of 1971, he announced delays and withholdings of salary increases for staff in FY 1972, with more layoffs in sight (layoffs also took place at other national labs; see, e.g., Holl 1997, 272). That summer, the union representing plant engineering and other groups called a nine-week strike, which also took a toll on morale.

During this period, the lab lost property as well as personnel. From the beginning, the lab had consisted of two tracts of the former Camp Upton, three thousand feet apart. All development had taken place on the southern tract, none on the 2,300 acre northern tract. In 1970, to curry political favor with local communities, Nixon directed the General Services Administration (GSA) to discover surplus government property and return it to local governments.[11] Brookhaven's north tract came to the attention of the GSA, which appointed a field study committee to see if its "current and extensive use" justified retention by the lab. "Some GSA New York people will be out on the 16th in this connection," wrote assistant director S. M. Tucker to George Vineyard, whom Goldhaber had made deputy director in 1967. "We don't know whether they are seriously interested or just going for a ride in the country." [12]

It was more than a ride in the country. On 24 June 1970, the field study recommended that "the entire North Tract consisting of approximately 2,299 acres be reported excess to General Services Administration." [13] The lab protested, but had little to offer in the way of current usage except for some of George Woodwell's ecology studies. More persuasively, it argued that the criterion of "current and extensive use" was inappropriate given the indefinite character of laboratory research; for example, it had a plan to build a future accelerator whose beam paths stretched onto the site. [14] Impressed neither by the existing ecology studies nor the future accelerator, the GSA declared the property surplus. In November 1971, in a ceremony at Berkner Hall attended by Governor Nelson A. Rockefeller and Julie Nixon Eisenhower, the land was turned over to the state, with five hundred acres going to Suffolk County. The county placed a prominent sign on the grounds of the former north tract advertising NIXON'S LEGACY OF PARKS. While in the 1968 presidential election, Suffolk County gave Nixon the largest margin of any county in the nation (96,000 votes), Nixon more than doubled that victory margin in 1972.

Yet another factor that, at the beginning of the 1970s, served to heighten sensitivities between the lab and the community was the Long Island Lighting Company's plan (mentioned in the previous chapter) to build a series of nuclear reactors, one in the nearby town of Shoreham. The hearings were nationally precedent setting and historically important for the U.S. nuclear power industry, with witnesses ranging from the country's most eminent scientists to, of course, Ernest Sternglass. The hearings over LILCO's proposed nuclear power plant increased local attention to possible dangers from Brookhaven's reactors, and—especially given Vance Sailor's active participation in a private group supporting the project—encouraged activists to attack the lab. The month the hearings began, one scientist received a phone call announcing that the Lloyd Harbor Study Group, an organization of private citizens opposed to the plant, would request an "investigation and survey of the records in the vicinity of BNL" after the Shoreham hearings as its "next move." [15]

Accusations that radiation emanating from Brookhaven was destroying the local environment continued to be lodged. Beekeepers blamed the destruction of a number of hives on radiation from the lab (a pesticide turned out to be the culprit). The declining osprey population on the east end of

Long Island was blamed on radiation the birds picked up flying over the lab (pesticides, again). A fish kill in a nearby lake was also attributed to radiation (studies showed deoxygenation the probable cause). The temptation of many who worked at Brookhaven was to blame public confusion on poor public relations. "I strongly feel that our Public Relations Department is exceptionally *incompetent*," wrote one frustrated lab employee, adding that "people on L.I. still believe we make bombs . . . we really do not have a P.R. Department." [16] But public relations was only a tiny part of a much bigger, culturewide failure to establish consensus on safety issues that would persist for decades. Nobody knew how to create a dialogue, neither at Brookhaven nor, it seemed, anywhere else.

In later years, Brookhaven's most serious problem with environmental contamination involved seepage of certain chemical pollutants (many of which were not considered hazardous at the time they were used and disposed of), a more genuine and less controllable danger than the small amount of additional radiation the lab contributed to the environment. But, as if to confirm Weart's thesis, the focus of the public, and the media, remained overridingly on radiation. The problem of establishing consensus on environmental safety grew worse: political passions and a certain amount of idealism subsided after the end of the Vietnam War, but antigovernment and antiauthority sentiment—as well as nuclear fear—remained. In 1997, discovery of a small leak from the HFBR's tritium-containing spent fuel pool triggered a series of events that included the shutting of the reactor—one of the nation's most valuable neutron-scattering instruments, and most dramatically, the DOE's cancellation of AUI's contract after half a century of operating the lab.

IN JULY 1972, Goldhaber informed the trustees that he was resigning as lab director, effective 31 December. He and the board had clashed on too many issues, and he was anxious to return to research. The board of trustees appointed a committee that produced a shortlist of candidates including Nevis Labs director Leon Lederman, SLAC director Wolfgang Panofsky, Yale physicist Jack Sandweiss, and George Vineyard. When the first two withdrew their names from consideration, the trustees considered the merits of a high-energy physicist (Sandweiss) versus a solid-state physicist (Vineyard) as director. While they ordinarily might have preferred a high-energy physicist, Vineyard (who had been chairman of the physics department from 1961 to 1966, associate director from 1966 to 1967, and deputy director since 1967) had strong internal support. Goldhaber had left to Vineyard the lion's share of responsibility for running the lab, and he had served effectively as the lab's executive officer. Vineyard was ultimately chosen.

In some ways, the lab had been a great success during the Goldhaber years. The success did not lie in any specific discovery or set of discoveries: associated production, the K_L, the omega minus, two neutrino discoveries, CP violation, or L-dopa as a therapy for Parkinson's, say. The success was rather that the lab had managed to remain vital and productive across many fields—not just physics—at the end of its first quarter-century of existence. It had fostered a wide variety of mutually supportive and continually developing experimental communities. Its internal structure remained flexible, and it was able to quickly establish interdisciplinary groups to capitalize on scientific developments—far more rapidly than would have been possible at a university, for instance. Its network of associations and collaborations had continued to grow, nationally and internationally, with new types of links to universities, industries, schools, museums, and other types of institutions. The experiment was working: Brookhaven, and its fellow laboratories, had shown that productive research could indeed happen in a special partnership involving federal funding and a university-type atmosphere. As science historian Robert Seidel wrote of the national labs:

> The government learned that it did not have to direct basic research to reap a harvest of applications. Scientists have found that research more productive thanks to the facilities placed at their disposal. Laboratory directors and scientific advisors learned to interpret agency missions in terms of scientific research programs that have real and permanent value in the quest for knowledge (Seidel 1986, 175).

Nonetheless, Brookhaven's future looked grim. Its relationship with the local community remained troubled. With the AEC's main original missions—weapons and power—no longer pressing, happy marriages between applied and basic interests were fewer. The lab's first-generation reactor and high-energy accelerator, the BGRR and the Cosmotron, had been shut off, while the second-generation instruments, the HFBR and AGS, had been surpassed by other instruments elsewhere. The AGS upgrade had been disappointing, taking longer, costing more, and resulting in less intensity than projected (fig. 13.4). Meanwhile, Wilson had built his 200 GeV machine, and in March 1972, Goldhaber would send him a telegram, "HEARTIEST CONGRATULATIONS FROM ALL YOUR FRIENDS AT BROOKHAVEN ON ACHIEVING 200 BEV." [17] Most ominous, Wilson was now talking about upgrading the Fermilab machine to 1,000 GeV, thus obviating the need for Brookhaven's long-planned next jump in accelerator energies to approximately the same energy. For most of the lab's early history, the diverse Brookhaven communities had experienced a focusing force in the effort to build or substantially renovate large machines—first the Cosmotron and the BGRR, then the AGS and the HFBR, then the AGS upgrade

13.4 Drift tubes in new AGS linear accelerator, the central feature of the upgrade.

and the Tandem Van de Graaff (completed in 1971). At the beginning of the 1970s, for the first time, that focus was lacking. Many of the lab's senior scientists were so concerned that the Brookhaven Council, Goldhaber's advisory committee, passed a resolution in April 1971 asserting that the trustees were not attending to the lab's future aggressively enough. Goldhaber had written inside his 1963 Christmas card:

> Speak of experiments in the past tense,
> of theories in the present tense,
> of machines in the future tense.

At the end of Goldhaber's directorship, the lab's new big machine remained in the far future tense.

The year 1972 was a transition point for the lab not only because of the change of directors, new competition from large facilities elsewhere, and lack of a new facility under construction. The politics both in and of science was changing.

The politics in science—the society inside the laboratory, supervised by the laboratory directorate—was affected by the increasing scale of scientific equipment and intricacy of techniques, which brought further subdivisions within experimental communities. The original vision of a university group able to bring its own detectors to the large facilities—clients feeding off a utility—was obsolete. A "user-type" atmosphere was retained for a time by transforming the detectors themselves into user facilities, incorporating members of Brookhaven staff into the group. But as the scale of the largest detectors kept growing, so did subdivisions and clashes of interest within these experimental groups. This happened even to experimental groups that weren't accelerator or reactor based, such as the interdisciplinary communities forming around superconductivity and PET scan research. Galison has written provocatively of "trading zones" that develop in such circumstances, intermediate domains between experimental subcultures "in which procedures could be coordinated locally even where broader meanings clashed" (Galison 1997, 46). Around 1972, with the transition between the (now) 33 GeV AGS and the 200+ GeV new accelerators elsewhere, and the new interdisciplinary groups, these trading zones grew more numerous, closer together, and more dependent on each other.

The politics of science—the relation between the laboratory and the rest of society, supervised by AUI—was also changing. If life at a reactor or accelerator was growing more complex, so was the politics of planning, approving, and building one.

Within a few years, the Atomic Energy Commission—which had supervised the national labs from the beginning, and whose commissioners always included at least one scientist—would be replaced first by the United

States Energy Research and Development Administration (ERDA) and then by the Department of Energy, a much larger and more bureaucratic institution. Pressure for applied research was as intense as ever—scientists were now firmly encamped behind the Yalu River. The increasing dependence on Washington was making it ever more difficult for the lab to encourage and support "grassroots" research programs that local scientists and administrators thought worthwhile without prior permission from Washington. Federal funding was the source of virtually all the lab's support, which threatened to make the lab more a collection of projects controlled in Washington than a community. This happened particularly in departments other than physics, where research depended less on the big facilities and more on individual initiatives. A particularly wrenching example was the fate of the medical department in the late 1960s. Shortly after Cronkite became department chairman in 1967, he was summoned to Washington to participate in a "Woolworth" exercise, or a discussion of what one would do with a 5 percent and a 10 percent budget cut. This marked the beginning of serious budgetary problems that took a severe toll on the department. Desperate, Cronkite pleaded for funds with many possible sources, including life insurance companies, pointing out that Brookhaven's work in hypertension and Parkinson's had made substantial contributions to improving the public health and welfare.

Some hope lay on the horizon. Two intriguing ideas for major new facilities were beginning to crystallize in 1972. One was for an accelerator to create and exploit synchrotron radiation, which eventually would become the National Synchrotron Light Source (NSLS). The impact of the NSLS on the lab would be huge: it was an applied and basic research machine that would bring industrial users into active collaboration with the laboratory, creating a vastly different kind of culture than the lab had known before. It would also be used by nearly all the departments at the laboratory and routinely create interdisciplinary collaborations of the sort that rarely existed at accelerators, which were more and more the exclusive province of high-energy physicists.

The second was an idea for a new high-energy accelerator. The events surrounding the decision to construct a 200 GeV machine in Illinois gave rise to a deep concern among the Brookhaven accelerator builders, for two reasons. One was the breakdown of the gentlemen's agreement to leapfrog forefront accelerators between Berkeley and Brookhaven, which created among some Brookhaven scientists "some feeling that our future might now be less assured." [18] Another reason was the spiraling cost of accelerators and the shrinking federal enthusiasm for basic research. Brookhaven's accelerator builders then began to focus on a higher step in energy, raising the energy of their planned future accelerator from 1,000 to 2,000 GeV.

Much of the enthusiasm was due to the promise shown by the work of a superconductivity group that had been established in the mid-1960s (an example of an interdisciplinary group, quickly established, having a major impact on the lab), for superconductors could vastly improve the capabilities of accelerator magnets. Goldhaber, however, was still unenthusiastic about a post-AGS accelerator. "We had just gotten money to improve the AGS, and I felt we couldn't do two things at once and that the projects would interfere with each other. I joked, 'Machines repel machines.' " [19]

Nonetheless, the board of trustees decided to take matters into its own hands. In July 1968 it appointed a High Energy Study Committee, chaired by Princeton physicist (and later AUI trustee) Val Fitch, to investigate "the question of what new major facilities should be planned and developed at BNL in the next 5 to 10 years." [20] At first the committee focused on plans for a 1,500–2,000 GeV accelerator. But in late 1970 and early 1971, the committee changed its mind. The looming completion of the 200 GeV Weston accelerator and the approval, at CERN, of a 300 GeV machine made the big accelerator seem less ambitious than previously. Moreover, in February 1971, CERN's 26 GeV Intersecting Storage Ring (ISR) accelerator had managed to store protons in both rings with unexpected ease; p-p collisions had been detected, demonstrating the potential of storage ring technology. Finally, the $SU(3) \times SU(2) \times U(1)$ gauge theory of particle physics, which would soon bear the awkward name "the standard model," was completed around that time, and it contained interesting predictions of phenomena at extremely high energies. Committee member John Blewett, an early supporter of storage rings at the AGS, again pressed for the idea, which soon became the focus of the committee's deliberations. [21] The idea became particularly interesting when the realization grew that there was no advantage to injecting particles into the storage ring at their final energies. Blewett and accelerator department chairman Frederick Mills baptized the machine ISABELLE (*I*ntersecting *S*torage *A*ccelerator + *belle* for beauty). At the end of 1971, the committee recommended to the board of trustees

> that BNL apply its pioneering development work in superconducting pulsed magnets to build two intersecting storage accelerating rings to operate in the neighborhood of 200 GeV. The rings would be supplied with 30-GeV protons from the AGS which would then be accelerated to the final energy in the storage rings. [22]

In January 1972, the trustees officially accepted the Fitch committee report and decided to go ahead in seeking approval of the machine. Goldhaber continued to resist, saying the lab should concentrate on getting the AGS running smoothly before seeking funds for ISABELLE design studies. [23] But he was effectively overridden, and a study group issued a

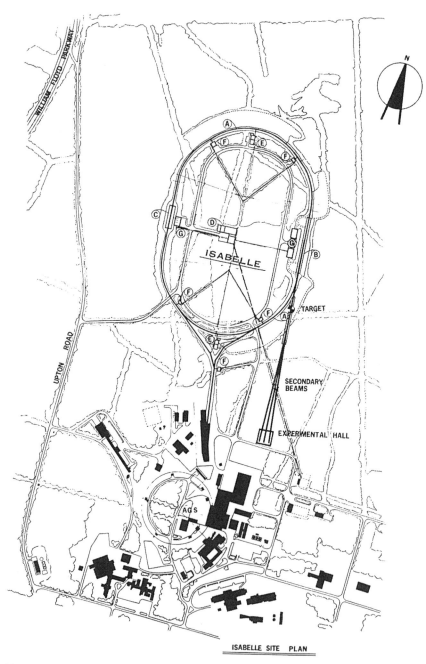

13.5 Drawing of ISABELLE, 1972.

preliminary design study, which became the focus of discussion at a three-week ISABELLE summer study in June 1972 (fig. 13.5).[24]

Amid the excitement, there were warning signs. In its FY 1972 authorization report, the JCAE applauded the six U.S. accelerator laboratories—four with proton accelerators (the 200–500 GeV synchrotron at Fermilab, the 33 GeV AGS at Brookhaven, the 12.5 GeV ZGS at Argonne, and the 6 GeV Bevatron at Berkeley) and two with electron accelerators (the 22 GeV linear accelerator at SLAC and the 6 GeV CEA), but expressed concern about their future due to the steady decline in the high-energy physics budget, and worried that support might be spread too thin:

> Therefore, the Joint Committee recommends that the AEC carefully examine the minimum level of support necessary to keep each of its high energy accelerator laboratories, including the NAL, viable and productive, and that it develop a priority listing of which accelerators should be kept operating should future funding be less than the minimum necessary to effectively support each of the six laboratories.[25]

But enthusiasm for ISABELLE mounted. Shortly after taking office, Goldhaber's replacement as director, George Vineyard, called it the most important single project facing the lab.[26]

Formidable challenges lay ahead, both for the lab and for AUI. For the lab, one challenge was to keep some influence in Washington over projects it thought scientifically valuable. These were what Smith has called the years of "policy disarray" (Smith 1990, chap. 4) when commitment to a sound scientific program crumbled. For many scientists and administrators at the lab, the lesson of the fourteen-foot bubble chamber was that, in Washington, political expediency could easily override scientific value. It was more important than ever for a laboratory director to be politically shrewd, and aggressive.

A second challenge was scientific-technological. In the past, accelerator building had provided a centripetal force to the lab community—but ISABELLE was a project on a *much* larger scale. It would introduce new kinds of strains into the lab community, and it would be more important than ever before to stay atop the problems that would inevitably appear in developing the requisite technology.

A third challenge was community relations. At the lab's founding, it enjoyed a measure of protection from the influence of the surrounding community, operated as it was by the AEC, and had little to fear from the concerns of the community. But the country was headed toward what Lewis Thomas called an "epidemic of apprehension" (Thomas 1983). And by the 1970s, big government-sponsored projects—and especially those involving reactors—were politically vulnerable in a way they never were before.

Meanwhile AUI, the organization that managed BNL and served as a buffer between it and Washington, also faced challenges. One involved making sure that the lab was in control of the immense and involved ISABELLE project, another that the lab make a priority of initiatives involving relations with the community rather than neglecting them. But a third key challenge had to do with science advocacy. AUI had been founded at a time when substantial, and rising, federal support of science could be taken for granted. Science was accepted, in Congress and among the public at large, as not one interest among others that competed for a slice of the federal pie, but as "a special interest," in Oppenheimer's phrase, support of which benefited everyone. In this historical context, the role of AUI was to ensure that the lab was well managed, and that the interests of northeastern scientists were well represented in the competition for resources with scientists from other parts of the country. The diversity of AUI-like contractors was thus a strength. By 1972, the situation had changed. The era of plentiful niches had come to an end. Basic research labs were hurt by the growing cost and size of accelerators and other forefront facilities. Competition among labs, always strong, grew fiercer in the face of huge layoffs, substantial program cuts, and declining federal interest in basic research. The diversity of contractors, formerly a strength, was now a liability. Not only had the pie shrunk in size, but science was now considered a special interest in a more current sense—an entitled elite with special privileges. The pulpits from which to advance the case for science were too small, and too numerous: the interests of science were represented by a number of different voices that had to battle each other to be heard and would sound to their audience like special pleading. Whether AUI could cope with that new challenge, and ensure that the case for science was strongly articulated, remained open.

In April 1972, accelerator department head Fred Mills gave a Brookhaven lecture that he entitled "Isabelle—A Crossroad for Physics." The word *crossroad* was, of course, a pun: physics at intersecting beams. But ISABELLE would be a crossroad for the lab in many more ways than Mills, or his audience, could have guessed. The trials and tribulations of this project, including the conflicts it brought about within the lab, AUI, and the entire high-energy physics community—in the course of which ISABELLE died and was reborn as a nuclear physics instrument—would be the single greatest force shaping the lab for the next quarter-century, and would take the lab in a new direction in a different world.

Some Key Personnel
of Brookhaven National
Laboratory and Associated
Universities, Inc.

BNL Directors

Philip M. Morse	September 1946–September 1948
Leland J. Haworth (acting)	July–October 1948
Leland J. Haworth	October 1948–March 1961
Gerald F. Tape (acting)	April–July 1961
Maurice Goldhaber	July 1961–December 1972

BNL Deputy Directors

Lincoln R. Thiesmeyer	Executive Assistant to the Director	October 1946–December 1949
Gerald F. Tape	Assistant to the Director	July 1950–September 1951
	Deputy Director	October 1951–March 1961
Clarke Williams	Deputy Director	May 1962–August 1966
George Vineyard	Deputy Director	July 1967–December 1972

BNL Assistant/Associate Directors (with planning or program responsibilities)

Robert D. Conrad	Assistant Director for Planning	June 1947–July 1949
Leland J. Haworth	Assistant Director for Projects	August 1947–July 1948
Donald D. Van Slyke	Assistant Director for Biology and Medicine	September 1948–June 1951
Rodney L. Cool	Assistant Director for High-Energy Physics Liaison	September 1965–June 1966
	Associate Director for High-Energy Physics	July 1966–September 1970

Victor P. Bond   Associate Director for Life Sciences and Chemistry
                                                    April 1967–October 1984
R. Ronald Rau   Associate Director for High-Energy Physics
                                                    October 1970–April 1978

AUI Presidents

Edward Reynolds   August 1946–November 1948
Frank D. Fackenthal   November 1948–February 1951
Lloyd V. Berkner   December 1951–November 1960
Leland J. Haworth   December 1960–March 1961
Isidor I. Rabi   April 1961–October 1962
Gerald F. Tape   October 1962–July 1963
Edward Reynolds   July 1963–November 1964
Theodore P. Wright   December 1964–September 1965
T. Keith Glennan   October 1965–June 1968
Franklin A. Long (acting)   July 1968–April 1969
Gerald F. Tape   May 1969–October 1980

AUI Officers Directly Involved with BNL Administration

Charles Dunbar   AUI Secretary and Legal Counsel   January 1949–
                                                             November 1969
N. Peter Rathvon   AUI Secretary and Legal Counsel   November 1969–1985
Eldon Shoup   Executive Vice President   October 1946–November 1950
John D. Jameson   Assistant Secretary   July 1947–June 1952
                      Assistant Treasurer   July 1947–June 1952
Lewis R. Burchill   Controller   February 1948–December 1973

Source: Needell 1978.

Chairmen of Key Brookhaven National Laboratory Departments

Accelerator Department (1960–84)

G. Kenneth Green, 1960–70

Frederick E. Mills, 1970–74

Biology Department (1947–)

Leslie Nims, 1947–50

Howard J. Curtis, 1950–65

C. H. W. Hirs, 1965–69

Howard Siegelman, 1969–74

Chemistry (1947–)

Richard W. Dodson, 1947–68

Gerhart Friedlander, 1968–77

Cosmotron Department (1954–60)

George B. Collins (1954–60)

Instrumentation and Health Physics Department (1948–70; this department then split into two divisions, instrumentation and health physics)

J. B. H. Kuper (1948–70)

Medical Department (1947–)

William Sunderman, 1947–48 (acting)

Robert D. Conrad, 1948–49 (acting)

Lee E. Farr, 1949–62

Victor P. Bond, 1962–67

Eugene Cronkite, 1967–79

Physics Department (1947–)

Norman Ramsey, 1947

Thomas H. Johnson, 1947–51

Samuel A. Goudsmit, 1951–60

Maurice Goldhaber, 1960–61

George H. Vineyard, 1961–66

Ronald Rau, 1966–71

Joseph Weneser, 1971–75

Reactor Science and Engineering Department (1949–52)

Lyle Borst (1949–51)

Clarke Williams and Marvin Fox (1951–52)

|

Nuclear Engineering Department (1952–67)

Clarke Williams, 1952–62

Warren E. Winsche, 1962–67

|

Department of Applied Science (1967–)

Warren E. Winsche (1967–75)

\

Reactor Department (1952–56)

Marvin Fox, 1952–56

\

Reactor Division (1956–)

Robert W. Powell, 1956–80

Notes

Introduction

1. In the recent monumental three-volume history, *Twentieth Century Physics* (Brown, Pais, and Pippard 1995), individual descriptions are confined to "biographical captions." One, devoted to British physicist Ernst Rutherford, discoverer of the atomic nucleus, quotes approvingly the following remark by a biographer: "People ask—what was the man like, did he really look like a farmer, what sort of accent did he have, did he believe in immortality? None of these things really matter." True, if one is narrowly interested in the product rather than process. But if this is *history*, then a sunset's a sextant setting and list of wavelengths. As for immortality, I. I. Rabi's approach to the divine (as a reader of this book will note in the first few pages) had much to do with his approach to science; and not to know that is not to know Rabi.

Chapter 1

1. Quotations in this paragraph are from "Reminiscences of I. I. Rabi," 11 January 1983, Columbia University Oral History Office. Biographical information here and in the following paragraphs is also drawn from Rigden (1987) and from Rabi's BNL Video Interview, 29 September 1982.

2. Rabi, "Reminiscences." "I've always said myself that for me and many of my colleagues, physics has essentially been a way of life, a point of departure from which you judged other activities, other disciplines, other ways of thinking, other ways of looking at things. It's a subject in which one expresses a large part of one's personality. One has a particular affection for certain sets of ideas. One has intuitive insights. It's not a closed subject. It's a growing, developing subject. It's a man-made thing."

3. Both quotations in this paragraph are from Rabi, BNL Video Interview, 29 September 1982. Rabi's equation of apprehension of the divine with the discovery of deeper forms of causation and interaction in nature is not unusual among scientists.

4. Rabi's quotation is from his BNL Video Interview, 29 September 1982; the insistence on becoming chair from the "Reminiscences."

5. At Los Alamos, Ramsey was in charge of converting the nuclear device put together by the scientists into a functional bomb. In 1945, he was chief scientist at the bomber base in the Marianas, from which the bombing missions over Hiroshima and Nagasaki were launched. In the movie *Enola Gay* (1980), about the pilot who flew the plane that dropped the first atomic bomb on Hiroshima, Ramsey briefly appears as a

character who is introduced to the hero Tibbets (the *Enola Gay's* pilot) by General Groves. Following the traditional Hollywood stereotype of scientists, Ramsey is portrayed as a superannuated nerd, balding seriously at the temples, who peers intently through steel wire-rims and smokes somberly on a pipe. "Dr. Ramsey's field is nuclear physics," the portly General Groves remarks, "His IQ equals my weight." The Ramsey character gets a handful of clichés to say for the rest of the movie. It would be difficult to get Ramsey more wrong.

6. Rabi, BNL Video Interview, 29 September 1982. For Pegram and the phone call, see Rhodes 1986, 293.

7. "Meeting of Representatives of Institutions and Organizations of New York City and Surrounding Region on the Subject of a Regional Nuclear Research Laboratory," 16 January 1946, D.O. Appendix 1.

8. "Draft of Proposed Letter," G. B. Pegram to General L. R. Groves, 17 January 1946; L.R. Groves to George B. Pegram, 22 January 1946, all in D.O. Appendix 1.

9. Rabi, BNL Video Interview, 29 September 1982; see also Ramsey 1966.

10. Letter, Karl Compton to George B. Pegram, 19 March 1946, D.O. Appendix 1.

11. J. C. Slater, "Proposal for Establishment of a Northeastern Regional Laboratory for Nuclear Science and Engineering," 9 February 1946, D.O. Appendix 1.

12. Norman Ramsey, "Memorandum" of meeting at Columbia University, 16 February 1946, D.O. Appendix 1. Reactors were then (rather unglamorously) called "piles" because they consisted of large piles of graphite bricks threaded with control rods and uranium fuel, the largest unbonded structures since the pyramids.

13. G. B. Pegram to General L. R. Groves, 3 March 1946, D.O. Appendix 1.

14. "Minutes of Meeting of Representatives of Universities Concerning a National Nuclear Science Laboratory," 26 March 1946, D.O. Appendix 1. The IUG arrangement preserved a distinction between scientists and science administrators that had been sharpened, and its importance underscored, by the war. "World War II taught us that neither group should look down their nose at the other group," Zacharias said (BNL Video Interview, 27 January 1983).

15. DuBridge, BNL Video Interview, October 1982. The first supplement was granted 3 April. At the end of the contract a report was issued describing how these funds were used, what activities had taken place, who was involved, etc. (Brookhaven National Laboratory 1948).

16. Rabi, "Reminiscences."

17. Letter, William Higinbotham to author, 4 February 1991.

18. She recounted the exchange in an unidentified article, dated 15 May 1952, in her file in Brookhaven's public relations office.

19. "Summary of Meeting of Initiatory Group," 30 March 1946, D.O. Appendix 1. The planners had debated asking Groves for a large particle accelerator as well as a pile at their meeting of 10 February, but tabled the proposal, in part because several of the universities already had plans for such machines, in part because accelerators were still inexpensive enough that building them did not yet require a cooperative venture. But their importance was becoming increasingly clear.

20. Livingston, BNL Video Interview, October 1982.

21. Zacharias, BNL Video Interview, 27 January 1983.

Chapter 2

1. DuBridge, BNL Video Interview, October 1982.

2. K. Compton to F. W. Loomis, 31 May 1946; Summary, Initiatory Group Planning Committee, 18 May 1946; Summary, Initiatory University Planning Committee, 31 May 1946; L. A. DuBridge to Wheeler Loomis, 24 May 1946; all in D.O. Appendix 1.

3. Zacharias, BNL Video Interview, 27 January 1983.

4. Philip M. Morse to Edward Reynolds, 22 July 1946, D.O. I 1.

5. Rabi, BNL Video Interview, 26 September 1982.

6. "Proposed Program for the New National Laboratory of Nuclear Science," 28 May 1946, with "Appendix A: Reactor Designs," June 1946, D.O. Appendix 1.

7. Report, Subcommittee on Personnel Policy, 16 April 1946, D.O. Appendix 1.

8. Smyth's suggestion that his subcommittee consist primarily of those in administrative departments of the universities was adopted, and it included Joseph R. Killian, Jr., the vice president of MIT, who had been involved with the management of the radlab and assistant to the president of MIT during the war, as well as Brakeley, Macaulay, Pegram, and Smyth himself.

9. DuBridge, BNL Video Interview, October 1982.

10. Minutes, planning committee, 3 August 1946. Associated Universities, Inc. (New Jersey), "Minutes of a Meeting of the Members, 10 July 1946," D.O. Appendix 1.

11. DuBridge, BNL Video Interview, October 1982.

12. Zacharias, BNL Video Interview, 27 January 1983.

13. Bacher, BNL Video Interview, October 1982.

14. Ibid.

15. Zacharias, BNL Video Interview, October 1982.

16. About Upton: When the United States declared war on Germany in April 1917, the War Department was suddenly faced with having to train some ten million eligible draftees across the country. For the New York region, almost ten thousand acres of mostly untouched land out on Long Island, previously used for cutting cordwood, was sold as a patriotic gesture by a scion of the Smith family for $1. An army camp was speedily built there. At first called Camp Long Island or Camp Yaphank, it was soon officially named for the Civil War general and West Point commandant Emory Upton, a native of nearby Batavia, Long Island. One inductee who trained at Upton was Irving Berlin, who wrote *Yip, Yip, Yaphank* while at the camp; the musical had a tryout at the camp theater before beginning a short run on Broadway. The musical's hit song, "Oh, How I Hate to Get Up in the Morning," is known to many an ex-G.I. today. Because World War I was to be the "war to end all wars," Camp Upton was dismantled after the armistice. During the Depression, part of the land was cleared by the government workfare agency, the Civilian Conservation Corps, and a few new structures were built. In the summer of 1940, after war broke out in Europe, Camp Upton was reactivated as an induction center, and the army hastily built a stockade for possibly interning German-Americans should trouble erupt. Though Japanese-Americans were relocated in internment camps in the west and midwest, no similar hysteria developed over German-Americans, and the stockade's eventual occupants were court-martialed G.I.s. As the war wound down, Camp Upton was rebuilt as a twenty-five-thousand-patient convales-

cent hospital in anticipation of a flood of casualties from the expected land invasion of Japan. To help rebuild the camp, the army moved almost a thousand German POWs into the stockade and set them to tearing down the tent frames and putting up barracks and concrete-block buildings (Donahue 1918; Bergreen 1990).

17. Bacher, BNL Video Interview, October 1982; Rabi, BNL Video Interview, 26 September 1982.

18. Mariette Kuper, BNL Oral Interview, 22 May 1990; Morse, BNL Video Interview, 26 January 1983.

19. The original trustees of Associated Universities, Inc., were: Arthur S. Adams (Cornell), Robert F. Bacher (Cornell), George A. Brakeley (Princeton), George B. Collins (Rochester), William H. DuBarry (Pennsylvania), Robert D. Fowler (Johns Hopkins), James R. Killian, Jr. (MIT), George B. Kistiakowsky (Harvard), Peter Stewart Macaulay (Johns Hopkins), George B. Pegram (Columbia), I. I. Rabi (Columbia), Edward Reynolds (Harvard), Louis H, Ridenour (Pennsylvania), Edmund W. Sinnott (Yale), Henry DeWolf Smyth (Princeton), Raymond L. Thompson (Rochester), William W. Watson (Yale), and Jerrold R. Zacharias (MIT). Bacher would resign in November 1946 and would be replaced by Franklin A. Long.

20. Mallory, BNL Video Interview, October 1982.

Chapter 3

1. Mallory, BNL Video Interview, October 1982.

2. Minutes, executive committee, 9 September 1946, D.O. Appendix 1.

3. Minutes, AUI board of trustees, 27 September 1946, D.O. Appendix 1.

4. Morse, BNL Video Interview, 26 January 1983.

5. Minutes, executive committee, 24 October 1946, D.O. Appendix 1.

6. Program of Brookhaven National Laboratory, 15 December 1946, D.O. Appendix 1.

7. Also at this meeting, Rabi was elected to take the place of Bacher, who had resigned after Truman selected him to be one of the first AEC commissioners. For two years Bacher would serve as the only scientist among the five, and would be eventually replaced on the commission by Smyth in 1949. Another AUI member to become an AEC commissioner was Campbell.

8. Groves to Reynolds, 24 December 1946, D.O. Appendix 1.

9. Edward Reynolds to Jerrold R. Zacharias and James R. Killian, 28 December 1946, D.O. Appendix 1.

10. Kuper to Morse, 27 December 1946, D.O. Appendix 5.

11. Kaplan, BNL Oral Interview, 1 August 1994.

12. On 1 January 1947, the Brookhaven site was still government property; formal transfer of ownership from the AEC to AUI would happen only in March, and signs across the middle of Long Island would point to Camp Upton for months.

Chapter 4

1. Minutes, AUI executive committee, 17 September 1948. Archives, Box 22, contains many files on early security issues.

2. Gurinsky, BNL Oral Interview, 18 February 1995.

3. Interview with Gerhart Friedlander, 18 November 1949, by the Personnel Security Board of the United States Atomic Energy Commission, New York Directed Operations. I am grateful to Gerhart Friedlander for supplying me with this transcript, executed by the AEC.

4. Letter, W. E. Kelley to Gerhart Friedlander, 29 March 1949, Friedlander file.

5. "I have read of cases where people have come under a sort of permanent 'cloud' through delay or refusal of FBI clearance, so that most other work possibilities also became closed to them," wrote Natalie to the lab's personnel officer in September 1947, "but it never seemed remotely possible that, through no fault of my own, such a shattering experience might ever come my way." Natalie Gurney to R. A. Patterson, 26 September 1947, Patterson correspondence file. A pair of scientists whose services were nearly lost was John and Hildred Blewett. During the war, they had both worked on top-secret radar countermeasures at General Electric's lab in Schenectady, New York; afterward John worked on accelerators and Hildred on the mathematical analysis of engineering problems. In fall 1946, Morse offered both of them positions. They accepted, and moved to Long Island. But on the last day of December, the AEC informed Morse that neither had received clearance and asked the laboratory to fire them. In the eyes of the AEC, the Blewetts were questionable security risks for several reasons. A former roommate of John had been jailed during a Canadian spy scare at the end of the war, and though the individual was eventually cleared, the incident cast a pall of suspicion over his friends. Moreover, the Blewetts had participated in a controversial strike at General Electric that had attracted national attention. At one point, left-wing actor Zero Mostel had been their houseguest when he visited Schenectady in support of the strike. In July 1947, the AEC commissioners, in their first nonunanimous decision, voted three to one, with one absence, to allow Brookhaven to employ the Blewetts (Lilienthal 1964, 189, 230).

6. Minutes, AUI executive committee, 19 December 1947.

7. Minutes, AUI executive committee, 15 July 1949, 21 July 1950.

8. The other Brookhaven scientist accused by McCarthy was William Higinbotham; the remaining five were Harold Urey, Philip Morrison, Linus Pauling, Edward U. Condon, and Kirtley Fletcher Mather. The accusations were made under immunity from lawsuit by virtue of being published in the *Congressional Record*. Leland J. Haworth to Members of the Policy and Program Committee, 5 September 1950, D.O. IV 16.

9. Morse to Dr. Lincoln R. Thiesmeyer, "Special Meeting of Laboratory Directors of Large Manhattan Projects," 17 October 1946. Archives, box 24, file 105.5.

10. Morse, BNL Video Interview, 26 January 1983.

11. Morse, "Special Meeting."

12. Morse, BNL Video Interview, 26 January 1983.

13. "Proposed Initial List of Unclassified Areas of Research for Brookhaven National Laboratory," undated, but evidently 1947 (D.O. III 8); a final list was agreed to and promulgated in October 1948; L. R. Thiesmeyer to Members of Policy and Program Committee, 5 October 1948, D.O. VI 40. For files on early classification struggles, see Archives, box 22.

14. Minutes, AUI board of trustees, 25 October 1947.

15. Memorandum, Eldon C. Shoup to Edward Reynolds, 10 November 1947, Reynolds correspondence file.

16. Minutes, AUI board of trustees, 16 January 1948.

17. AEC Contract No. AT-30-2-GEN-16.

18. *AEC Contract Policy and Operations,* U.S. Atomic Energy Commission, January 1951, 64. For contrast with typical previous cost plus fixed fee contracts, see "Manual for Cost-Plus-a-Fixed-Fee Supply Contracts," War Department Technical Manual TM14-910, December 1944. See also "History of Management Contracts," put together by Allen Moseley of the Department of Energy's Oak Ridge office, 9 May 1983.

19. Minutes, AUI board of trustees, 21 October 1949.

20. Minutes, AUI board of trustees, 15 April 1949 and 21 October 1949.

21. "Foreword" to "The Brookhaven Concept," material prepared for Reynolds Committee, 6 October 1949. H.O drawer 1949.

22. Minutes, AUI board of trustees, 21 April 1950.

23. "A modus vivendi must be contrived," one trustee observed at an early meeting, "which is effective for the present . . . promises that creative research of the highest order flourishes . . . [and avoids] those pitfalls and blind alleys which lie in wait for a new enterprise." "Policy Areas Requiring Examination and Discussion by Trustees," 27 May 1948 (a summary of the meeting of 16 April).

24. E. L. Van Horn to L. Haworth, 1 November 1960, Haworth correspondence file. BERA's constitution was adopted in May 1948; minutes of policy and program committee, 18 May 1948, D.O. IV 16.

25. Material on these subjects is found in D.O. VI 36 and 37.

26. Arland Carsten to author, 3 February 1997.

27. Eldon G. Shoup to E. L. Van Horn, 18 September 1947, Shoup correspondence file.

28. David H. Gurinsky, *Memoirs: Après la Guerre,* unpublished manuscript.

29. On the founding of Berkeley's Division of Medical Physics, by contrast, see Westwick 1996.

30. Long, BNL Video Interview, 25 June 1986.

31. Sunderman, "Remarks at the Opening of the Medical Department and Hospital Laboratories at the Brookhaven National Laboratory"; the document bears the date 1 May, but that is apparently erroneous. Sunderman file.

32. Sunderman to policy and program committee, 23 July 1947; minutes, AUI executive committee, 17 September 1948.

33. Hastings, interview by Peter D. Olch, History of Medicine Division, National Library of Medicine, Bethesda, Maryland, 1969; Van Slyke, ibid.

34. Ibid.

35. Memo, Kelley and Wilson to BNL, 31 March 1948; Warren to Van Slyke, 8 March 1948; Shoup to Reynolds, 15 April 1948, D.O. V 46.

36. Minutes, AUI executive committee, 14 October 1948.

37. Van Slyke, interview by Olch. Some idea of how well known Van Slyke was throughout the lab is clear from the following story. It seems that one protein chemistry symposium early in the lab's history ended late at night in a boisterous party involving heavy drinking and a craps game on the floor. Attracted by the commotion, a security

guard appeared, confiscated the dice, and said he'd have to report the incident and take down everyone's name. The first person asked to identify himself said Louis Pasteur. The guard wrote that down. The second person gave his name as Charles Darwin, and after him came Michael Faraday, Gregor Mendel, and a few others, and the guard copied down those names. Then someone identified himself as "D. D. Van Slyke." The guard straightened up angrily. "You're not telling the truth," he said. "I *know* D. D. Van Slyke!" Van Slyke was called and asked to try to enforce discipline. Daniel Koshland, Jr., Reminiscence Seminars, Biology Department, 13 May 1997.

38. Minutes, AUI executive committee, 19 December 1947.

39. Long, BNL Video Interview, 24 June 1986.

Chapter 5

1. The CP-1 was rebuilt as CP-2 in March 1943; the CP-3 was built the following year.

2. Kouts, BNL Video Interview, 15 May 1986.

3. Borst, BNL Video Interview, 26 June 1986. The basic components of a reactor were a fissionable fuel, a moderator, a cooling medium, and control rods. Possible fuels were natural uranium, enriched uranium (uranium with a higher concentration of the fissionable isotope ^{235}U), and plutonium. The moderator, which slowed neutrons to the energy at which they would be most efficiently absorbed by the fuel, could be light water, heavy water, or graphite. The chief possibilities at the time for a coolant, to transfer heat out of the reactor core, were air and water. The control rods, used to slow or stop the reaction, were made largely of neutron absorbers such as boron or cadmium. In principle, these components could be combined in many ways; in practice, Brookhaven's reactor group had little flexibility. Borst and others were interested in using enriched uranium, but the MED did not want to approve it for a nonmilitary reactor (Morse to Kelley, 16 May 1947, Morse correspondence file). Thus natural uranium would be the fuel, for which graphite was the moderator of choice. Air rather than water would have to be the coolant partly because of the lab's location on Long Island, which drew its water supply from a single-source aquifer with limited groundwater. The CP-1, X-10, Chalk River, and Hanford reactors ran on natural uranium. The tiny Los Alamos "water-boiler," which went critical in 1944, ran on enriched uranium, which was advantageous because the more powerful fuel would allow more neutrons to be produced by a reactor of smaller size; later, Clementine, a reactor that ran on plutonium, was built at that lab. The Hanford reactors were graphite moderated and cooled by light water, while the Chalk River reactor was moderated by heavy water.

4. Proposed Scientific Program, December 1946, H.O. drawer 1946.

5. A safety feature of the horizontally positioned control rods was that each had its own flywheel, kept constantly spinning by an electric motor, with enough stored energy to drive in the rod. When the reactor shut down or "scrammed," whether deliberately or due to sudden failure of electricity, the hydraulic motor automatically engaged to the flywheel, driving it in place in about two seconds.

6. The protective metal casing had to be conductive enough to transfer heat from the uranium to the air coolant in the fuel channel rapidly enough to keep the uranium from swelling. It was not easy to find a protective metal that was at once easy to weld, strong

enough to protect the uranium from corrosion, conductive enough to transfer heat efficiently, and not so absorbent that it would siphon off an excessive number of neutrons, causing what was called parasitic neutron loss. And failure of a single weld could shut down the reactor. This difficulty was known as the canning problem during the war. In the Smyth report, the author commented that "on periodic visits to Chicago the writer could roughly estimate the state of the canning problem by the atmosphere of gloom or joy to be found around the laboratory" (p. 146). Powell, who had worked on the X-10, recalled of that machine:

> To protect against ruptured welds, the exhaust gas was monitored for fission products; if they appeared, you shut down the reactor to look for swollen fuel elements. We had a primitive method for this: we shone a light down into the channel at one end of the reactor and eyeballed the other end—we figured we wouldn't receive enough radiation in the eye to give us any trouble, though they did stop the practice later on. We simply looked through the channel at the profile of the fuel elements, and if there was any irregularity, we took the whole channel out and discarded it. But if a fuel element swelled enough to block the channel, there's no air flow to transmit those fission products to the detector and you don't know it's leaking. It gets by you, you continue to run, there's no cooling for the rest of the fuel in the channel, and they sequentially fail. (Powell to author, 1 August 1997)

7. Materials on this project are found in the "Heat Recovery Correspondence" and "Reactor—Power Recovery" folders, H.O. drawer BGRR. Minutes, scientific steering committee, 10 September 1948, D.O. IV 19. Independent study: Gibbs & Cox, Inc., "Investigation of the Conversion of Waste Heat to Electric Power for H. K. Ferguson Company, Inc.," 30 June 1948, H.O. drawer BGRR.

8. In preparation for studying how the gap would affect the flow of neutrons in the reactor, Kaplan consulted an important paper on the general theory of gaps in reactors that had been written by a trio of physicists: a husband and wife team, Marvin and Mildred Goldberger, and a black physicist named J. Ernest Wilkins (report CP-3443). Another problem was to deter leakage of neutrons from the gap. Kaplan's solution was to "step" the gap, making it turn a quick series of corners at the edges of the graphite cube, so neutrons could not flow out directly.

9. Correspondence to and from Borst in H.O. drawer BGRR; Borst's BNL Video Interview, 26 June 1986.

10. The reasons are outlined in a draft memo from Beeler to Reynolds dated 10 January 1947, D.O. V 78.

11. Minutes, AUI board of trustees, 20 October 1950.

12. They asked that AUI be assigned "full responsibility" for "the design, engineering and construction," which, they said, would improve the chances for a quick success of the process. Shoup to Reynolds, 27 June 1947, Shoup correspondence files.

13. Van Horn: "Conference with Area Manager, Brookhaven Area, U.S. Atomic Energy Commission," 14 May 1947, D.O. V 78.

14. "Report on Reactor Duct Repairs, Evaluation and History," 23 June 1949, D.O. V 78; Ferguson contract: Letter Contract AT-30-2-GEN-3, 14 July 1947. Management arrangement: "Meeting on July 17, 1947 with Representatives of H. K. Ferguson Company," 17 July 1947, D.O. V 78.

15. Kaplan, interview with author, 1 August 1994.

16. Borst, BNL Video Interview, 26 June 1986.

17. Borst, Reminiscence Seminars, Graphite Reactor, 7 May 1997, BNL Video.

18. *Isotopics* 2 (Jan.-Feb. 1948).

19. Van Horn to Shoup, 10 May 1948, D.O. III 8.

20. Early heat removal discussion meetings: Janice Cutler to Morse, "Meeting with Hydrocarbon to discuss the problem of heat removal of pile #1," 24 December 1946, History of Reactor Construction file. "Report on Duct Repairs," D.O. V 78.

21. G. M. Inman to Borst, "Pile Operation Div. Construction Progress Report #6," 1 June 1948, D.O. V 78. Shoup to Van Horn, 13 September 1950, Shoup correspondence files.

22. Minutes, AUI board of trustees, 23 July 1948. Also, Ferguson had begun to deal directly with the New York office of the AEC on a number of technical matters, bypassing and sometimes not even informing the laboratory—a development Haworth protested as "unsound."

23. Ferguson's bafflement: Shoup to Van Horn, 13 September 1950, Shoup correspondence files.

24. Borst, BNL Video Interview, 26 June 1986.

25. The start-up group's first task, in the fall of 1949, was to measure the thermal conductivity of the reactor's individual graphite blocks, and then of the entire pile. That meant heating the entire pile of hundreds of tons of graphite with "cal rods," electric stoves in the shape of long rods, which were inserted into the reactor's fuel channels and connected to several thousand volts of electricity (a household stove uses 220 volts). Several hundred thermocouples (electrical instruments that measure temperature) had been installed in other fuel channels. To read each thermocouple, an operator cranked a handle on a device attached to it, and the devices were set up on the balconies all around the reactor. Dozens of people were enlisted to help take these readings, "not only technicians," Sailor recalled, "but guys like Borst and Clarke Williams, and even Lee Haworth took a shift." The measurements, completed by the end of 1949, proved valuable in B&W's final assessment of the thermal stress problem.

26. To count the flow of neutrons under various conditions, Sailor's start-up crew inserted thin sheets of indium or gold for a specific period of time into channels in the reactor, measured the amount of radioactivity that had been produced in the foil, and calculated how many atoms had been activated. With this information, and the cross-section of the metal (a cross-section is a measure of the rate of interaction between neutrons and a nucleus), they could calculate how many neutrons had been flowing at that particular point and place. It was, Sailor said, "a torturous but well-beaten path to determining the flux," torturous not least because the calculations had to be done on hand-operated calculating machines. See J. Chernick, J. J. Floyd, J. W. Kunstadter, and V. L. Sailor, "Subcritical Experiments on BNL Reactor," in "Classified Progress Report of the Reactor Science and Engineering Department, January 1–March 31, 1950," BNL 52. See also J. Chernick, I. Kaplan, J. W. Kunstadter, V. L. Sailor, and C. Williams, "An Experimental and Theoretical Study of the Subcritical BNL Reactor," BNL-60 (15 June 1950).

27. Minutes, AUI board of trustees, 17 February 1950; minutes, AUI executive committee, 19 January 1950. Milbank, Tweed had several reasons for its pessimism. First, Delner and not Ferguson had done much of the original design under contract

with the AEC. Second, a reactor was a highly uncommon kind of project for which no standard building procedures existed. Third, there had been no operational tests, and in their absence, the Brookhaven staff and AEC had approved plans. Fourth, the nature of liability in a contract of the sort AUI had with the AEC was uncertain. Fifth, much potential evidence was classified and could not be introduced in court. Therefore, litigation was likely to result in a "mud slinging match" followed by a congressional inquiry and staff dissension. Edgar Baker (of Milbank, Tweed) to Associated Universities, Inc., 17 March 1950.

28. The first hot lab had been built at Oak Ridge, and had consisted of large open cells with remote-controlled manipulators in what was called the RaLa facility, used for making radiolanthanum for implosion studies at Los Alamos, the first time manipulators were developed for handling radioactive materials.

29. Much material on early waste disposal is in D.O. IV 17.

30. Manowitz to W. E. Winsche, 19 April 1948, BNL document C-1698.

31. By the time the liquid was discharged into the Peconic River, the average radiation level of the effluent was well below the limit set by the AEC. The Peconic was not used for drinking and was considerably diluted by other streams. For an extensive assessment of the liquid waste disposal practices of the early years, see Hull 1969.

32. Many BNL reports were issued on computed and measured dose rates from the BGRR. The work of the meteorology group consisted of "stack meteorology," or the study of the patterns of advection and eddy diffusion of effluents emerging from stacks, and had both a research and applied component. The research involved the development of stack meteorology for use at other reactor facilities, while the applied work required the development of forecasts that would be used to schedule the operation of the pile. Before the reactor went into operation, the meteorological group conducted an environmental survey. Monitoring instruments were installed off-site and on-site. Two towers with monitors were built, one 420 feet and the other 160 feet high, to permit the study of wind velocity and temperature fluctuations in the atmosphere and atmospheric turbulence. The leading figure in the field, O. Sutton, had done the key theoretical treatment of the problem of diffusion from raised sources such as smokestacks, and Sutton himself visited the lab in May 1950. The Brookhaven meteorology group also worked with the Cornell Aeronautical Laboratory, which performed wind tunnel tests on models of the reactor building and studied the air currents likely to develop around it.

33. Borst, BNL Video Interview, 26 June 1986.

34. Borst, "The Brookhaven Nuclear Reactor," AECU-1089, 1951.

35. Sailor to author, 4 May 1996.

36. Manowitz, BNL Oral Interview, 5 January 1995

37. Borst, BNL Video Interview, 26 June 1986.

38. Shoup: James M. Knox to files, "Meeting with HKF and AEC on December 13, 1948," 14 December 1948, Knox correspondence files.

39. Borst, BNL Video Interview, 26 June 1986.

40. Kaplan to author, 8 March 1995.

41. Kaplan, BNL Oral Interview, 1 August 1994.

42. One other nonmilitary reactor was completed before the BGRR; ZOE, built in France under the direction of Frederic Joliot-Curie, whose membership in the Com-

munist Party eventually caused the government to remove him (ZOE was nicknamed by some the "Communist Pile"). But ZOE was more of engineering exercise (on ZOE, see Weart 1979, 247–48, 253–56). Also, a number of reactors built during the Manhattan Project, such as CP-3 (completed in 1944) were used after the war for experimental research. The BGRR was the first designed and built in peacetime for experimental research.

43. Borst, BNL Video Interview, 26 June 1986.

Chapter 6

1. Jameson to R. O. Niehoff, 22 April 1948, D.O. VII 37.

2. Hartzell, "Education in Atomic Energy," 5 February 1952, H.O. drawer Public Relations

3. Rabi, BNL Video Interview, 26 September 1982.

4. A general discussion of the possible role of the corporation in public education took place on 23 September 1949. Opinion about their value was divided, but the sense of the meeting was that "the assumption of any major responsibility for production or direct program action should be avoided" (Minutes, AUI executive committee, 23 September 1949). The following April, Hartzell presented his outline of a possible program on public education to the trustees, who judged it only remotely related to Brookhaven's goals; the AEC, they felt, should be induced to divorce this program from the laboratory's mission. Though Hartzell continued to press for an effort, the trustees' interest steadily declined. After his departure in 1952, early attempts to find a public education role for the lab all but ended.

5. Brookhaven National Laboratory Record of Press Conference, 28 February 1947, H.O. drawer 1947.

6. John D. Jameson to G. Edward Pendray, 4 March 1949, D.O. VII 37.

7. Morse, BNL Video Interview, 26 January 1983.

8. Long, BNL Video Interview, 24 June 1986.

9. P. E. Kraght to Norman Beers, 24 January 1948, D.O. VII 37.

10. *Newsday*, 7 June 1948, p. 3.

11. *Hartford Times*, 23 August 1948, p. 14.

12. Handwritten note accompanying a copy of the article in H.O. drawer BGRR, Power Recovery folder.

13. The AP story was dated 13 November 1948; see for instance, "Smog Is New Death Peril," *Tulsa World*, 14 November 1948, p. 13.

14. Pearson, "Washington Merry-Go-Round," 3 January 1948. Donora and Brookhaven were vastly different in numerous ways. The short-lived argon at Brookhaven was very different from Donora's smog. Brookhaven was located in a flat area, while Donora was deep in a river valley that collected stagnant air. Brookhaven had a single smokestack, while Donora, one of the most heavily industrialized areas in the country, had thousands. Not least, Brookhaven had meteorological monitoring facilities and safeguards, while Donora had none.

15. *Islip Press*, 13 January 1949.

16. Anderson, BNL Oral Interview, 7 December 1994.

17. *New York Post,* 16 October 1956, 4.

18 "Report to Dr. C. L. Dunham from Dr. John E. Rose," Argonne National Laboratory, 16 November 1957, Koerber file.

19. Graham DuShane, "Canard Corrected," *Science* 125 (12 April 1957).

20. *New York Times,* 16 November 1960, 26, 43. On 19 November, the *Times* ran a small correction (p. 11).

21. Kuper et al., "Radiation Protection at Brookhaven National Laboratory, Presentation to the Board of Trustees," April 1955.

22. Higinbotham recounts:

> The lab had several analog computers, which came with books telling you how to display things like bouncing balls, etc. I looked at them and said, "Well, obviously, with this machine, I can fix it so that, instead of having it preprogrammed, people can control it." That was the "great invention," if you want to call it that! It didn't strike me as the least bit novel. All the circuits I used were circuits used by people before, except for putting these hand controls in, and a game which would go with that. (Higinbotham, BNL Oral Interview, 28 June 1991)

23. D.O. II 28.

24. Rabi, BNL Video Interview, 26 September 1982.

Chapter 7

1. A happy by-product of the slightly bowed shape of the magnetic field between the poles of a cylindrical magnet (a "fringing field") was that the particles tended to remain focused in the midplane, a phenomenon Lawrence in particular came to appreciate and use to advantage.

2. Livingston, BNL Video Interview, October 1982.

3. Ibid.

4. Ibid.

5. Livingston: "Memorandum of Conference," 9 May 1947, History of Reactor Construction file; Shoup to Morse, 12 May 1947; "Happy windfall": minutes, scientific steering committee, 22 May 1947, D.O. IV 19. Proposal and Specifications, 3.5 MeV Electrostatic Positive Ion Accelerator, Specification No. SPA-64. General Electric Company, revised 22 May 1947.

6. Hoey to Livingston, 25 June 1947, D.O. V 7.

7. Hafner, BNL Oral Interview, 24 February 1995.

8. The accelerating tube consists of a series of metal rings or electrodes. Each electrode is separated from its neighbors by an insulating ring but connected via resistors, so that the voltage is gradually and evenly stepped down from the extremely high voltage on the electrodes at the terminal end of the column to a low voltage at the base (creating what is known as a series of "equipotential planes"). When charged particles are released at the terminal, they are sucked down the accelerating tube by the voltage differential and emerge through a hole in the base plate.

9. Van de Graaff Record Book #1, entries for 10 and 11 October.

10. Hafner, BNL Oral Interview, 24 February 1995.

11. The reference is in Shoup to Morse, 12 May 1947, Morse correspondence files.

12. G. B. Collins to Haworth, 31 August 1950, D.O. V 21.

13. Haworth to Van Horn, 10 September 1950, D.O. V 14.

14. Haworth to Van Horn, 16 November 1950, D.O. V 14.

15. Livingston, BNL Video Interview, October 1982.

16. Morse to Shoup, "Division of Responsibilities on the Cyclotron Project," 6 March 1947, Shoup correspondence files, 1947.

17. Report of the Director, Status of the Pile Project and the High Voltage Machine Project, 20 May 1947, D.O. V 7.

18. "Specifications for 60″ Cyclotron for Brookhaven National Laboratory," approved by Salisbury and Livingston, D.O. V 3.

19. Van Horn to Shoup, 14 July 1948, D.O. V 2; Van Horn to Shoup, 27 July 1949, D.O. V 2.

20. "Notes on the Meeting with Collins Radio Company," 3 April 1951, D.O. V 3.

21. Van Horn to Haworth, 25 October 1949, D.O. V 2.

22. James M. Knox to files, 20 March 1951, D.O. V 3.

23. Salisbury's departure: "Notes on Argonne 60″ Cyclotron," W. W. Merkle, 18 January 1952. Ending the contract: Gerald F. Tape to Arthur Collins, 22 April 1952, D.O. V 3.

24. D.O. V 3. "L.I.L." is Long Island Lighting Company, which supplied the electricity.

25. Plotkin, BNL Technical Report #1, 11 January 1947.

26. Synchrocyclotrons are themselves limited by the fact that, as in all cyclotrons, the magnet covers the entire diameter of the vacuum chamber. Each small increase in magnet radius entails a huge additional cost: the rule of thumb was that the cost of a cyclotron was roughly proportional to the cube of the radius. According to conventional wisdom, the effective limit of synchrocyclotrons was about one billion electron volts (abbreviated BeV, but later changed to GeV when "giga" was substituted for "billion"; I use both as appropriate to the time described). "Entirely new acceleration schemes may have to be devised" to reach a level, say, of 10 BeV, says the Initial Program Report, and "the Laboratory must actively engage in the development of super high energy machines, more powerful than any now being designed, in order to solve the mysteries of nuclear forces and the phenomena associated with cosmic rays." 1946 Program Report, appendix B, part 2.

27. Rabi, BNL Video Interview, 26 September 1982.

28. Livingston, BNL Video Interview, October 1982.

29. Minutes, planning committee, 2 October 1946, D.O. Appendix 1. In a survey of accelerator designs made for a scientific advisory committee meeting in April of 1947, Livingston admitted that "each machine has its strong supporters for first priority at Brookhaven," but stated his own case for the synchrocyclotron: "The cyclotron offers very high intensities at medium energy and with poor directionality; its design is practical and it could be completed soonest with minimum demands on scientific personnel. . . . The synchrotron is the only machine capable of really high energy, but at the expense of exceedingly low intensity; its development will take many years and involve major investments of money and scientific talent." D.O. IV 16.

30. Minutes, scientific advisory committee, 21 April 1947, D.O. IV 16.

31. G. K. Green, "Notes on Conference," 8 August 1947, D.O. IV 16.

32. The possibility of increasing that to 1 BeV with 288-inch poles was discussed and rejected; it would add a full 75 percent to the cost. A third possible intermediate step would be to work on a smaller proton synchrotron of the sort Oliphant was building of about 1 to 2 BeV.

33. Green, "Notes on Conference."

34. Minutes, policy and program committee, 21 October 1947, D.O. IV 16.

35. Memo, Livingston, 20 October 1947, D.O. IV 16.

36. The accelerator physicists involved in "ultra" or "super" high-energy design had been working on two versions, one a 25-foot radius, 2.5 BeV design, the other a more ambitious 10 BeV design. A few conservative-minded physicists thought the lab should build the smaller one first, arguing that only such experience could clarify important design decisions and make it possible to plan an efficient big machine. Others preferred the small machine in part because it might interfere less with the pile project, demonstrating the continued priority the lab still put on reactors. Rabi disagreed; his go-for-broke philosophy called for shooting at 10 BeV right away. He received more ammunition in February, when Berkeley reported it had produced pions with its new 184-inch synchrocyclotron. Brookhaven, he felt, risked being left in a scientific backwater if it did not aim higher than Berkeley. Rabi bent Morse to his views, and the two pressed for 10 BeV. "Memorandum on the Present State of Particle Accelerators," J. C. Slater, 15 December 1947, D.O. V 21. Minutes, executive committee, 19 December 1947; minutes, executive committee, 20 February 1948.

37. Livingston to Haworth, 27 February 1948, D.O. V 21.

38. Livingston, BNL Video Interview, October 1982.

39. Ibid.

40. Rabi, BNL Video Interview, 26 September 1982. Heilbron, Berkeley's historian, has said that Brookhaven was "hoodwinked" in this decision, but in view of the arguments, this is probably a rare instance of a bad call on Heilbron's part. For more on the politics in Washington of this decision, see Seidel 1982, 395-96.

41. Minutes, accelerator project staff meeting, 16 March 1948, D.O. V 21.

42. Minutes, accelerator project staff meeting, 22 March 1948, D.O. V 21.

43. James M. Knox to Morse, 4 May 1948, D.O. V 21.

44. Blewett, BNL Video Interview, 17 March 1986.

45. Rabi, BNL Video Interview, 26 September 1982.

46. Blewett, BNL Video Interview, 17 March 1986.

47. Livingston, BNL Video Interview, October 1982.

48. Plotkin, BNL Oral Interview, 13 March 1986.

49. The oscillations were of two principal kinds: betatron oscillations (the wobble, or set of wavelike motions, that particles naturally make about a perfect, center-line orbit), and synchrotron oscillations. The latter arise because the particles do not all arrive at the accelerating cavity in a precise dot but in a small bunch in which the faster particles are in front and the laggards in back—which means that they get slightly different "kicks," sending them in slightly different orbits. Perturbations, or displacements of the particles from their orbits, could be produced by various errors in the injection, by acceleration or magnets, or by gas scattering. How big were the errors that could not be avoided, and how small could the aperture be so that it would still be big enough to take account of these errors? And because most focusing would be in the

vertical direction with little focusing horizontally, the aperture needed to be wider than it was high.

50. Blewett, BNL Video Interview, 17 March 1986.

51. Ibid.

52. The top and bottom of the chamber consisted of ribs of 2″-wide steel bars with a 1/16″ separation between them. These ribs were covered by sheets of Myvaseal, a low-vapor-pressure rubber, leaving a vertical gap of about 6¼″ for the beam. The inner and outer walls were made from solid, inch-thick stainless steel; the strength of eddy currents goes by area at right angles to the magnetic field, and since the walls showed only a small area at right angles to the field, eddy currents in them would be insignificant.

53. Memo, Graph of Accelerator History, J. P. Blewett, 9 June 1950, H.O. drawer Cosmotron. This graph has become known as the Livingston plot, evidently because Livingston kept using it and updating it in his book on accelerators, but it really should be called the Blewett plot.

54. Memo, M. H. Blewett, 14 November 1951, D.O. V 22.

55. Dexter, BNL Oral Interview, 30 April 1991.

56. J. Blewett, BNL Video Interview, 17 March 1986.

57. CERN's governing council had met for only the second time in Copenhagen at the end of June 1952, the gathering at which the delegates officially chose the name CERN, but had not yet decided on a site. Geneva was selected in October, and work on the site began in May 1954.

58. If a field has zero gradient, for instance, it means that the field is constant and does not change, while a positive gradient means that the field grows stronger and a negative gradient means that the field grows weaker as one moves outward from the center. Livingston's proposal would mean "alternating" the gradient, or arranging the magnets so that the gradients would alternate positive and negative.

59. The butterfly concept sought to increase the mean field in a cyclotron through alternating ridges of strong and weak fields, keeping the frequency constant with azimuthal variations of field strength. A fundamental problem of so-called Alvarez-type linear accelerators was that the accelerating fields, if stable for acceleration, defocused radially. As a solution to this problem, focusing grids had been used in the drift tubes, but they scattered a major portion of the beam and interfered with the focusing. The use of quadrupoles in their stead, Blewett showed, would eliminate the scattering and vastly improve focusing.

60. Haworth to Robert L. Thornton, 29 August 1952, Van Horn to Haworth, 19 September 1952, D.O. III 8; Courant, BNL Video Interview, 3 April 1986.

61. Minutes, AUI executive committee, 16 September 1952.

62. Minutes, AUI executive committee, 20 February 1953.

63. Minutes, sixth weekly meeting on Multi-Billion Volt Synchrotron, 18 November 1952, D.O. V 22.

64. This development was reported in the RSI's Cosmotron issue, and discussed more thoroughly in a later issue.

65. Letter, T. H. Johnson to L. V. Berkner and L. J. Haworth, 29 August 1952, D.O. VI 5.

66. The letter suggesting the engraving was Collins to Smyth, 9 March 1953, D.O. V 22.

67. Courant, BNL Oral Interview, 10 July 1991.

68. Reminiscences of the Cosmotron dedication ceremony are from BNL Oral Interviews with Blewett, Courant, Dexter, Serber, and others.

Chapter 8

1. Sailor to the author, 4 May 1996.

2. Ibid.

3. E.g., Haworth to Van Horn, 2 January 1958; Haworth to scientific department chairmen, 17 April 1957. Haworth correspondence files.

4. The unit of choice of size in the submicroscopic world is the *barn,* 10^{-24} square centimeters, based on the ironic jest that, to a neutron, a nucleus seems as big and easy to hit as the side of a barn; it was a natural choice of a unit since a typical cross-sectional area is between 1 and 10 barns. The unit was coined by Charles Baker, later to become head of Brookhaven's cyclotron project (Holloway and Baker 1972). Excited states of nuclei may consist of both single-particle excitations or collective excitations of various kinds. If one weighed the excited state of the compound nucleus and its normal or "ground" state, the difference between the two is known as the excitation or "binding" energy. And just as the study of the excited states of atoms through analysis of the spectral lines they emit is called spectroscopy, so the study of the excited states of nuclei is known as nuclear spectroscopy. The most important prewar cross-section research had been carried out at Columbia, in an extensive series of measurements by Rainwater and Havens, while during the war Hy Goldsmith, an instructor at City College of New York before the war, and H. W. Ibser compiled cross-sections at the metallurgical laboratory of the University of Chicago. After the war, cross-section research was carried on at Argonne, and Brookhaven was early in the compilation business. Brookhaven had hired Goldsmith to head up an information section, and he promptly published a revised compilation of neutron cross-sections with Brookhaven as his institutional affiliation (Goldsmith, Ibser, and Feld 1947). In his new job at Brookhaven, he planned to update that compilation, and arranged with Heinz Barschall at Wisconsin for a graduate student (Robert Adair) to complete the task. Adair came to Brookhaven in summer 1947 and finished and published the compilation back at Wisconsin that fall (Adair 1950), with institutional affiliations at Brookhaven and Wisconsin.

5. G. A. Kolstad to P. W. McDaniel, "Neutron Cross-Section Measurements for the Division of Reactor Development," 8 August 1951, D.O. V 56.

6. The AEC's reactor development division had established the neutron cross-sections committee in 1948 to study the cross-section needs of the weapons, reactor design, and nuclear submarine programs, rank the desired cross-sections by category in order of priority, and coordinate research. Hughes was interested in cross-sections from the standpoint of basic research, and though lacking any real interest in reactor design he was willing to coordinate the program as the price for bountiful support for his own interests. The cross-section committee uncovered much confusion: some research was duplicated, and other important work was neglected. Early in 1950, Hughes proposed a project to prepare an "authoritative compilation" of cross-section data for the cross-section committee, whose "logical headquarters" would be Brookhaven, and Hughes was soon asked to head the project. See for instance A. H. Snell to Hughes, 4 April

1950; Hughes to members of the cross-section committee, 9 June 1950. G. A. Kolstad to P. W. McDaniel, 6 August 1951, D.O. V 56.

7. Minutes, meeting of AEC neutron cross-sections advisory group, 8–10 October 1951, D.O. V 56.

8. "Thermal Cross Sections," AEC cross-sections committee, 20 August 1951. *Neutron Cross Sections,* AECU 2040 (Office of Technical Services, Department of Commerce, Washington, D.C., 1952).

9. Sailor to the author, 4 May 1996.

10. The fast chopper, which spun about 10,000 rpm and the principal facility for determining nuclear cross-sections with neutrons of 3 to 1,000 electron volts, was installed at the north face, where a single beam hole looked deep into the core and provided higher-energy neutrons than the other beam holes. A detector measured the neutron counts of a time-of-flight beam with and without the target in place. The difference between the two measurements revealed how many neutrons at that energy had been absorbed, and the cross-section of the target. Data taking was (until 1957) manual. Detectors recorded the number of neutrons in each time channel (i.e., of a given energy) with the same type of mechanical register used by traffic controllers to count vehicles passing a certain location, and graduate students stopped by periodically to read off the information.

11. Theorists were able to take such information and relate it to a statistical model of the nucleus called the "theory of orthogonal ensembles." Individually, in this theory, resonances are an example of nuclear chaos, and cannot be predicted, for they are formed of random superpositions of matrix elements in a classic example of chaos theory. But patterns emerge even in the chaotic behavior, and these can be related to statistical models. At high energies, nuclei had too many resonances to study individually. But the total activity induced in a sample by a spectrum of fission-energy neutrons can be measured, by which means one can obtain a "nuclear level density," which is related to a kind of average cross-section. The neutron group created a facility for this work atop the reactor, using "converter plates" to convert the reactor-supplied thermal neutron spectrum into a fission-energy spectrum, and their results proved valuable in fundamental nuclear theories (Carter, Hughes, and Palevsky 1954).

12. All that remains of the imposing-looking security booth is the outline on the floor in green tile in the reactor building, looking like a mosaic of a champagne cork. The reason for the Soviet close copying of the U.S. work no doubt was that they were receiving regular information from Klaus Fuchs relating to the U.S. nuclear weapons program. Another important publication was BNL 400, *Angular Distributions.*

13. Neutrons, like all particles, have a wavelike nature, and when they pass through the ordered, latticelike array of atoms in solids they fan out, or scatter, giving rise to interference patterns among the waves. These patterns may be likened to those created by waves passing through a set of pilings; the patterns can in principle be examined to reveal the relative positions of the pilings. The patterns are easiest to interpret when the sizes of the waves are roughly the same as those of the passages, and by a happy coincidence of nature the so-called thermal neutrons produced by reactors have just the right wavelength for examining interatomic structure. The scattering is said to be *elastic* when the neutrons bounce off, without change of energy, from essentially rigid forms. But the lattice itself vibrates, the pilings themselves are joined and undulate together as

it were, and a coherent, normal motion of a solid material is called a *phonon*. Some fraction of the neutrons therefore pick up and lose energy in the scattering process, called *inelastic* scattering. While elastic scattering can reveal the structure of the lattice, inelastic scattering can reveal its dynamics. To study inelastic scattering, experimenters must know the precise energy of the neutrons they use, and must use neutrons whose energies are low enough so that any gains or losses are detectable. The slow chopper was designed to use the time-of-flight technique with "cold" neutrons, neutrons of low energy ($<$ 0.005 eV) and consequently long wavelength. The energy of such neutrons is so small compared to lattice vibrational energies that whatever energy they gained or lost could be attributed to the low-energy collective motions they were being used to study. To obtain cold neutrons, Hughes installed on the chopper a beryllium filter that scattered away neutrons with an energy $>$ 0.005 eV, a wavelength $<$ 4 Å (Johnson et al. 1951). Elastic scattering had been done at Oak Ridge by Shull and Wollan, inelastic scattering by Brockhouse, both of whom spent time at Brookhaven during the 1950s. When Shull and Brockhouse shared the Nobel Prize in 1995, the citation remarked that Shull taught us where atoms were, Brockhouse what they did. The slow chopper was first installed at the BGRR's east face, then moved to the top of the reactor. An early report was Seidl et al. 1951. Hughes also exploited his knowledge of neutron optics to develop other kinds of instruments besides choppers. These included neutron mirrors, surfaces that reflected neutrons, with which he made key measurements in nuclear physics with a new sensitivity. One was of the neutron-electron interaction: The neutron was envisioned not as a point particle of zero charge but rather as having regions of opposite charge that canceled each other. Neutrons and electrons can therefore interact, as Enrico Fermi and Leona Marshall had already shown while Hughes was at Argonne. Hughes used neutron mirrors to make some of the most reliable early measurements of that interaction, an important test of nuclear theories of the day (Hughes et al. 1953).

14. They made an early observation of the neutron-phonon interaction (Johnson, Palevsky, and Hughes 1951); at around the same time, however, Brockhouse published a more definitive treatment of phonon scattering. Hughes and company measured the temperature dependence of inelastic scattering, or how the atomic undulations change with temperature. They found that inelastic scattering in iron was twice that predicted by theory, and attributed it to the fact that neutrons act like tiny magnets and that the contributions of magnetic scattering must also be taken into account; this, in turn, suggested a new way to study the magnetic properties of matter (Palevsky and Hughes 1952). Hughes, Palevsky, and company also studied helium II, the only substance that turns into a "superfluid" with peculiar properties close to absolute zero. Soviet physicist Lev Landau had attributed helium II's peculiar properties to low-energy collective excitations called rotons, and Palevsky and company were able to detect these rotons (Carter et al. 1955; Palevsky et al. 1957; Palevsky, Otnes, and Larsson 1958). In 1959, Landau was awarded a Nobel Prize for the theory of He II; definitive measurements of collective excitations in superfluid helium were later made by John Yarnell at Los Alamos. The neutron physics group studied critical magnetic scattering. A critical point is a temperature at which an unusual type of phase transition occurs, such as liquid helium's transition superfluidity. Near a critical point, magnetic systems begin to form locally ordered regions that scatter neutrons very effectively and whose lifetimes and spatial extent increase closer to the critical point. Palevsky and collaborators managed

to observe critical point scattering in iron, noting that small temperature variations around the critical point produced large differences in scattered intensity. Much of this work had been pioneered elsewhere, or was done more definitively elsewhere; researchers at Oak Ridge had done the earliest work, and the first publication of a magnetic neutron diffraction pattern was in 1949 (Shull and Smart 1949).

15. The paper on magnetic scattering gives a general formula for the energy and angular distribution of magnetically scattered neutrons, using a correlation function that describes how the particles move around in the system (expressing the probability that if you have a particle that experiences some disturbance at time t you have another one at distance r disturbed at t').

16. Kato, BNL Oral Interview, 29 November 1995.

17. Maurice Goldhaber, BNL Oral Interview, 30 August 1995.

18. Gertrude Goldhaber, unpublished reminiscences.

19. After the discovery of fission, Goldhaber drew up some early schemes for reactors and went to Washington to offer his services to Gregory Breit, the person initially in charge of the uranium project. Breit said that Fermi and Szilard had the same ideas, and gave Maurice the feeling he was one foreigner too many. When Maurice approached Robert Oppenheimer, he rebuffed Goldhaber curtly, and Goldhaber finally gave up. As a result, he stayed in Urbana during the war, measuring cross-sections.

20. Der Mateosian, BNL Oral Interview, 30 August 1995.

21. Maurice Goldhaber, BNL Oral Interview, 11 December 1990.

22. Ibid.

23. The generic term *spectrometer* describes instruments that resolve incoming radiation into discrete wavelengths (energies). An optical spectrometer, for example, uses a prism to separate the wavelengths of light in an incident beam, taking advantage of the way light of different wavelengths refracts by different amounts when passing though glass. Neutron spectrometers employ single crystals that when properly aligned will selectively reflect a beam of neutrons of a given energy at a given angle, thus allowing the experimenter to produce beams of neutrons with well-defined wavelengths (energies). Because of their wavelike properties, neutrons (like X rays) are selectively reflected by the regularly spaced atoms in the crystal volume. Constructive interference takes place between neutron waves of a given wavelength at specific angles, called (by analogy with what happens with X rays) "Bragg angles" after Sir William Lawrence Bragg, who first described the "Bragg condition," or at what angles of a given wavelength X rays will be diffracted by a crystal. By selecting neutrons reflected by a crystal at a given angle, it is thus possible to create monochromatic beams of neutrons, i.e., neutrons with specific wavelengths.

24. Sailor, BNL Oral Interview, 1 March 1989.

25. Ibid.

26. Records of early computing are in D.O. V 19.

27. Gerald Tape to Van Horn, 26 June 1953, D.O. V 19.

28. By October, Ralph Shutt's cloud chamber group (see chapter 9) was using it. When design work started on the lab's next major accelerator, the Alternating Gradient Synchrotron (the AGS), accelerator physicists found the computer indispensable. The department of nuclear engineering, which was trying to develop models to determine how parameters like neutron flux and temperature changed over time inside a reactor,

also discovered the value of computers. But the 409-2R was capable only of simple mathematical calculations. By 1955 it was clear that demands on computer time and ability were soaring and that the 409-2R was woefully inadequate. Palevsky's group built a nonprogrammable, multipurpose analyzer with a Williams-tube memory able to handle 1,024 channels; this was an early type of memory system in which spots on the face of tubes would be read out continuously by an electron beam. It was built by Marty Graham, was working by 1957, and operated for five years. For more on the background of AEC computers, see Seidel 1996.

29. The first cross-sections Sailor measured were of isotopes of indium, an element easily activated by neutrons. Sailor and Borst had used indium foils to measure the neutron flux distribution during the reactor's start-up, and had a lot of very pure indium samples conveniently on hand. After indium, he turned to the rare earths and ^{235}U. Rare earth cross-sections were of vital importance in reactor engineering. Reactors are loaded with enough fuel for an operating cycle of a certain period of time, and at the beginning of the cycle they have excess reactivity, or more neutrons than needed to sustain the reaction at the given power. Reactor engineers therefore incorporate "burnable poisons," neutron-absorbing elements with high cross-sections, with the fuel to control the reaction. Burnable poisons have to have a high cross-section over a wide range of energies, and on capturing neutrons have to convert to elements or isotopes of the same element all with very nearly the same chemical and mechanical properties as the original poison so that the structural material in which they were incorporated would not deteriorate. Rare earth oxides, such as EuO_2 and Dy_2O_3, were found to be good for this purpose, and cross-section studies were required to establish which and how much of these burnable poisons to use. This application was especially important in special reactors where refueling would be infrequent, as in nuclear submarines.

30. The refrigerator Sailor commenced to build was in technical language a "two-step adiabatic demagnetization refrigerator." *Adiabatic* simply means that the process is kept isolated from heat transfer with the environment, while *demagnetization refrigerator* refers to a method developed by the U.S. Nobel laureate in chemistry William Giauque. The method is easily described by analogy with the operation of a conventional household refrigerator, in which a gas is compressed, heating up in the process, and a heat exchanger then used to cool the compressed gas back down to room temperature. When this gas is abruptly expanded, it lowers the temperature to create the refrigerating effect. In Giauque's technique, a paramagnetic material (one in which the atomic spins are not aligned but point any which way) is subject to a magnetic field to line up the spins. The work done by the electromagnetic fields in aligning the spins is analogous to the mechanical work involved in compressing a gas, and the material heats up. The paramagnetic material is then cooled to a temperature of a few degrees above absolute zero in the aligned state, and the field removed. The disordering of the spins produces an additional cooling effect sufficient to reduce the temperature of the target, suspended below the paramagnet, to a few hundredths of a degree above absolute zero. It was "two-step" in that two paramagnets were used, one serving as a low-temperature reservoir to keep the other cool.

31. Lawrence Passell to author, 25 February 1997.

32. "Proposal for Annealing of Graphite Structure in the Brookhaven National Laboratory Reactor," 4 September 1953, D.O. V 71.

33. As, for instance, Haworth to Lloyd Berkner, 13 November 1952, BNL C-6624.

34. One of McReynolds most important experiments concerned an experiment to measure the fall of neutrons under the force of gravity, with a beam of neutrons created at the reactor and with detectors set up in the hot lab some fifty meters away. He found that the neutrons did indeed fall a few millimeters in traversing that distance under the influence of gravity, though the gravitational constant for neutrons was the same as that of ordinary materials, as far as he could tell; if it had been different, the result would have been a shock to the world of physics.

35. Dienes, BNL Oral Interview, 1 August 1995. In the summer of 1952, Dienes organized the first of a series of summer programs at Brookhaven in solid-state physics. One of the attendees was George Vineyard, a solid-state physicist from Missouri, who soon joined the lab and became one of Dienes's most valuable collaborators. In 1957, Dienes and Vineyard wrote *Radiation Damage in Solids,* which treated on a basic level the effect of radiation on solids and helped to bolster the image of solid-state physics as a basic branch of physics. Vineyard would rise up the hierarchy of the lab until 1973, when he became its director.

36. Kouts, BNL Video Interview, 15 May 1986.

37. Ibid.

38. Several elements have a high enough thermal neutron cross-section to become candidates for target elements, including uranium-235 (549 barns), lithium-6 (950 barns), and boron-10 (3,990 barns). Ordinary boron in the earth's crust consists of 80 percent ^{11}B and 20 percent ^{10}B. On absorbing a neutron, a ^{10}B nucleus flies apart in two fragments, an alpha particle and a lithium nucleus, with the two pieces dividing the 2.4 million electron volts of energy released. These two pieces would travel a short distance, localizing the damage they cause to the cell containing the original ^{10}B nucleus. Perhaps the earliest suggestion of this trick is Locher 1936. As suggested about 1950 by William Sweet, of the Massachusetts General Hospital and Harvard Medical School, this is particularly attractive for brain tumor radiation therapy, for while the blood-brain barrier retards the uptake of many compounds by the normal brain, no such barrier exists in tumors.

39. Lee Edward Farr, "Neutron Capture Therapy: Years of Experimentation—Years of Reflection," BNL 47087.

40. Minutes, AUI executive committee, 19 April 1951; minutes, AUI board of trustees, 20 April 1951; Lear 1951. Bonestell had worked in Hollywood as a special effects matte painter for movies that included *Citizen Kane, Destination Moon,* and *War of the Worlds,* and later turned his talents to painting dramatic pictures of future space flight missions and fanciful images of craggy landscapes on planets and satellites of the solar system, collaborating with space writers including Willy Ley and Werner van Braun. Bonestell's paintings captured the imaginations of scientists, politicians, and the public, and the impact of his work in inspiring space research was such that an asteroid was named after him.

41. Farr to Paul Teng, 7 November 1957.

42. Teng to Farr, 14 November 1957.

43. An interesting exchange between Gerald L. Geison and M. F. Perutz raising exactly this issue in connection with Pasteur's use of the rabies vaccination appears in the letters column of *The New York Review of Books,* 4 April 1966, 68–69.

44. Some files on this are in D.O. VII 10.

45. Tape to Van Horn, 27 August 1953, D.O. VII 22. See also "Research Reactor Charges," Brookhaven National Laboratory, 27 January 1955.

46. "Reactor Irradiation Services and Radioisotopes," Brookhaven National Laboratory, January 1951.

47. The nuclear engineering department's many activities during the 1950s cannot be described here. I cannot resist, however, recounting an episode that exhibits the cleverness and high spirits of its members in this period. In 1958, the department hosted a symposium on the fissionable material thorium 233, with an eye to its use as a possible reactor fuel. The conference organizer was Bernard Manowitz, who had worked during the war at Oak Ridge and had acquired a reputation as one of the lab's most entertaining pranksters. While planning the conference, he remembered that, during the Manhattan Project, fissile materials had been assigned code names: ^{238}U was "Tube Alloy," while ^{235}U was "Ore Alloy." Thorium's code name was "Myrna Loy," simultaneously carrying on the pattern and making a gag of it. That gave Manowitz an idea. Discovering that the eponymous actress lived in New York, he convinced her to drop in at the start of the conference. In his welcoming remarks, after recounting the story of thorium's code name to the nuclear physicists present, he announced, with a dramatic flourish of his hand, "And here's Myrna Loy!"—and contentedly watched the dropping jaws and popping eyes as she emerged from the wings of the old Camp Upton theater, an entrance surely as grand and startling as any she had made on-screen.

48. Fox to S. G. English, 24 August 1953, D.O. V 42.

49. Fox to Spofford English, 16 February 1953, D.O. V 42.

50. R. W. Powell, "Reactor Operations Division Monthly Report," January 1952; F. P. Cowan, "Health Physics Summary for January 1952."

51. R. W. Powell, "Reactor Operations Division Monthly Report," March 1952, August, November 1953. A good source of information about radiation releases by leakers as well as by any other source at the Brookhaven site is the monthly "Health Physics Summary," which was issued by health physics.

52. M. W. Weiss to J. B. H. Kuper, "Summary Report of Data from Stations O-7, O-8, O-9, and O-10 for Period April 1949 thru March 1952," 2 June 1952, D.O. III 18. A study comparing the fallout from a slug rupture and from a single atomic bomb test in Nevada showed a huge increase in background in the region after the latter. "[C]ontamination of the BNL site and environs due to the slug rupture of 8.24.53 was minor in comparison with that resulting from one of the test detonations in Nevada last spring [6 April 1953]. Ground contamination was not detectable in the former case but was severe and long lasting in the latter." L. Gemmell and F. P. Cowan to M. Fox, 24 November 1953. Defective tubes in the evaporation plant, allowing active wastes to leak into the condensate, caused a high spike in the effluent in the late 1950s; see "Health Physics Summary for July 1959."

53. Minutes, meeting of the visiting committee of the nuclear engineering department, 5 January 1954.

54. Van Horn to Haworth, 2 February 1954; BNL Log No. C-8110, 21 June 1954. The new flux was 2×10^{13} cm^{-2} s^{-1}. The change also meant a change in the neutron energy profile, in the shutdown schedules, and in the fuel management system. "An-

nouncing a Change in the Shutdown Schedule of the BNL Reactor," 20 February 1957, D.O. V 78.

55. M. M. Weiss, "Area Survey Manual," BNL 344, 15 June 1955. A network of monitoring stations continuously measured the radiation levels.

56. A good summary of practices at this time is in F. P. Cowan and L. Gemmell, "Waste Management Operations at Brookhaven National Laboratory," prepared for hearings of the Special Sub-committee on Radiation of the Joint Committee on Atomic Energy, 28 January–3 February 1959, D.O. V 66. Many environmental reports were compiled at the time and made publicly available. One good source, surveying Brookhaven's liquid radioactive releases to the environment in the period of the BGRR's lifetime, is Hull 1969. A list of BNL reports on atmospheric radiation and fallout is found in D.O. VI 8.

57. Haworth to Hughes et al., 7 August 1956, Reactor file.

58. Hastings, BNL Oral Interview, 18 August 1995.

59. U.S. Atomic Energy Commission, Construction Project Description FY 1960, Project 562-60, "High Flux Research Reactor," July 1958.

Chapter 9

1. Haworth, "Organizational Changes," 1 July 1953, D.O. V 29. Also, staff members of the former accelerator department working with the cyclotrons and the Van de Graaff were transferred to the physics department, while an accelerator development division was created to work on future accelerator projects. But initially the distinction between users and operators was not neat, and some Cosmotron department members were experimenters, while many physics department members devoted a significant amount of time to working on the machine.

2. Swartz, BNL Oral Interview, 26 February 1991.

3. "Summary of July 3rd [1952] Meeting of Experimental Program Committee," in "Cosmotron BNL" file, 1951–1953, Cosmotron drawer.

4. Adair, BNL Oral Interview, 15 February 1995.

5. Shutt, BNL Oral Interview, 16 January 1995.

6. Because strange particles did not follow the rules of strong interactions, Pais thought their decay might be an instance of the Fermi force governing beta decay, implying a longer decay time because the force was correspondingly weaker. Thus he sought "selection rules which would hold for the strong and electromagnetic but not for weak processes," and came up with one he called the even-odd rule. If one assigned the number 0 to all the traditional particles and 1 to strange particles, and called the sum of these values for the initial state n_i and for the final state n_f, then in all strong and electromagnetic interactions n_i and n_f are both even or both odd, while in weak decays one is odd and the other even. As Pais soon discovered, similar selection rules were being considered simultaneously by a number of Japanese physicists (Pais 1986, 518).

7. The principle is as follows: the gas in the chamber is thoroughly saturated until it can hold no more vapor at that temperature and pressure. A piston is then released, expanding the chamber and thus lowering the pressure inside so the gas supersaturates with vapor. The slightest disturbance inside the chamber—including the passage of

charged particles—will then set off condensation. The degree of expansion has to be carefully adjusted; with too little expansion no tracks appear, while too much creates a dense fog, obliterating the tracks.

8. Thorndike, BNL Oral Interview, 15 February 1989.

9. Shutt, BNL Oral Interview, 16 January 1995. Langsdorf had made a small prototype, but was unsure if it could become a useful particle detection device. Shutt introduced an important improvement: he substituted high-pressure hydrogen for ordinary air at normal pressure. This was desirable because hydrogen gas consists of protons, and collisions between single particles were of more interest than collisions between particles and the nuclei of the gases that make up ordinary air; the physicists would be able to understand the particle collisions in much more detail. Moreover, the high-pressure hydrogen would slow the particles, allowing more precise measurements. After building and testing a small model, the members of the cloud chamber group described their work at an APS meeting in November 1950 and published a paper on it in 1951 (Miller, Fowler, and Shutt 1951). The Shutt group first reported tracks of a diffusion chamber with high-pressure hydrogen using pion beams from the Nevis cyclotron in Shutt et al. 1951 before moving the device to the Cosmotron when it came into operation.

10. Thorndike, BNL Oral Interview, 15 February 1989.

11. The lab's first professional scanner was Mary Burns, a local housewife. At the beginning of 1952, after her husband had died leaving her with two young children to support, Burns looked for a job at Brookhaven. Personnel sent her to Shutt, who assigned her a room in the cloud chamber building along with a film projector. He and others taught her how to recognize the signatures of particles: pions, strange particles, electron-positron pairs. Burns quickly learned to scan pictures—at first, those taken at Nevis—and cull the interesting events to give to the physicists to analyze.

12. Morris to author, 15 April 1995.

13. Lederman to author, 23 October 1995.

14. Ibid. Lederman, too, was interested in strange particles, and in 1954 began some careful measurements of strange particle lifetimes that were considerably more precise than those made by cosmic ray experimenters. Meanwhile, groups from Princeton and Yale with their own chambers also set up shop on the floor of building 902, while groups from Cornell, Duke, Fordham, and Purdue collaborated with the members of Shutt's team on Brookhaven-built chambers.

15. Friedlander, BNL Oral Interview, 3 January 1993.

16. Ibid.

17. R. P. Risch to D. H. Gurinsky, "Cosmotron Magnet Coil Cooling Tube Failure: Work to Date," 21 April 1955.

18. "Progress Report on Cosmotron Failure of November 5, 1954," 20 January 1955, D.O. V 29.

19. Swartz, "A Shielding Manifesto," 5 October 1956, D.O. V 28.

20. Swartz, BNL Oral Interview, 26 February 1991.

21. The Christofilos story is a saga in itself. He was born in Boston in 1916 but moved seven years later to Athens with his father. An inquisitive and voluble youth, he built radio sets and was a devoted ham radio operator until 1936 when the Greek government outlawed amateur radio transmission. In 1938 he received a degree in electrical and mechanical engineering at the National Technical University in Athens. That

year, he joined a company that installed and maintained elevators in apartment and office buildings. When Germany invaded Greece, the elevator installation factory was commandeered as a truck repair shop. As repairing trucks put no serious drain on Christofilos's time, he read German science and technology books, which were cheap and widely available, including a book on particle accelerator design. At war's end, Christofilos founded his own elevator installation company, but began to work out designs for synchrotrons, thinking of ingenious ways to combine the varying magnetic field of the betatron with the radio frequency acceleration method of the cyclotron. In 1946, he applied for a patent in Greece and in the U.S., for what was essentially a synchrotron, but soon discovered the earlier work of McMillan and Veksler. He continued to work, and sent a paper on his ideas to Berkeley. Accelerator physicists there wrote him back pointing out that it was unsound. Christofilos learned from the comments and revised his ideas, applying for U.S. patents in 1949 and 1950. He went back to work, to invent what was a version of strong focusing, and sent a letter back to Berkeley describing his new scheme. At Berkeley this was treated as another crank invention and the document was filed without study.

22. License and Agreement, 29 May 1953. Acknowledging that Christofilos "is the sole owner of U.S. Letters Patent 2,531,028 and 2,567,904 and U.S. Patent Application S. N. 148,920 entitled 'Focusing System for Ions and Electrons,'" the agreement gave the government the right to use the strong focusing and other methods, for which Brookhaven would pay Christofilos ten thousand dollars.

23. *Washington Post,* 14 February 1958.

24. G. Kenneth Green to Burton H. Wolfe, 21 March 1958, Green correspondence files.

25. Minutes of the Alternating Gradient Synchrotron group, 27 January 1953; 3 February 1953; 10 February 1953; H.O. drawer AGS.

26. M. Hildred Blewett, "Notes on the Joint Meeting of the Alternating-Gradient Design Groups Held at Brookhaven," 10–11 March 1953, H.O. drawer AGS.

27. Letter, Haworth to T. H. Johnson, 21 August 1953. This and other important documents concerning the electron analogue have been collected and published by Martin Plotkin: "The Brookhaven Electron Analogue 1953–1957," BNL 45058, 18 December 1991. To give an idea of how feared was the phase transition problem: When strong focusing was discovered, the USSR was building a weak-focusing 10 GeV "synchrophasotron" (essentially an enlarged copy of the Bevatron) at its Joint Institute for Nuclear Research in Dubna (itself loosely patterned on CERN, only for the Soviet republics), which had its first beam in 1957 to become the most powerful accelerator in the world. After the discovery of strong focusing, Vladimirsky and Tarasov came up with a scheme to eliminate the need for phase transition by incorporating negative bends in the ring, and proposed and planned a 50 GeV machine to be built at Protvino (Serpukhov) with this feature. Parameters for these and other machines appear in the Appendix of the Proceedings of the 1967 Accelerator Conference, Cambridge, Mass.

28. Haworth to Johnson, 9 September 1953, Haworth correspondence files.

29. Adams to Haworth, 3 June 1954, D.O. VII 2.

30. John and Hildred Blewett to Haworth, 5 November 1953, Haworth correspondence files.

31. Letter, Haworth to Hildred and John Blewett, 19 October 1953, Haworth correspondence files.

32. The electron analogue was the only synchrotron ever built that used electrostatic lenses rather than magnetic devices. Although this type of design was completely different from anything done before, it could be built quickly. The longest procurement time for conventional accelerators is usually for the magnets and coils, and the analogue did not have these. Instead of a large, high-current power source, as a motor-flywheel arrangement, the analogue used electronic power supplies to provide linear sawtooth voltages to the electrostatic lenses. Tolerances on all components were critical: the exact shape of the pole tips of the electrostatic lenses, mechanical alignment, the beam injection system, the low-level and high-level radiofrequency systems, and the beam observation devices. Under Green's tutelage, all problems were worked out, and the analogue worked almost immediately. One of the worries was possible residual magnetic fields in the stainless steel electrostatic lenses. To preclude this, the lenses were degaussed in a setup in the Cosmotron building and then carried to the test shack. To prevent remagnetization by the earth's magnetic field, the people carrying the lenses walked along rotating their bodies to average out the earth's effect.

33. Numerous people were present who recollect this incident, including Martin Plotkin. The Serpukhov machine began operation in 1967.

34. Collins to Haworth, 8 February 1955, D.O. V 29.

35. Friedlander, Piccioni, Ed Salant, and Shutt to Haworth, 23 March 1955, D.O. V 29.

36. "Minutes of 1st Meeting of High Energy Policy Committee," 27 June 1955, D.O. V 37.

37. Goldhaber to author, 30 March 1998.

38. Memo from Goudsmit to Cool, et al., 18 May 1956. Galison also cited this memo as illustrating how "experimenters were being reconstructed by the machines on which they worked" (1997, 307), though he misidentifies Goudsmit as the Cosmotron's director. It is significant that Goudsmit, the physics department chair, had to step in authoritatively in place of the less forceful Collins, chair of the Cosmotron department.

39. Collins to Tape, 29 January 1957, D.O. V 29.

40. The standardization took place across several conferences, which began at a conference in December 1951 in Bristol, England, and extended through a conference in summer 1953 (see Proceedings of the Congrès International sur le Rayonnement Cosmique 1953), with results later published in Amaldi et al. 1954.

41. Looking for symmetries and other conserved properties in the comings and goings of the subatomic world was an old and profitable game; for instance, it had guided Gell-Mann and Pais in their work on strangeness. One important set of symmetries and conservation laws had to do with the behavior of wave functions, the equations used to describe particle interactions. Wave functions include variables relating to the spatial location, charge, and temporal direction of particles. Three key operations can be performed on wave functions: charge conjugation (C), parity, and time reversal. Charge conjugation is the operation of changing all the particles in an interaction to their antiparticles. When any state (particle or set of particles) changes to itself under the C operation, it is said to have "even" or "odd" charge-parity depending on whether the amplitude stays the same or changes sign. Gell-Mann and Pais had been exploring the applications of symmetries and conservation laws to strange particles, and were

the first to realize that the neutral K mesons known as thetas were not behaving as they ought with respect to charge conjugation symmetry, and were seemingly able to turn into anti-thetas in defiance of the presumed conservation law according to which a particle and antiparticle were quite distinct entities.

42. Adair, BNL Oral Interview, 31 January 1996.

43. Chinowsky to author, 31 March 1995. The Steinberger paper is Budde et al. 1956.

44. In subsequent years, particle mixing between the K_L and the K_S, as they are now called (or "the K-system") was extensively studied. The phenomenon of particle mixing has not been observed elsewhere (though it is expected to take place in the b-quark system) and some ideas of particle mixing are also involved in neutrino oscillation theory. Even in the b-quark system, however, the mixing is not as dramatic as in the K-system. For while the two K's have strikingly different decay properties—two vs. three pions, and lifetimes that differ by a factor of five hundred—b-quark particles have many different decay channels, making for a less dramatic difference in decay modes and lifetime.

45. Searching the scientific literature for experiments with superheated liquids, Glaser came across a few articles by some Canadian chemists on superheated ether written in the early 1920s and promptly reduplicated their apparatus in work he reported in 1952. Later that year, he and his first graduate student, David Rahm, built a bigger chamber, with a volume of about two cubic centimeters, which was expanded by a hand crank. Using a high-speed motion picture camera, they managed to take pictures, for the first time, of a cosmic ray track in liquid, which were published in Glaser 1953.

46. Adair, BNL Oral Interview, 31 January 1996.

47. A. M. Thorndike to R. R. Rau, 16 February 1954, D.O. V 12.

48. Shutt, "Proposal and Cost Estimate for a Large Liquid Hydrogen Bubble Chamber Provided with Magnetic Field," June 1955, D.O. V 12.

49. Adair, BNL Oral Interview, 31 January 1996.

50. Leipuner to author, 1 June 1995.

51. Schwartz, BNL Oral Interview, 20 June 1995.

52. Van Horn to Haworth, 29 June 1953, Haworth correspondence files.

53. Samios, BNL Oral Interview, 20 November 1989.

54. Ibid.

55. Allan Thorndike, BNL Oral Interview, 15 February 1989.

56. Schwartz, BNL Oral Interview, 20 June 1995.

57. Charge conjugation (C) has already been mentioned, while time reversal (T) is the operation of reversing the direction of time in the wave equation, which is tantamount to saying that if you took a film of a subatomic particle interaction and ran it backward, you still have a film of a possible interaction. According to a fundamental theorem of quantum mechanics established in 1953, wave functions are invariant when all three operations—C, P, and T—are performed at once. It was also assumed (without proof) that wave functions were invariant under each individual operation; that when one performed the operation of switching the charges or parities or temporal direction, the wave equation would be unaffected. Physicists often refer to invariant properties as reflecting symmetries; in this case, they refer to charge conjugation, parity, and time reversal symmetry.

58. Dalitz also developed a second kind of plot for identifying resonances in which the dots are equally distributed if there is no resonance and are clustered if there is.

59. Under gauge symmetry, the value of a quantity can be arbitrarily changed from point to point. Gauge symmetry was based on an analogy with electromagnetism. In quantum electrodynamics, the phase of the wave function can be changed arbitrarily in space and time without affecting the electric charge, because the action of the electromagnetic field cancels out the alteration. Yang and Mills set out to construct a theory in which one could analogously shift isotopic spin arbitrarily, because what they called a *B-field* would counteract the change. Just as the electromagnetic field maintains the gauge symmetry of electromagnetic interactions about the phase by particles called photons, this B-field would maintain the gauge symmetry of strong interactions about the orientation of isotopic spin by particles called *vector bosons*. Though Yang and Mills hoped that their equations would show that vector bosons would convey a force identical to the strong force, they were unable to do so, and lost interest in the idea. This theory of Yang and Mills would transform the role of symmetry in quantum physics at the end of the next decade, by demonstrating that gauge invariance is a principle that can actively generate forces; indeed, it is the *only* principle that can generate forces. Alas, the authors tried to make their theory stick on the wrong phenomenon, and its proper place would not be found for a decade (Yang and Mills 1954; Yang 1983, 19–21, 172–76).

60. Adair, BNL Oral Interview, 31 January 1996.

61. Rau, BNL Oral Interview, 13 November 1990.

62. A number of philosophers and sociologists of science, including Collins, Pinch, Schaffer, and Shapin, have argued that replication, supposedly the basis of science, is at least very hard and nearly always requires direct contact with the original experimenters, therefore hardly ever occurs—and without it, there is no empirical base to science, which is instead a feat of construction. But it is an error to suppose *that* kind of replication to be the basis of science. One needs to draw a distinction between an experiment, which as a performance gives one profile of a phenomenon, and the phenomenon itself, which can be viewed through an infinite number of profiles. Parity violation in the weak interaction was a phenomenon with a number of different profiles (Crease 1993; Eger 1997).

63. Adair, BNL Oral Interview, 31 January 1996.

64. Smith to S. Winter, 11 November 1957, D.O. V 23.

65. "Investigation of Cosmotron Coil Failure of Nov. 5, 1957," 25 November 1957, D.O. V 23.

66. Adair, BNL Oral Interview, 31 January 1996.

67. Goldhaber, BNL Oral Interview, 31 August 1995.

68. Ibid.

69. Ibid. Robert Adair once remarked that, while in most scientific discoveries one feels that had the actual discoverers missed the boat the discoveries would still have been made in a few years, this one was different. It was so extraordinarily clever that had Maurice Goldhaber not existed, Adair said, "I am not sure that the helicity of the neutrino would ever have been measured." Adair, "Landmarks in Particle Physics at Brookhaven National Laboratory," BNL-52129.

70. Adams, "Estimate of the Staff Required for the CERN Proton Synchrotron," CERN-PS/Admin. 16/JRA, 12 August 1955.

71. Weinberg to author, 11 July 1995.

72. As the project neared completion, Haworth began to realize that the lab was not well prepared for the complexity of problems that would accompany the new scale of research. The already constructed experimental area was now "completely inadequate," Haworth wrote AEC area manager Van Horn. "Particle beams have grown in length and complexity. Thus, modern high energy experiments usually involve beam layouts up to several hundred feet of length which include large numbers of analyzing, bending and focusing magnets, electrostatic beam separators, vast arrays of counters and very large bubble chambers." A series of small movable houses for the experimenters, which would have been plush for cloud chambers, were woefully inadequate, and Haworth requested close to $2 million to build a new target area and equip it with overhead cranes and hydrogen safety equipment for use with liquid hydrogen bubble chambers. Haworth recalled Rodney Cool, away on leave at Berkeley, to undertake primary responsibility for the program of preparing to use the AGS. Letter, Haworth to Van Horn, 2 November 1959; Haworth to Cool, 23 May 1958, Haworth correspondence files.

73. Green's recollections of that week are in the *Brookhaven Bulletin*, 30 July 1960.

74. AUI executive committee, 14 October 1959. M. Q. Barton to G. B. Collins, 3 February 1960, D.O. V 29.

75. Collins to Haworth, 11 January 1960; Collins to Haworth, 11 February 1960, D.O. V 29.

76. Haworth to Smith, 4 April 1960, D.O. V 29.

77. Haworth, "Organizational Announcement," 6 May 1960, D.O. V 29.

Chapter 10

1. Haworth to Donald Quarles, 1 December 1953; Haworth to Lewis L. Strauss, 20 June 1958, Haworth correspondence files.

2. Kouts, BNL Video Interview, 15 May 1986.

3. Weinberg to Haworth, 6 November 1957, Haworth correspondence files.

4. Brookhaven was one of eight "multiprogram laboratories"; there were also five project engineering laboratories (including Bettis, Sandia, and Knolls Atomic Power Laboratory), two production plant laboratories (Hanford and Savannah River), and five university laboratories (including the Cambridge Electron Accelerator and Princeton-Pennsylvania proton accelerator labs). The multiprogram laboratories were further subdivided into national laboratories (Argonne, Brookhaven, and Oak Ridge); weapons laboratories (Los Alamos and Livermore); and on-campus laboratories (Ames Laboratory at the University of Iowa, and Berkeley). U. S. Congress, Joint Committee on Atomic Energy, "The Future Role of the Atomic Energy Commission Laboratories," October 1960, pp. 9, 11, 12, 33.

5. On AUI's participation in the National Radio Astronomy Observatory, located in West Virginia, see Needell 1987.

6. Seaborg to author, 8 January 1992.

7. Minutes, AUI executive committee, 20–21 July 1961.

8. See, for instance, Morris Glasoe to Haworth, 15 March 1961, Haworth correspondence files.

9. Goldhaber, BNL Oral Interview, 30 November 1995.

10. Goldhaber, BNL Video Interview, 25 March 1986.

11. Previous PSAC/GAC panels had met in November 1958 and December 1960, to discuss ten-year programs; frequent review of long-range plans seemed appropriate in such a rapidly moving field. The Ramsey committee's report is entitled "Report of the Panel on High Energy Accelerator Physics of the General Advisory Committee to the Atomic Energy Commission and the President's Science Advisory Committee," 26 April 1963. The reasons for Brookhaven's decision against the storage ring idea are articulated in the report submitted to the board of trustees on 2 January 1964. Storage rings were not as versatile an instrument as conventional fixed-target machines: they permitted collisions of just one kind of particle at a small number of locations, and collided them with just one kind of target. For the increase in energy they sacrificed a great deal of luminosity (a measure of the collisions/second). Equipment for studying such collisions was not well advanced. Finally, Schwartz, Lederman, and other Brookhaven experimenters interested in neutrino beams were vigorously opposed to storage rings at the AGS, preferring that resources be devoted to upgrading the AGS's intensity, which would boost prospects for their own neutrino beam research.

12. The phrase "big science" had been coined the previous year by Oak Ridge National Laboratory director Alvin Weinberg to refer to the recent increase in scale of large scientific projects.

13. Yuan, BNL Oral Interview, 21 October 1994.

14. "Current efforts to keep government-owned research and production facilities operating at artificially high levels, long after completion of their missions, can establish precedents that will be felt by many business and educational institutions," the article began. The author, who had just returned from a tour of the Soviet Union, where he had examined its academic institutions, was an advocate for sharply restricting the scope of national labs (Parsegian 1962). Similar attacks followed.

15. *Washington Evening Star,* 9 March 1965.

16. Letter, Goldhaber to Van Horn, 5 May 1965, H.O. drawer 200 BEV.

17. Lee E. Koppelman to Goldhaber, 28 May 1965, H.O. drawer 200 BEV.

18. Goldhaber to McMillan, 16 June 1965, H.O. drawer 200 BEV.

19. Samuel Devons, "Comments on the High Energy Physics Program in the United States," 13 October 1965, H.O. drawer 200 BEV.

20. Wilson to McMillan, 27 September 1965, H.O. drawer 200 BEV.

21. Wilson to Tape, 19 October 1965; Wilson, "The 200 Bev Synchrotron," 22 December 1965, H.O. drawer 200 BEV.

22. "Since it is no longer clear that a site near the Radiation Laboratory will be chosen or that Berkeley will be assigned the responsibility for design and construction of the machine, we consider ourselves a contender for the proposed accelerator." Memo, Clarke Williams to Charles R. Dominy, 10 January 1965, H.O. drawer 200 BEV.

23. Unattributed letter to Goldhaber, 21 January 1966, H.O. drawer 200 BEV.

24. Goldwasser to McDaniel, 18 January 1966, H.O. drawer 200 BEV.

25. Rockefeller to Glennan, 22 March 1966, H.O. drawer 200 BEV.

26. Glennan to Seaborg, 4 May 1966, with an attached appendix, "Suitability of the Brookhaven National Laboratory as a Site for a Super-Energy Accelerator," May 1966, H.O. drawer 200 BEV.

27. In retrospect, many ironies are apparent in the episode, especially in the various arguments made about the virtues of consolidating facilities versus spreading them around. Brookhaven's arguments for having the machine at its lab so that the AGS could

be used as an injector, rather than starting over from scratch, were good ones—and they would be the same arguments Fermilab would use years later, unsuccessfully, in trying to attract the Superconducting Supercollider to Illinois.

28. "Notes of Coordination Meeting on Construction of Large Liquid Hydrogen Bubble Chambers," 28 April 1965, "H" Street, Washington, D.C. H.O. drawer 14 Foot BC.

29. W. B. Fowler to C. Falk, 26 May 1965. H.O. drawer 14 Foot BC.

30. Shutt to Goldhaber, "Proposal for 16-Foot Chamber," and "Proposal for Construction of a 14-Foot Diameter Liquid Hydrogen Bubble Chamber for Use at the Brookhaven Alternating Gradient Synchrotron," 1 June 1964, BNL 8266. 46,000 liters of liquid hydrogen would be visible to three cameras, at a total project cost of $16.3 million. A decision to use superconducting magnetic coils with no iron return yoke eliminated about three thousand tons of iron from the magnet and allowed for a higher magnetic field, BNL 9695.

31. Shutt to Goldhaber, 23 June 1964. H.O. drawer 14 Foot BC.

32. Goldhaber to Paul W. McDaniel, 27 July 1964. H.O. drawer 14 Foot BC.

33. Shutt and Fowler viewed the cancellation as indicating a new lack of interest on the part of the government in scientific justifications in favor of political concerns. They complained to Goldhaber about the AEC's actions, although the letter seemed equally directed at the lab administration and its budgeting process:

> [W]e are treated here just like a manufacturing concern; our project is discarded by those responsible, as if we had asked for it for our own personal profit: there has been no apparent concern for the fact that a very large bubble chamber is needed in some years for a logical continuation of our work, for the fact that Brookhaven is one of the very few centers in the world where now the necessary experience and staff exists for such a project, and for the fact that most of the ideas for the project have originated here. If there had been such concern, before an important scientific project is just disregarded, we might have expected some discussion of questions such as "What would happen if the project were discarded or delayed?" or "What money might be needed to give the project a chance to get started?" Instead we are merely given the opportunity to stare into a vacuum. Those ultimately responsible indeed have a lot to learn; so far their treatment of scientific projects in the same manner as government manufacturing contracts reveals an appalling incompetence.

Fowler and Shutt to Goldhaber, 13 January 1965, H.O. drawer 14 Foot BC.

34. Letter, Edwin Goldwasser to Shutt, 26 August 1970, H.O. drawer 14 Foot BC.

35. The liability problem had not been addressed in the 1946 Atomic Energy Act because all nuclear reactors at that time were government owned and operated and thus self-insured. The architects of the 1954 bill revising the original act also bypassed the liability issue, for two reasons. One was that no information was available on potential liability, and sponsors were eager to push the bill through and did not want to wait for results of a study. Another was that operation of privately owned nuclear power plants was years in the future and the question did not seem urgent (Mazuzan and Walker 1984).

36. Letter, Anderson to K. E. Fields, 6 July 1956, D.O. III, 17. At the beginning of 1956, Anderson and Price had introduced a bill into Congress limiting company liability, with the government providing additional coverage, but it never came to a vote that session. Knowing that Congress would not write the utilities a blank check, Anderson sought to know how much was involved. A report on "runaways" or what might happen

in event of an accident at a reactor, had been made at the Atoms for Peace meeting in Geneva the year before; Anderson wanted a more thorough study.

37. Telegram, Louis H. Roddis, Jr., to Haworth, 29 August 1956, D.O. III 17.

38. Kuper to N. Dernbach, 30 April 1970, D.O. III 17.

39. "Theoretical Consequences of Major Accidents in Large Nuclear Power Plants: A Study of Possible Consequences if Certain Assumed Accidents, Theoretically Possible but Highly Improbable, Were to Occur in Large Nuclear Power Plants." United States Atomic Energy Commission, March 1957, WASH-740, p. 2.

40. Ibid., p. 23. It should be noted that, although the Chernobyl accident was very bad news, there was a core meltdown, but no China syndrome phenomenon occurred.

41. It required reactor operators to purchase the maximum amount of coverage offered by private insurance carriers: about $60 million. The operators then become eligible for a federal insurance program created by the act providing additional coverage of up to $500 million. The coverage extended to contractors and subcontractors who worked on the reactor, who otherwise might be held liable. If damages of an accident exceeded $560 million, claimants were out of luck.

42. *Business Week,* 24 August 1957, 124.

43. "A major reason for reconsidering [WASH-740]," states an AEC memo, "was that many people feel that new estimates will be lower." Albert P. Kenneke to steering committee, 9 February 1965, H.O. drawer WASH.

44. Minutes, steering committee, 21 October 1964, H.O. drawer WASH.

45. "The 1957 Brookhaven Report has been a serious public relations liability in the promotion of nuclear power installations" (Pesonen 1965, 245), and the report has been brought up again and again in debates about nuclear power. In 1969, an article in *Natural History* (Curtis and Hogan 1969) attacked the WASH report for vagueness. In an unpublished response, Kouts concurred, citing the ground rules laid down by the AEC. Letter, H. Kouts to editor, *Natural History Magazine,* 23 July 1969. But he complained that while WASH-740 had been written as a laboratory report and meant to be read as such, selected ideas in it were regularly excerpted and used out of context in support of preconceived ideas. One of the most extreme examples of such excerption occurred in the 1979 movie *The China Syndrome,* in which the character Dr. Eliott Lowell, a "physics professor," first explains to the characters played by Jane Fonda and Michael Douglas the meaning of the phrase providing the title of the movie. When Dr. Lowell tells them that a meltdown accident could "render an area the size of Pennsylvania permanently uninhabitable," he is paraphrasing part of the WASH-740 update.

46. Rau to Samios, 9 October 1963.

Chapter 11

1. Green, "A Brief History of Accelerators and Accelerator Management," undated but clearly around 1961, Archives box 3, file "Accelerator History."

2. Ibid.

3. Lederman to author, 23 October 1995.

4. J.-M. Gaillard, L. Lederman, M. Schwartz, and J. Steinberger, "A Proposal for the Study of High Energy Neutrino Interactions at the A.G.S.," 26 May 1960, AGS proposal files.

5. Lederman and Schwartz to HEAC, 22 September 1960, "BNL-Columbia Neutrino Collaboration," AGS proposal files.

6. Goldhaber, introduction to "This Was the Particle Physics That Was: The Years from P and C Violation to CP Violation," BNL 52214.

7. Shutt to Cool, 9 June 1961, D.O. V 64.

8. Lederman to author, 23 October 1995.

9. To hunt for the W, the neutrino group constructed a sixty-ton spark chamber in a specially built area southwest of the AGS between that machine and the Cosmotron. The pion-producing target was removed and a separate beam line built to carry protons to the shielding, vastly improving the number of pions and thus neutrinos. The idea was as follows: If a neutrino flying through their detector happened to collide with something and create a W plus a muon, the W would decay immediately to a muon or electron. The experimenters then would see *another* muon or electron at large angles from the first; because the W is massive, its decay would be able to send the much lighter particle off at a sharp angle. This "dilepton" signature would become a standard way of looking for W's in the next two decades.

10. Adair, BNL Oral Interview, 31 January 1996.

11. J. W. Cronin, V. L. Fitch, R. Turlay, AGS #181, 10 April 1963, AGS proposal files.

12. Adair, "Early Experiments," BNL 52129.

13. Ibid.

14. Adair's experiment is an interesting lesson in the history of science that reveals shortcomings of both the current realist and constructivist approaches to scientific practice. It is a mistake to treat his result as experimental error, as an "artifact," as something due to the experimenters rather than to what was "really there" (as some historians of science have done, e.g. Franklin 1986, 86). In noticing forbidden K decays, Adair was not reading something into the performance of the equipment that wasn't there. The decays were in fact happening, as was confirmed when another experimenter borrowed the equipment and obtained a similar result. To call misapprehension of a new phenomenon of nature the same as a signal artificially produced by a machine (an artifact) is a gross misunderstanding of the experimental process. That Adair could not pick out the explanation that was to become universally accepted did not mean the performance of his equipment was "unnatural" or due to poor experimental practices. But neither did it mean that this subsequently accepted explanation was entirely an artifact of discourse, of talk, the outcome of a negotiation of interests. Rather, it is an example of how new experimental performances are needed to understand the enigmas of the present. In the light of the new experiment by Fitch and Cronin, existing concepts were able to be developed and extended, and applied not only to their experiment, but also to Adair's. Another such prediscovery was made by a team from the University of Illinois, who found a hint of CP violation in a spark chamber also positioned in Inner Mongolia.

15. Adair, "Early Experiments," BNL 52129.

16. Limon, AGS Experimental Proposal #549, "Proposal to use the Multiparticle Magnetic Spectrometer to Detect Lepton Pair Production," received 22 September 1970, AGS proposal files.

17. Limon to author, 10 January 1997.

18. These input particles would be produced in the thicket of secondary particles produced in collisions between AGS protons and a target, and part of a group's work

would be to develop particle beam separators to select out the desired input particles. Beam separators worked with crossed electric and magnetic fields, arranged so that the desired input particle moving at a certain velocity would sail straight forward, with all others bent to one side. The limitation on a beam separator was the maximum electric field that could be produced, and experimenters used glass plates and filled the space in between with inert gas to get as high a voltage as one could without sparking. Beam separators had been introduced at Berkeley, and the first ones had been for pi beams. Pis are produced one hundred times more frequently than K's, so to get K's one needed an intense beam. Schwartz and others had tried to create a K beam separator at the Cosmotron, without success, but at the AGS it was possible. The first beam separator at the AGS was built by a collaboration between the BNL-Yale and BNL-Syracuse groups for use at the 20″, and it could be used to select either antiprotons or K⁻s around a 2.24 GeV momentum limit; all one had to do was change the tuning of the electric and magnetic fields. Meanwhile, at Berkeley, Alvarez was developing a technology to obviate the need for scanners with a device to read bubble chamber photographs electronically; at Brookhaven, scientists built their own device, a Hough-Powell flying spot digitizer; Powell was at CERN, Hough at Brookhaven.

19. Gell-Mann, Caltech report CTSL-20. A similar scheme was simultaneously developed by Israeli physicist Yuval Ne'eman.

20. Samios, BNL Interview, 20 November 1989.

21. Ibid.

22. Samios and Leitner to Rau, 5 September 1962, in AGS proposal files.

23. This information was vital for placing resonances in families, but it was more difficult to obtain for rarer particles. If a particle had 0 spin, the decay particles come out isotropically; isotropic decay is, in fact, the signature of 0 spin. If a particle has a spin, the decay products do not come out isotropically. Determining the spin of a particle therefore required looking at enough events to determine the angular distribution of the decay. When Gell-Mann made his omega minus prediction, he knew the J^P value only of the delta in the decimet (3/2+), and the J^P of the sigma star wasn't known yet. It was happy news when the BNL-Syracuse group determined that it was 3/2+, like that of the delta.

24. Rau to Samios, 31 October 1963, Samios correspondence.

25. Samios, BNL Oral Interview, 12 April 1995.

26. Ibid.

27. Rau to Samios, 11 December 1963, Samios correspondence.

28. Samios, BNL Oral Interview, 12 April 1995.

29. Neale to author, 25 March 1997.

30. Samios, BNL Oral Interview, 12 April 1995.

31. Matthews admitted to knowing of the discovery in a letter sent to Goldhaber (21 February 1964), in Archives, box 4, folder "Information Division: Publicity—Omega Minus."

32. Bubble Chamber Film Analysis Groups (Birmingham, Glasgow, Imperial College, Munich, Oxford and Rutherford Laboratory), "A Possible Ω−," no date.

33. More specifically, they showed that the size of the effect was linked with the presence of types of quark that existed—and they showed that the size of the effect could not be explained on the basis of four quarks alone. CP violation thus played in

some ways an analogous role in the discovery of the third generation of quarks that the discovery of the deviations in the orbit of Uranus did in the discovery of the planet Neptune; these deviations could not be adequately explained on the basis of a seven-planet solar system, and were taken to indicate the presence of an eighth planet, still further out from the sun, which was then hunted for and found.

34. Samios, BNL Oral Interview, 12 April 1995.

35. Lederman to author, 23 October 1995.

Chapter 12

1. Hendrie, BNL Oral Interview, 11 November 1992.

2. One discussion of this issue, following completion of the BMRR, is in AUI, minutes of the executive committee, 16 April 1959.

3. Hendrie, BNL Oral Interview, 11 November 1992.

4. Private communication, Hendrie to author, 20 May 1998.

5. Hendrie, BNL Oral Interview, 11 November 1992.

6. "Final Safety Analysis Report on the Brookhaven High Flux Beam Research Reactor," BNL 7661 (2 vols).

7. Hendrie, BNL Oral Interview, 11 November 1992.

8. Sweet inquired whether a cave for medical research might be built into the HFBR design, specifically for BNCT, which was not yet dead, but it was decided that that would be too costly and not provide enough advantages. Charles Falk to William Sweet, 28 July 1961, Falk correspondence files.

9. Herbert Kouts to Charles Falk, 28 January 1963, Falk correspondence files.

10. Goldhaber, "Research at the HFBR," April 1963; Goldhaber to scientific staff, 6 June 1963, Goldhaber correspondence files.

11. Sailor also installed a subsidiary beam at H-1 where a conventional spectrometer could be placed for use by visiting scientists. As at the BGRR, the group played host to a range of visitors, from Richard Wilson, a physics professor from Harvard, to Joseph Roberge, a high-school physics teacher from Bellport (where the school board president was Emery Van Horn) who worked with Sailor during the summertime.

12. Sailor, BNL Oral Interview, 1 March 1989.

13. Report—Associated Universities, Inc., November 1971, "Assessment of Energy Technologies: Report on Step 1," AET-7, submitted to Office of Science and Technology, contract OST-30. No authors listed.

14. Chrien, BNL Oral Interview, 8 February 1989.

15. More specifically, Chrien wanted to use the capture gamma method to identify the direct capture mechanism in neutron absorption by nuclei. Nuclei can absorb neutrons in two ways: by the resonance mechanism, which is a collective behavior of the nucleus involving creation of a compound nucleus; and by what is variously called *direct, hard sphere,* or *channel capture,* in which no compound nucleus is formed and a direct (and fast) reaction takes place between the incoming neutron and certain nucleons; direct capture takes place at energies between resonances. The existence of this latter type of nuclear reaction had been predicted, but as its cross-section was extremely low it had not been observed.

16. The group first looked at cobalt, known to have a strong resonance at 132 eV, in

a good range for the HFBR. First they examined the precise shape of the resonance, and then what the shape looked like off the resonance. They were able to detect an interference effect between the resonance mechanism and direct capture, the first experimental evidence for the direct capture mechanism. They reported on the result in fall 1966 at a conference at Argonne, and when the work was published it was the first experimental paper from the HFBR. Conference of Slow-Neutron-Capture Gamma-Ray Spectroscopy, Argonne National Laboratory, 2–4 November 1966; Wasson et al. 1966.

17. Kane kept nominal charge of the other still unused port, at which he later attempted to build an electron conversion spectrometer, yet another tool with which to study high-energy resonance states.

18. Schoenborn, BNL Oral Interview, 27 June 1995.

19. "Symposium on Inelastic Scattering of Neutrons by Condensed Systems," BNL 940, 20–22 September 1965.

20. This reorientation continued the following year with the addition of Vic Emery, a theorist also interested in many-body problems.

21. Kevey, BNL Oral Interview, 16 September 1995.

22. The computer engineering for this was done by Robert Spinrad of the instrumentation department, the program developed by Walter Hamilton of chemistry.

23. This and subsequent quotes are from Gen Shirane, BNL Oral Interview, 30 April 1994.

24. For an informal discussion of these systems, see Birgeneau and Shirane 1978.

25. Two of the magnetic systems they explored involved Ising magnets and Heisenberg magnets. Ernst Ising was a theorist who had developed a model of the ordering of atoms in magnets that assumed that the spins were pointed either directly up or down. While not a realistic model, as spins could be pointed in any direction, it had the tremendous advantage that exact calculations and predictions could be made with it, which could be compared with experimental results. While Ising's original model was in only one dimension, a two-dimensional model with these assumptions was developed; but for the much more difficult three-dimensional Ising problem, experimental input would be required. Corliss and Hastings produced that input, in a study of dysprosium aluminum garnet: $3\text{-}Dy_2O_3 * 5Al_2O_3$, known as DAG. With a contribution by Blume, this work also pointed the way toward the discovery of a new kind of magnetic field behavior, in a classic Brookhaven interplay between theory and experiment (Blume et al. 1974). The elaboration and completion was by Nick Giordano (1977). A Heisenberg ferromagnet owes its magnetism to *super exchange,* which is one of the two basic processes (the other is via *conduction electrons*) by which spins of ions in a crystal couple with each other. In super exchange, two ions sense each other via the outer electrons of an intermediary atom. In most ferromagnets, including iron, a combination of these two processes is at work—but theorists required accurate models of each. Corliss and Hastings examined the system in detail, helping to elaborate the theory (Tucciarone et al. 1971).

26. Oak Ridge had originally studied normal, disordered ice, ice-I, by neutron diffraction. Barclay Kamb, a Caltech crystallographer and Pauling's son-in-law, and Hamilton were the first to detect proton ordering in high-pressure ice.

27. "An Introduction to the Protein Data Bank," September 1972.

28. Koetzle to author, 1 July 1997.

29. Following the Bragg condition, scattering depends on the wavelength of the

neutron and the size of the object from which the scattering is taking place; the larger the scattering object, the smaller the angle at which peaks occur and the more difficult it is to collect information.

30. Sailor to users' committee, 9 February 1960, H.O. drawer HFBR.

31. A slow chopper would cut the beam into short bursts of slow neutrons for the study of the dynamical response of solids and liquids—a very different objective from the fast chopper program, which was aimed at nuclei. The slow chopper, moreover, would be complementary to a triple-axis spectrometer. First of all, it could do experiments at very low energies such as millivolts. Second, while a triple-axis spectrometer can examine a given point in energy-momentum space (reciprocal space) at a time, with choppers and time-of-flight measurements one can look everywhere in reciprocal space at once. As a consequence, choppers and time of flight are useful in situations where one is interested in nondispersive excitations, ones whose energy is independent of the direction of propagation. Triple-axis spectrometers are the instruments of choice when the situation is just the opposite—when one is studying a particular excitation traveling in a particular direction.

32. Nathans, BNL Oral Interview, 25 April 1995.

33. J. W. Dean and J. E. Jensen, "Direct Cooling of the Cold Neutron Moderator by Helium," 5 January 1973; revised 13 January 1973, Cold Neutron Facility.

34. Hendrie and Kouts, "A Program for the Development and Design of a Pulsed Fast Research Reactor," 1966; Kouts, "Repetitively-Pulsed Fast Research Reactors," BNL 12084, January 1968.

35. Memo, Paul W. McDaniel, "Close Out of Pulsed Reactor Development Work," 18 October 1968.

36. Hendrie, BNL Oral Interview, 11 November 1992.

36. Ibid.

Chapter 13

1. Quoted in a letter from Bond to Goldhaber, 14 April 1969, Goldhaber correspondence files.

2. Rau to David, 4 March 1971.

3. V. P. Bond to department chairmen and division heads, 8 August 1972; the relevant documents are AECM 0510, AECM 0513, and AECM 0524.

4. Ron Peierls, introductory note, "About 'Brookhaven/Facade,'" H.O. drawer public relations. *Brookhaven/Facade*'s full title was *An Examination and Historification of BROOKHAVEN as Performed by the Inmates of the Site at Upton under the Marquee of the Director's FACADE.* That was a takeoff on the full title of a play by Peter Weiss of a few years earlier, about a play put on by inmates of a mental institution outside Paris, *The Persecution and Assassination of Jean-Paul Marat as Performed by the Inmates of the Asylum of Charenton under the Direction of the Marquis de Sade*, a title universally shortened to *Marat/Sade*.

5. In 1969, Brookhaven's Theatre Group chose to produce *The Crucible,* Arthur Miller's 1953 play about innocent people condemned for absurd reasons on the basis of fabricated evidence, in which those who attempt to introduce reasonable discourse into the controversy bring upon themselves ridicule, contempt, vituperation, and persecution.

6. Memo, Goldhaber, 15 October 1969, Goldhaber correspondence files. Alterna-

tive organizations and newsletters had sprung up at several national labs; Argonne had the Concerned Argonne Scientists and Berkeley *The Real Lab News,* one of whose contributors, Charles Schwartz, was one of the four cofounders of Scientists and Engineers for Social and Political Action (SESPA) and one of the few important radical figures in the scientific community. See Gladstone 1987.

7. Director's office to department chairmen and administrative heads, "Distribution Brookhaven Free Press," 13 May 1970.

8. Boyce Rensberger, "Bellport: Friction with the 'Lab People,'" *The New York Times,* 13 June 1971, 100.

9. *Main Street Press,* 31 July 1970.

10. Letter, H.O. drawer public relations.

11. "Providing for the Identification of Unneeded Federal Real Property," Executive Order 11508, 10 February 1970, *Federal Register,* 35, no. 30 (12 February 1970): 2855–56.

12. Memo, Tucker to Vineyard, 11 June 1970, Vineyard correspondence files.

13. Report of GSA Field Survey of Brookhaven National Laboratory, 24 June 1970, H.O. drawer North Fork.

14. John A. Erlewine, H.O. drawer North Fork.

15. Minutes, department chairmen's meeting, 19 September 1970, H.O. Archives box 14.

16. Memo, Icarus Pyros to Vineyard, 26 February 1971, Vineyard correspondence files.

17. Telegram, Goldhaber and Rau to Wilson, 2 March 1972.

18. "First Report of an Ad-Hoc Committee on New Directions for Brookhaven," 20 March 1968, H.O. drawer ISABELLE.

19. Goldhaber to author, 30 March 1998.

20. Fitch committee report, H.O. drawer ISABELLE.

21. Much of the enthusiasm was due to promise shown by the work of a superconductivity group that had been established in the mid-1960s. Superconductors could carry large currents even in high magnetic fields, making them valuable in places, such as magnets, where power dissipation is a significant problem. In 1964, a number of people began to agitate for a group whose business it would be to develop and build superconducting magnets. In 1965, Goldhaber asked the AEC to support an expanded superconducting materials program at the lab, and the following year the group completed a superconducting magnet that made the cover of the lab's Christmas card. By the end of the decade, the superconducting group appeared to have solved all the important scientific problems associated with superconducting magnets.

22. "Report to the Trustees of Associated Universities, Inc., by the High Energy Study Committee," November 1971.

23. Minutes, department chairmen's meeting, 21 January 1972, H.O. Archives box 14.

24. J. P. Blewett and H. Hahn, eds., "200-GeV Intersecting Storage Accelerators ISABELLE—a Preliminary Design Study," BNL Report 16716 (1972).

25. Paul W. McDaniel to Goldhaber, 16 July 1971.

26. Minutes, department chairmen's meeting, 19 January 1973, H.O. Archives box 14.

A Note on Sources

The early historical archives at Brookhaven National Laboratory were surveyed and organized—it would not be too strong to say "rescued"—by Allan Needell (with the help of Joan Warnow) in the late 1970s and early 1980s, and an inventory of these archives is found in the pamphlet *Sources for the History of Physics at Brookhaven National Laboratory,* by Allan Needell (New York: American Institute of Physics, 1978). I refer principally to two sets of these records. The first is a group of records from the Director's Office dated 1946–1961, which are catalogued by series and folder; thus "D.O. V 15" refers to an item in the Director's Office set of records, series V, folder 15. These are also available on microfilm in the BNL research library. The second is a group of assorted archives, which include everything from memos and correspondence to blueprints and log books, which are catalogued by box number and name of the folder; thus "Archives, box 23 C. Swartz" refers to an item in a folder of box 23 in this set. Finally, a number of items are collected in drawers organized by subject in the Historian's Office; "H.O. drawer BGRR 'Pile Costs and Schedules,'" refers to an item in a folder of the BGRR drawer. I also had access to the minutes of the Board of Trustees and Executive Committee of Associated Universities, Inc., which I was allowed to quote with the kind courtesy of the AUI Board of Trustees. I also drew material from a set of videotaped interviews, made in the early to mid-1980s, with individuals who were involved in the founding of Brookhaven. This project was the idea of Lou Harson, the lab's principal architect (1976–1986) and an amateur historian. He and senior technical photographer Douglas Humphrey drew up a list of individuals involved in the founding of Brookhaven, packed up a camera and equipment, hauled them around in a rented station wagon, and conducted the interviews, which contain much material of interest. These are called "BNL Video Interviews," and identified by individual interviewed and date. I also conducted a number of tape-recorded interviews myself, which are identified as "BNL Oral Interviews" and identified by individual interviewed and date. Copies of both sets of interviews are in the Historian's Office. Quotations were checked with the original sources for grammar and content, and have occasionally been corrected. Other materials in the Historian's Office are located in personal files and correspondence files, as identified. Brookhaven National Laboratory publishes a series of reports (available in its library), which are identified by number; thus BNL 283 refers to Brookhaven Report # 283.

Other key sources on Brookhaven's early history include: Allan Needell, "Nuclear Reactors and the Founding of Brookhaven National Laboratory," in *Historical Studies in the Physical Sciences* 14:93–122 (1983); Norman F. Ramsey, "Early History of Associated Universities and Brookhaven National Laboratory," BNL 992; *Final*

Report of Work by Columbia University under Contracts W-7405-eng-50, W-31-109-eng-15, and AT-30-1-GEN-71, Upton, New York, 15 January 1948; and items in D.O. Appendix I.

To fully appreciate Brookhaven's scientific contributions, this book would need to be read alongside histories of each field in which it participated. Many marvelous histories already exist of the remarkable postwar developments in these fields. In physics (the field most extensively covered in this book), these include Laurie M. Brown, Abraham Pais, Brian Pippard, eds., *Twentieth Century Physics*, 3 vols. (London: Institute of Physics Publishing; New York: American Institute of Physics Press, 1996); Abraham Pais, *Inward Bound: Of Matter and Forces in the Physical World* (New York: Oxford University Press, 1986); Donald Kevles, *The Physicists: The History of a Scientific Community in Modern America* (New York: Knopf, 1987).

On the U.S. national laboratory system specifically, see Robert W. Seidel, "A Home for Big Science: The Atomic Energy Commission's Laboratory System," *Historical Studies in the Physical and Biological Sciences* 16 (1986): 135–75; and "Golden Anniversaries: The 50th Anniversaries of National Labs," forthcoming. For histories of Brookhaven's "sibling" laboratories at the birth of the national laboratory system, Argonne National Laboratory and Oak Ridge National Laboratory, see: Jack M. Holl, *Argonne National Laboratory 1946–1996* (Urbana: University of Illinois Press, 1997), and Leonard Greenbaum, *A Special Interest: The Atomic Energy Commission, Argonne National Laboratory, and the Midwestern Universities* (Ann Arbor: University of Michigan Press, 1971); Charles W. Johnson and Charles O. Jackson, *City Behind a Fence: Oak Ridge, Tennessee, 1942–1946* (Knoxville: University of Tennessee Press, 1961).

On the history of the Atomic Energy Commission, which funded the national labs, see the official history of the AEC: R. G. Hewlett and O. E. Anderson, Jr., *The New World, 1939–1946* (University Park: Pennsylvania State University Press, 1962); R. G. Hewlett and Francis Duncan, *Atomic Shield, 1947–1952* (Berkeley and Los Angeles: University of California Press, 1991); Richard G. Hewlett and Jack M. Holl, *Atoms for Peace and War, 1953–1961* (Berkeley and Los Angeles: University of California Press, 1989).

Bibliography

Abashian, A., et al. 1957. "Angular Distributions of Positrons from π^+-μ^+-e^+ Decays Observed in a Liquid Hydrogen Bubble Chamber." *Physical Review* 105:1927–28.

Adair, R. K. 1950. "Neutron Cross Sections of the Elements." *Reviews of Modern Physics* 22:249–89.

Adair, R. K. 1990. "CP-Nonconservation: The Early Experiments." In *CP Violation in Particle Physics and Astrophysics,* ed. J. Tran Thanh Van, 37–54. Paris: Editions Frontières.

Adams, J. B. . M. G. N. Hine, and J. D. Lawson. 1953. "Effect of Magnet Inhomogeneity in the Strong-Focusing Synchrotron." *Nature* 171:926–27.

Alvarez, L. 1989. "The Hydrogen Bubble Chamber and the Strange Resonances." In *Pions to Quarks: Particle Physics in the 1950s,* ed. L. Brown, M. Dresden, and L. Hoddeson, 299–306. Cambridge: Cambridge University Press.

Amaldi, E., et al. 1954. "Symbols for Fundamental Particles." *Nature* 173:123.

Anderson, H. L., et al. 1952. "Total Cross Sections of Positive Pions in Hydrogen." *Physical Review* 85:936.

Auerbach, T., et al. 1958. "Preliminary Design of a High-Flux Epithermal Research Reactor." Proceedings of the Second United Nations International Conference on the Peaceful Uses of Atomic Energy, Geneva, 10:60.

Bardon, M., K. Landé, and L. M. Lederman. 1958. "Long-Lived Neutral K Mesons." *Annals of Physics* 5:156–81.

Barnes, V., et al. 1964. "Observation of a Hyperon with Strangeness Minus Three." *Physical Review Letters* 12:204–6.

Bergreen, L. 1990. "Oh! How He Hated to Get Up in the Morning." *MHQ: The Quarterly Journal of Military History* 2:4.

Bertanza, L., et al. 1963. "Spin of the Y_1^*." *Physical Review Letters* 10:1976–79.

Bethe, H. 1937. "Nuclear Physics." *Reviews of Modern Physics* 9:69–224.

Birgeneau, R. J., and G. Shirane. 1978. "Magnetism in One Dimension." *Physics Today* 31:32–43.

Blewett, J. P. 1952. "Radial Focusing in the Linear Accelerator." *Physical Review* 88:1197–99.

Bloor, D. 1991. [1976]. *Knowledge and Social Imagery.* 2d ed. Chicago: University of Chicago Press.

Blume, M., et al. 1974. "Observation of an Antiferromagnet in an Induced Staggered Magnetic Field: Dysprosium Aluminum Garnet near the Tricritical Point." *Physical Review Letters* 32:544–47.

Boffey, P. M. 1969. "Ernest J. Sternglass: Controversial Prophet of Doom." *Science* 166:195–200.

413

Bohr, N., and J. Wheeler. 1939. "The Mechanism of Nuclear Fission." *Physical Review* 56:426–50.

Borst, L. B. 1949. "Nuclear Reactor at Brookhaven National Laboratory" (abstract). *Physical Review* 75:330.

Borst, L. B., et al. 1948. "Nuclear Reactor at Brookhaven National Laboratory." *Physical Review* 74:1883–84.

Borst, L. B., and V. L. Sailor. 1953. "Neutron Measurements with the Brookhaven Crystal Spectrometer." *Review of Scientific Instruments* 24:141–48.

Brockhouse, B. N. 1986. "A Childhood of Slow Neutron Spectroscopy." In *Fifty Years of Neutron Diffraction,* ed. G. E. Bacon, 35–46. Bristol: Adam Hilger.

Bromley, D. A. 1974. "The Development of Electrostatic Accelerators." *Nuclear Instruments and Methods* 122:1–34.

Bromley, D. A., and E. W. Vogt, eds. 1960. *Proceedings of the International Conference on Nuclear Structure.* Toronto: University of Toronto Press.

Brookhaven National Laboratory. 1948. "The Founding of the Brookhaven National Laboratory by Associated Universities, Inc." Final Report of Work by Columbia University under Contracts W-7405-eng-50, W-31-109-eng-15 and AT-30-1-GEN-71, Upton, New York.

Brown, L. M., A. Pais, and B. Pippard, eds. 1995. *Twentieth Century Physics.* 3 vols. London: Institute of Physics Publishing; New York: American Institute of Physics Press.

Brueckner, K. A. 1952. "Meson-Nucleon Scattering and Nucleon Isobars." *Physical Review* 86:106–9.

Budde, R. M., et al. 1956. "Properties of Heavy Unstable Particles Produced by 1.3-Bev π^- Mesons." *Physical Review* 103:1827–36.

Burns, R., et al. 1965. "Search for Intermediate Boson in Proton-Nucleon Collisions." *Physical Review Letters* 15:830–34.

Cabibbo, N. 1963. "Unitary Symmetry and the Leptonic Decays." *Physical Review Letters* 10:531–33.

Carson, R. 1962. *Silent Spring.* Boston: Houghton Mifflin.

Carter, R. S., D. J. Hughes, and H. Palevsky. 1955. "Energy Distribution of Cold Neutrons Scattered by Lattice Vibrations" (abstract). *Physical Review* 99:611.

Carter, R. S., et al. 1954. "Ratio of r/D for Slow Neutron Resonances." *Physical Review* 96:113–14.

Catton, W. R., and R. E. Dunlap. 1980. "A New Ecological Paradigm for Post-Exuberant Sociology." *Animal Behavioral Scientist* 24:15–47.

Chrien, R. 1980. "Investigating the Nucleus by Neutron Radiative Capture." *Transactions of the New York Academy of Sciences,* 2d ser., 40:40–64.

Christenson, J., et al. 1964. "Evidence for the 2π Decay of the K_2^0 Meson." *Physical Review Letters* 13:138–40.

Christenson, J., et al. 1970. "Observation of Massive Muon Pairs in Hadron Collisions." *Physical Review Letters* 25:1523–26.

Cohen, M. D., and J. G. March. 1974. *Leadership and Ambiguity.* Boston: Harvard University Press.

Collins, H. M., and T. Pinch. 1993. *The Golem.* Cambridge: Cambridge University Press.

Condon, E. U., and P. M. Morse. 1929. *Quantum Mechanics.* New York: McGraw-Hill.

Cool, R., O. Piccioni, and D. Clark. 1956. "Pion-Proton Total Cross Sections from 0.45 to 1.9 BeV." *Physical Review* 103:1082–97.

Corliss, L., and J. Hastings. 1953. "Neutron Diffraction Studies of Zinc Ferrite and Nickel Ferrite." *Reviews of Modern Physics* 25:114–21.

Courant, E. D. 1953. "Field Inhomogeneities in Alternating Gradient Synchrotron" (abstract). *Physical Review* 91:456.

Courant, E. D. 1997. "M. Stanley Livingston." *Biographical Memoirs of the National Academy of Sciences* 72:264–86.

Courant, E. D., M. S. Livingston, and H. S. Snyder. 1952. "The Strong-Focusing Synchrotron: A New High Energy Accelerator." *Physical Review* 88:1190–96.

Courant, E. D., M. S. Livingston, H. S. Snyder, and J. P. Blewett. 1953. "Origin of the 'Strong-Focusing' Principle." *Physical Review* 91:202–3.

Crease, R. P. 1993. *The Play of Nature: Experimentation as Performance.* Bloomington: Indiana University Press.

Crease, R. P., ed. 1997. *Hermeneutics and the Natural Sciences.* Dordrecht: Kluwer.

Crease, R. P., and C. C. Mann. 1996. *The Second Creation: Makers of the Revolution in 20th Century Physics.* New Brunswick: Rutgers University Press.

Crease, R. P., and N. P. Samios. 1991. "Managing the Unmanageable." *Atlantic,* January, 80–88.

Cumming, J. 1963. "Monitor Reactions for High Energy Proton Beams." *Annual Review of Nuclear Science* 13:261–86.

Cumming, J. B., G. Friedlander, and C. E. Swartz. 1958. "$C^{12}(p, pn)C^{11}$ Cross Section at 2 and 3 Bev." *Physical Review* 111:1386–94.

Cumming, J. B., and R. Hoffmann. 1958. "Efficient Low-level Counting System for C^{11}." *Review of Scientific Instruments* 29:1104–7.

Curtis, R., and E. Hogan. 1969. "The Myth of the Peaceful Atom." *Natural History,* March, 6–16, 71–76.

Dalitz, R. 1953. "On the Analysis of τ-Meson Data and the Nature of the τ-Meson." *Philosophical Magazine* 44:1068–80.

Dalitz, R. 1954. "Decay of τ Mesons of Known Charge." *Physical Review* 44:1046–51.

Danby, G., et al. 1962. "Observation of High-Energy Neutrino Reactions and the Existence of Two Kinds of Neutrinos." *Physical Review Letters* 9:36–44.

Davis, N. P. 1968. *Lawrence and Oppenheimer.* New York: Simon and Schuster.

Day, J. S., A. Krisch, and L. Ratner, eds. 1980. *History of the ZGS.* New York: American Institute of Physics.

de Solla Price, D. J. 1963. *Little Science, Big Science.* New York: Columbia University Press.

Dienes, G. J. 1952. "Mechanism for Self-Diffusion in Graphite." *Journal of Applied Physics* 23:1194–200.

Dienes, G. J., and G. Vineyard. 1957. *Radiation Damage in Solids.* New York: Interscience Publishers.

Donahue, J. J. 1918. "The Building of Camp Upton." *Public Service Record* 5:2.

Durant, F. C., III, and R. Miller. 1983. *Worlds Beyond: The Art of Chesley Bonestell.* Virginia Beach, Va.: Donning.

Eger, M. 1997. "Achievements of the Hermeneutic-Phenomenological Approach to Natural Science." In *Hermeneutics and the Natural Sciences,* ed. R. P. Crease. Dordrecht: Kluwer, 85–109.

Farr, L. E., and T. Konikowski. 1967. "Transplantable Mouse Neoplasm Control by Neutron Capture Therapy." *Nature* 215:550–52.

Feinberg, G. 1958. "Decays of the μ Meson in the Intermediate-Meson Theory." *Physical Review* 110:1482–83

Feinberg, G. 1988. "This Was the Particle Physics That Was: The Years from P and C Violation to CP Violation." BNL 52214.

Fermi, E. 1950. "High Energy Nuclear Events." *Progress of Theoretical Physics* 5: 570–83.

Fermi, E. 1953. "Multiple Production of Pions in Nucleon-Nucleon Collisions at Cosmotron Energies." *Physical Review* 92:452–53.

Feshbach, H., C. E. Porter, and V. F. Weisskopf. 1954. "Model for Nuclear Reactions with Neutrons." *Physical Review* 96:448–64.

Feynman, R. 1961. *The Theory of Fundamental Processes.* Reading, Mass.: W. A. Benjamin.

Fitch, V. 1981. "The Discovery of Charge-Conjugation Parity Asymmetry." In *Les Prix Nobel 1980,* 84–92. Stockholm: Nobel Foundation.

Fowler, W. B., et al. 1953a. "A Diffusion Cloud Chamber Study of Pion Interactions in Hydrogen and Helium." *Physical Review* 91:135–49.

Fowler, W. B., et al. 1953b. "Observation of V⁰ Particles Produced at the Cosmotron." *Physical Review* 90:1126–27.

Fowler, W. B., et al. 1953c. "Production of V_1^0 Particles by Negative Pions in Hydrogen." *Physical Review* 91:1287.

Fowler, W. B., et al. 1954a. "Production of Heavy Unstable Particles by Negative Pions." *Physical Review* 93:861–67.

Fowler, W. B., et al. 1954b. "Diffusion Cloud Chambers for Cosmotron Experiments." *Review of Scientific Instruments* 25:996–1003.

Fowler, W. B., et al. 1955. "Production of Heavy Unstable Particles by 1.37-Bev Pions." *Physical Review* 98:121–30.

Franklin, A. 1986. *The Neglect of Experiment.* Cambridge: Cambridge University Press.

Friedlander, G., et al. 1954. "Nuclear Reactions of Copper with 2.2 BeV Protons." *Physical Review* 94:727–28.

Fuller, S. 1993. *Philosophy, Rhetoric, and the End of Knowledge.* Madison: University of Wisconsin Press.

Furman, N. S. 1990. *Sandia National Laboratories: The Postwar Decade.* Albuquerque: University of New Mexico Press.

Galison, P. 1985. "Bubble Chambers and the Experimental Workplace." In *Observation, Experiment, and Hypothesis in Modern Physical Science,* ed. P. Achenstein and O. Hannaway, 309–73. Cambridge: MIT Press.

Galison, P. 1997. *Image and Logic: A Material Culture of Microphysics.* Chicago: University of Chicago Press.

Galison, P., and B. Hevly, eds. 1992. *Big Science: The Growth of Large-Scale Research.* Stanford: Stanford University Press.

Gell-Mann, M., and A. Pais. 1955. "Behavior of Neutral Particles under Charge Conjugation." *Physical Review* 97:1387–89.

Giordano, N. 1977. "Magnetic Phase Transitions in Dysprosium Aluminum Garnet." Ph.D. diss., Yale University.

Gladstone, K. 1987. "The Unfolding of the Nuclear Age: A Psychohistorical Investigation into the Lives of Ten Men." Ph.D. diss., Department of Psychology, University of California, Berkeley.

Glaser, D. 1953. "Bubble Chamber Tracks of Penetrating Cosmic-Ray Particles." *Physical Review* 91:762–63.

Goldhaber, M. 1993. "Reminiscences from the Cavendish Laboratory." *Annual Review of Nuclear and Particle Science* 43:1–25.

Goldhaber, M., L. Grodzins, and A. W. Sunyar. 1958. "Helicity of Neutrinos." *Physical Review* 109:1015–17.

Goldhaber, M., and R. D. Hill. 1952. "Nuclear Isomerism and Shell Structure." *Reviews of Modern Physics* 24:179–239.

Goldhaber, M., and G. Scharff-Goldhaber. 1948. "Identification of Beta-Rays with Atomic Electrons." *Physical Review* 73:1472–73.

Goldhaber, M., and A. W. Sunyar. 1951. "Classification of Nuclear Isomers." *Physical Review* 83:906–18.

Goldhaber, M., and J. Weneser. 1955. "Electromagnetic Transitions in Nuclei." *Annual Review of Nuclear Science* 5:1–24.

Goldsmith, H. H. W. Ibser, and B. T. Feld. 1947. "Neutron Cross Sections of the Elements: A Compilation." *Reviews of Modern Physics* 19:259–97.

Greenbaum, L. 1971. *A Special Interest: The Atomic Energy Commission, Argonne National Laboratory, and the Midwestern Universities.* Ann Arbor: University of Michigan Press.

Gurinsky, D. H. [N.d.]a. "Memoirs: Après la Guerre." Unpublished manuscript.

Gurinsky, D. H. [N.d.]b. "The Brookhaven Graphite Research Reactor: A Personal Memoir." Unpublished manuscript.

Hamilton, W. 1965. "Significance Tests on the Crystallographic R Factor." *Acta Crystallographica* 18:502–10.

Hawkins, C. J. B. 1967. "K_2° Interactions, Decays, and Regenerative Properties at 590 MeV/c in Liquid Hydrogen." *Physical Review* 156:1444–50.

Haworth, L. J. 1961. "Dedicatory Address." *Physics Today* 14, no. 12 (December): 23–30.

Heilbron, J. L., and R. W. Seidel. 1989. *Lawrence and His Laboratory: A History of the Lawrence Berkeley Laboratory.* Vol. 1. Berkeley and Los Angeles: University of California Press.

Herb, R. G. 1959. "Van de Graaff Generators." *Handbuch der Physik* 44, pp. 64–104. Berlin: Springer-Verlag.

Hermann, A., et al. 1987. *History of CERN.* Vol. 1. New York: North-Holland.

Hewlett, R. G., and O. E. Anderson, Jr. 1962. *The New World, 1939–1946.* University Park: Pennsylvania State University Press.

Hewlett, R. G., and F. Duncan. 1991. *Atomic Shield, 1947–1952.* Berkeley and Los Angeles: University of California Press.

Hewlett, R. G., and J. M. Holl. 1989. *Atoms for Peace and War, 1953–1961.* Berkeley and Los Angeles: University of California Press.

Hillaby, J. 1964. "Key Particle Found in the Atom, Ending Nuclear Physics 'Chaos.'" *New York Times,* 20 February, 1, 12.

Holl, J. M. 1997. *Argonne National Laboratory 1946–1996.* Urbana: University of Illinois Press.

Hollinger, D. A. 1996. *Science, Jews, and Secular Culture: Studies in Mid-Twentieth-Century American Intellectual History.* Princeton: Princeton University Press.

Holloway, M., and C. Baker. 1972. "How the Barn Was Born." *Physics Today,* July, 9.

Holton, G. 1973. *Thematic Origins of Scientific Thought.* Cambridge: Harvard University Press.

Hughes, D. J. 1953a. "The Nuclear Reactor as a Research Instrument." *Scientific American,* August, 23–29.

Hughes, D. J. 1953b. *Pile Neutron Research.* Cambridge: Addison-Wesley.

Hughes, D. J. 1954a. *Neutron Optics.* New York: Interscience Publishers.

Hughes, D. J. 1954b. "Statistical Factors in Radiation Widths of Nuclear Energy Levels." *Physical Review* 94:740–41.

Hughes, D. J., R. C. Garth, and J. A. Levin. 1953. "Fast Neutron Cross Sections and Nuclear Level Density." *Physical Review* 91:1423–29.

Hughes, D. J., and J. A. Harvey. 1954. "Radiation-Widths of Nuclear Energy-Levels." *Nature* 173:942–43.

Hughes, D. J., et al. 1953. "The Neutron-Electron Interaction." *Physical Review* 90:497–98.

Hull, A. P. 1969. "Liquid Radioactive Waste Disposal and Related Environmental Concentrations at Brookhaven National Laboratory." BNL 14797.

Jachim, A. G. 1967. *Science Policy Making in the United States and the Batavia Accelerator.* Carbondale: Southern Illinois University Press.

Johnson, C. W., and C. O. Jackson. 1961. *City behind a Fence: Oak Ridge, Tennessee, 1942–1946.* Knoxville: University of Tennessee Press.

Johnson, G. W., H. Palevsky, and D. J. Hughes. 1951. "Inelastic Scattering of Low Energy Neutrons in Graphite" (abstract). *Physical Review* 82:345.

Jungk, R. 1968. *The Big Machine.* New York: Scribner's.

Kamb, B. 1983. "Crystallography of Water Structures." In *Crystallography in North America,* ed. D. McLachlan and J. P. Blusker, 336–42. New York: American Crystallographic Association.

Kamb, B., et al. 1971. "Ordered Proton Configuration in Ice II, from Single-Crystal Neutron Diffraction." *Journal of Chemical Physics* 55:1934–45.

Kendrew, J. C., et al. 1960. "Structure of Myoglobin." *Nature* 185:422–27.

Kevey, A. 1991. *Béla Keredy: A Hungarian Odyssey.* Santa Barbara: Fithian Press, 1991.

Kevey, A. [N.d.] "The Plastic Duck." Unpublished memoir.

Kevles, D. 1978. *The Physicists: The History of a Scientific Community in Modern America.* New York: Knopf.

Kosiba, W. L., G. J. Dienes, and D. H. Gurinsky. 1956. "Some Effects Produced in Graphite by Neutron Irradiation in the BNL Reactor." In *Proceedings of the Conferences on Carbon, University of Buffalo,* 143–48.

Krige, J., and D. Pestre, eds. 1997. *Science in the Twentieth Century.* Amsterdam: Harwood Academic.

Landé, K., et al. 1956. "Observation of Long-Lived Neutral V Particles." *Physical Review* 103:1901–4.

Landé, K., L. M. Lederman, and W. Chinowsky. 1957. "Report on Long-Lived K Mesons." *Physical Review* 105:1925–27.

Lang, D. 1947. "A Reporter at Large: The Long Island Atoms." *New Yorker,* 20 December, 33–43.

La Placa, S., et al. 1969. "Nature of the Metal-Hydrogen Bond in Transition Metal-Hydrogen Complexes: Neutron and X-Ray Diffraction Studies of β-Pentacarbonyl-manganese Hydride." *Inorganic Chemistry* 8:1928–35.

Latour, B. 1987. *Science in Action.* Cambridge: Harvard University Press.

Latour, B., and S. Woolgar. 1979. *Laboratory Life.* Princeton: Princeton University Press.

Lear, J. 1951. "Atomic Miracle: Science Explodes an Atom in a Woman's Brain." *Collier's,* 21 April, 15–17.

Lederman, L., M. Schwartz, and J.-M. Gaillard. 1960. "A Neutrino Detector for Use at the Brookhaven AGS." In *Proceedings of the International Conference on Instrumentation, Berkeley 1960,* 201. New York: Wiley-Interscience.

Lee, T. D., and C. N. Yang. 1957. "Question of Parity Conservation in Weak Interactions." *Physical Review* 104:254–58.

Lee, T. D., and C. N. Yang. 1960. "Theoretical Discussions on Possible High-Energy Neutrino Experiments." *Physical Review Letters* 4:307–11.

Leipuner, L. B., et al. 1963. "Anomalous Regeneration of K_1^0 Mesons From K_2^0 Mesons." *Physical Review* 132:2285–90.

Lilienthal, D. E. 1964. *The Journals of David E. Lilienthal.* Vol. 2. *The Atomic Energy Years.* New York: Harper and Row, 1964.

Lindenbaum, S. J., and L. C. L. Yuan. 1953. "The Interaction Cross Section at Hydrogen and Heavier Elements for 450-Mev Negative and 340-Mev Positive Pions." *Physical Review* 92:1578–79.

Lindenbaum, S. J., and L. C. L. Yuan. 1954. "Total Interaction Cross Section of Hydrogen for Positive and Negative 200–700 Mev Pions" (abstract). *Physical Review* 93:917–18.

Locher, G. L. 1936. "Biological Effects and Therapeutic Possibilities of Neutrons." *American Journal of Roentgenology* 36:1–13.

Marshall, W., and S. W. Lovesy. 1971. *Theory of Thermal Neutron Scattering.* London: Oxford University Press.

Matthews, P. T. 1964. "Order out of Sub-nuclear Chaos." *New Scientist,* 20 February, 458–60.

Mazuzan, G. T., and J. S. Walker. 1984. *Controlling the Atom: The Beginnings of Nuclear Regulation 1946–1962.* Berkeley and Los Angeles: University of California Press.

McCaffrey, D. P. 1990. *The Politics of Nuclear Power: A History of the Shoreham Nuclear Power Plant.* Dordrecht, the Netherlands: Kluwer.

McMillan, E. 1945. "The Synchrotron: A Proposed High Energy Particle Accelerator." *Physical Review* 68:143–44.

Miller, D. H., E. C. Fowler, and R. P. Shutt. 1951. "Operation of a Diffusion Cloud Chamber with Hydrogen at Pressures up to 15 Atmospheres." *Review of Scientific Instruments* 22:280.

Morse, P. M. 1936. *Vibration and Sound.* New York: McGraw-Hill.

Morse, P. M. 1977. *In at the Beginnings: A Physicist's Life.* Cambridge: MIT Press.

Musolino, S., et al. 1964. "Comments on 'Breast Cancer: Evidence for a Relation to Fission Products in the Diet.'" *International Journal of Health Services* 25: 475–80.

Neagu, D., et al. 1961. "Decay Properties of $K_2^°$ Mesons." *Physical Review Letters* 6: 552–53.

Needell, A. 1978. *Sources for the History of Physics at Brookhaven National Laboratory.* Report under NSF Grant SOC 77-02247. New York: American Institute of Physics.

Needell, A. 1983. "Nuclear Reactors and the Founding of Brookhaven National Laboratory." *Historical Studies in the Physical Sciences* 14:93–122.

Needell, A. 1987. "Lloyd Berkner, Merle Tuve, and the Federal Role in Radio Astronomy." *Osiris* 3:261–88.

Néel, L. 1948. "Propriétés magnétiques des ferrites: ferrimagnétisme et antiferromagnétisme." *Annales de Physique* 3:139.

New Yorker. 1964. "Omega Minus." *New Yorker,* 7 March, 39–40.

Nichols, K. D. 1987. *The Road to Trinity.* New York: Morrow.

Noyes, H. P., et al., eds. 1952. *Proceedings of the Third Annual Rochester Conference.* Rochester: University of Rochester.

Orlans, H. 1967. *Contracting for Atoms: A Study of Public Policy Issues Posed by the Atomic Energy Commission's Contracting for Research, Development, and Managerial Services.* Washington: Brookings Institution.

Pais, A. 1952. "Some Remarks on the V-Particles." *Physical Review* 86:663–72.

Pais, A. 1986. *Inward Bound: Of Matter and Forces in the Physical World.* New York: Oxford University Press.

Pais, A., and O. Piccioni 1955. "Note on the Decay and Absorption of the Θ^0." *Physical Review* 100:1487–88.

Palevsky, H. 1962. "The Contributions to Pile Neutron Research of the Late Donald J. Hughes." In *Pile Neutron Research in Physics,* 3–7. Vienna: International Atomic Energy Agency.

Palevsky, H., and D. J. Hughes. 1952. "Cross Section of Fe for Cold Neutrons" (abstract). *Physical Review* 87:221.

Palevsky, H. K. Otnes, and K. E. Larsson. 1958. "Excitation of Rotons in Helium II by Cold Neutrons." *Physical Review* 112:11–18.

Palevsky, H. K., et al. 1957. "Excitation of Rotons in Helium II by Cold Neutrons." *Physical Review* 108:1346–48.

Parsegian, V. L. 1962. "Makework Projects Waste U.S. Brain Power: Unneeded Scientists Kept by Federal Facilities." *Nation's Business,* September, 106–7.

Pearson, D. 1948. "Washington Merry-Go-Round." *New Kensington (Pa.) Dispatch,* 3 January.

Pesonen, D. 1965. "Atomic Insurance: The Ticklish Statistics." *The Nation,* 18 October, 242–45.

Piccioni, O., et al. 1955. "External Proton Beam of the Cosmotron." *Review of Scientific Instruments* 26 : 232–33.

Pinch, T. 1982. "The Development of Solar-Neutrino Astronomy." Ph.D. diss., University of Bath.

Pinch, T. 1986. *Confronting Nature: The Sociology of Solar-Neutrino Detection.* Dordrecht: Reidel.

Polkinghorne, J. 1989. *Rochester Roundabout: The Story of High-Energy Physics.* New York: W. H. Freeman.

Postma, H., et al. 1962. "Transmission of Polarized Neutrons through Samples of Polarized Ho^{185} Nuclei." *Physical Review* 126 : 979–85.

Prentki, J., ed. 1962. *Eleventh International Conference on High Energy Physics.* Geneva: CERN.

"Proceedings of the Congrès International sur le Rayonnement Cosmique," Bagnères de Bigorre, 1953. Unpublished typescript.

Ramsey, N. 1967. "Early History of Associated Universities and Brookhaven National Laboratory," BNL #992.

Rensberger, B. 1971. "Bellport: Friction with the 'Lab People.'" *New York Times,* 13 June, 100.

Rhodes, R. 1986. *The Making of the Atomic Bomb.* New York: Simon and Schuster.

Richards, P., W. Tucker, and S. Srivastava. 1982. "Technetium-99*m:* An Historical Perspective." *International Journal of Applied Radiation and Isotopes,* 33 : 793–99.

Rigden, J. S. 1987. *Rabi: Scientist and Citizen.* New York: Basic Books.

Rochester, G. D., and C. C. Butler. 1947. "Evidence for the Existence of New Unstable Elementary Particles." *Nature* 160 : 855–57.

Rogers, M. 1990. *Acorn Days: The Environmental Defense Fund and How It Grew.* New York: Environmental Defense Fund.

Sailor, V. 1953. "Parameters for the Slow Neutron Resonance in Rhodium." *Physical Review* 91 : 53–57.

Sakurai, J. J. 1964. *Invariance Principles and Elementary Particles,* Princeton: Princeton University Press.

Scharff-Goldhaber, G. 1953. "Excited States of Even-Even Nuclei." *Physical Review* 90 : 587–602.

Schirmer, R. I., et al. 1968. "Spin Dependence of the U^{235} Low-Energy Neutron Cross Section." *Physical Review* 167 : 1121–30.

Schwartz, M. 1960. "Feasibility of Using High-Energy Neutrinos to Study the Weak Interactions." *Physical Review Letters* 4 : 306–7.

Schwartz, M. 1972. "Discovery of Two Kinds of Neutrinos: One Researcher's Personal Account." *Adventures in Experimental Physics* 1 : 82.

Schwartz, M. 1986. "Jack Steinberger: The Best Teacher I Have Known." In *Particles and Detectors: Festschrift for Jack Steinberger,* ed. K. Kleinknecht and T. D. Lee, 269–74. New York: Springer.

Schweitzer, D. G. 1993. "Experimental Results of Air Ingress in Heated Graphite Channels: A Summary of American Analyses of the Windscale and Chernobyl Accidents." IAEA Technical Committee Meeting on Response of Fuels, Fuel Elements and Gas-Cooled Reactor Cores under Accidental Air or Water Ingress Conditions, Institute for Nuclear Energy Technology, Tsinghua University, Beijing, China.

Seidel, R. W. 1982. "Accelerator Science: The Postwar Transformation of the Lawrence Radiation Laboratory." *Historical Studies in the Physical Sciences* 13, pt. 2: 375–400.

Seidel, R. W. 1986. "A Home for Big Science: The Atomic Energy Commission's Laboratory System." *Historical Studies in the Physical and Biological Sciences* 16: 135–75.

Seidel, R. W. 1996. "From Mars to Minerva: The Origins of Scientific Computing in the AEC Labs." *Physics Today,* October, 33–39.

Seidl, F. G. P., et al. 1951. "Thermal Neutron Time-of-Flight Velocity Selector." *Physical Review* 82:345.

Serber, R. 1947. "Nuclear Reactions at High Energies." *Physical Review* 72:1114–15.

Serber, R., with R. P. Crease. 1998. *Peace and War: Reflections of a Life on the Frontiers of Science.* New York: Columbia University Press.

Shull, C. G., and J. S. Smart. 1949. "Detection of Antiferromagnetism by Neutron Diffraction." *Physical Review* 76:1256–57.

Shutt et al. 1951. "π^--p Scattering Observed in a Diffusion Cloud Chamber." *Physical Review* 84:1247–48.

Slatkin, D. N. 1991. "A History of Boron Neutron Capture Therapy of Brain Tumors." *Brain* 114:1609–29.

Slatkin, D. N., et al. 1986. "A Retrospective Study of 457 Neurosurgical Patients with Cerebral Malignant Glioma at the Massachusetts General Hospital 1952–1981: Implications for Sequential Trials of Postoperative Therapy." In *Neutron Capture Therapy,* ed. H. Hatanaka, 434–46. Niigata: Nishimura.

Smith, A. K. 1970. *A Peril and a Hope.* Cambridge: MIT Press.

Smith, B. L. R. 1990. *American Science Policy since World War II.* Washington: Brookings Institution.

Smyth, H. D. 1945. *Atomic Energy for Military Purposes: The Official Report on the Development of the Atomic Bomb under the Auspices of the United States Government, 1940–45.* Princeton: Princeton University Press.

Steiner, G. 1989. *Real Presences.* Chicago: University of Chicago Press.

Thomas, L. 1983. "An Epidemic of Apprehension." *Discover,* November, 78–80.

Tucciarone, A., et al. 1971. "Quantitative Analysis of Inelastic Scattering in Two-Crystal and Three-Crystal Neutron Spectrometry: Critical Scattering from $RbMnF_3$." *Physical Review B* 4:3206–45.

Turner, C. M. 1951. "Electron Loading in Ion Accelerating Tubes" (abstract). *Physical Review* 81:305.

U.S. Atomic Energy Commission. 1951. *AEC Contract Policy and Operations.* Washington: U.S. Atomic Energy Commission.

U.S. Atomic Energy Commission. 1952. *Neutron Cross Sections.* AECU 2040. Washington: Office of Technical Services, Department of Commerce.

U.S. Atomic Energy Commission. 1957. *Theoretical Consequences of Major Accidents in Large Nuclear Power Plants: A Study of Possible Consequences If Certain Assumed Accidents, Theoretically Possible but Highly Improbable, Were to Occur in Large Nuclear Power Plants.* WASH-740. Washington: U.S. Atomic Energy Commission.

U.S. Atomic Energy Commission. 1965. *Policy for National Action in the Field of High Energy Physics."* Washington: U.S. Atomic Energy Commission.

U.S. Congress, Joint Committee on Atomic Energy. 1960. *The Future Role of the Atomic Energy Commission Laboratories*. 86th Cong. 2d sess., 60475.

Van Allen, J. 1997. "Energetic Particles in the Earth's External Magnetic Field." In *Discovery of the Magnetosphere*, ed. C. Gillmor and J. Spreiter, 235–51. Washington: American Geophysical Union.

Van Hove, L. 1954. "Time-Dependent Correlations between Spins and Neutron Scattering in Ferromagnetic Crystals." *Physical Review* 95: 1374–84.

Veksler, V. 1944. "A New Method of Acceleration of Relativistic Particles" [in Russian]. *Comptes Rendus (Doklady) de l'Académie des Sciences de l'URSS* 43: 329–31; 44: 365–68. (An abbreviated English translation appeared in *Journal of Physics USSR* 9 [1945]: 153–58.)

Vonnegut, K. 1963. *Cat's Cradle*. New York: Delacorte Press.

Walker, J. A. 1992. *Containing the Atom: Nuclear Regulation in a Changing Environment 1963–1971*. Berkeley and Los Angeles: University of California Press.

Walker, R. L. 1989. "Learning about Nuclear Resonances with Pion Photoproduction." In *Pions to Quarks: Particle Physics in the 1950s*, ed. L. M. Brown, et al., 111–25. Cambridge: Cambridge University Press.

Wang, Z. 1995. "Politics of Big Science in the Cold War: PSAC and the Funding of SLAC." *Historical Studies in the Physical and Biological Sciences* 85: 329–56.

Wasson, O. A., et al. 1966. "Direct Neutron Capture in Co^{59} (n, γ) Co^{60}." *Physical Review Letters* 17: 1220.

Watson, W. W. 1946. "Associated Universities, Inc." *Yale Scientific Magazine*, December, 7–8, 16, 18, 20.

Weart, S. R. 1979. *Scientists in Power*. Cambridge: Harvard University Press.

Weart, S. R. 1988. *Nuclear Fear: A History of Images*. Cambridge: Harvard University Press.

Weisskopf, V. 1960. *Proceedings of the International Conference on Nuclear Structure*, ed. D. A. Bromley and E. W. Vogt, 890–905. Toronto: University of Toronto Press.

Westfall, C. 1988. "The First 'Truly National Laboratory': The Birth of Fermilab." Ph.D. diss., Michigan State University.

Westwick, P. J. 1996. "Abraded from Several Corners: Medical Physics and Biophysics at Berkeley." *Historical Studies in the Physical Sciences* 2: 131–62.

Wood, R. E., H. H. Landon, and V. L. Sailor. 1955. "Parameters for the Neutron Resonance in Gold." *Physical Review* 98: 639–45.

Wright, B. T. 1954. "Magnetic Deflector for the Bevatron." *Review of Scientific Instruments* 25: 429–31.

Wu, C. S., et al. 1957. "Experimental Test of Parity Conservation in Beta Decay." *Physical Review* 105: 1413–15.

Yang, C. N. 1983. *Selected Papers, 1945–1980, with Commentary*. San Francisco: W. H. Freeman.

Yang, C. N., and R. L. Mills. 1954. "Conservation of Isotopic Spin and Isotopic Gauge Invariance." *Physical Review* 96: 191–95.

Yuan, L., ed. 1964. *Nature of Matter: Purposes of High Energy Physics*. BNL 888.

Zahl, P. A., F. S. Cooper, and J. R. Dunning. 1940. "Some *in vivo* Effects of Localized Nuclear Disintegration Products on a Transplantable Mouse Sarcoma." *Proceedings of the National Academy of Sciences of the USA* 26: 589–98.

Index

accelerators
  alternating gradient synchrotron
    (AGS), 365
    construction, 251–52
    design, 219–25
    experiments, 281–315
    naming, 220–21
    upgrade, 270, 272, 315, 361, 363
  Bevatron (Berkeley), 123–27, 145–46,
    150, 225, 367
  butterfly cyclotron, 145–46, 220,
    387 n. 59
  Cambridge Electron Accelerator, 255,
    344
  Cosmotron, 125–51, 200–256
  cyclotrons, general, 108, 122, 143
  eighteen-inch cyclotron, 159
  electrostatic accelerator (Van de
    Graaff), 111–18, 121, 129
  ISABELLE, 364–68
  ISR (CERN), 315, 365
  project, 20, 25, 43, 49, 61, 68, 108–51
  Proton Synchrotron (CERN), 145, 147,
    223, 241, 251–52, 282–83, 287–
    90, 303–12
  Relativistic Heavy Ion Collider, 256
  600–1,000 GeV (BNL), 262–65, 267
  sixty-inch cyclotron, 118–21, 129, 213
  Stanford Linear Accelerator Center,
    315, 367
  storage rings, 263, 365
  synchrocyclotrons, 121, 385 n. 26
  synchrotrons, 123
  200 GeV (Fermilab), 262–64, 266–70,
    315, 361, 363–65, 367
  240-inch synchrocyclotron (BNL),
    121–25
  zero gradient synchrotron (Argonne),
    251, 272, 303–6, 312, 367
Adair, Robert, 201, 231, 234–35, 244–
  47, 254–55, 294–98, 388 n. 4
Adams, J. B., 223, 251
Alburger, David, 113
Acta Crystallographica, 341
Alfred I. duPont Institute, 65
Alperin, Harvey, 332
alternating gradient synchrotron. *See*
  accelerators
Alvarez, Luis, 117, 147, 150, 233–35,
  239–41, 270, 301, 313
American Crystallographic Association,
  341–42
American Institute of Physics, 311
American Museum of Natural History, 192
American Nuclear Society, 191, 199
American Physical Society, 14, 19, 67,
  166, 246, 248, 350
Anderson, Clinton, 274
Anderson, R. Christian, 60, 101, 102, 105
annealing (of BGRR), 177–78
applied science, department of, 350
Argonne Cancer Research Hospital, 191
Argonne National Laboratory, 1, 25–26,
  31, 45–46, 72, 93, 102, 153, 157–
  59, 167, 182, 206, 251, 257, 272,
  304–6, 312, 346, 354, 388 n. 4
associated production, 205, 210, 238,
  302, 361
Associated Universities, Inc.
  contract, 55–58
  formation of, 33
  future challenges of, 368
Association of Oak Ridge Scientists, 70
Aston, Francis, 165
Atomic Energy Acts, 40, 50, 258, 274,
  403, 346
Atomic Energy Commission
  classification of research, 54–55
  contract negotiations, 55–58
  division of biology and medicine, 65, 184
  security clearances, 49–58
  WASH-740 report, 274–77

Atomics International, 346
Atoms for Peace, 104, 160–61, 186

Babcock and Wilcox Co., 80–81
Bacher, Robert, 20, 25–26, 28–32, 125, 260
Baker, Charles, 388 n. 4
"Barn Book," 159, 162, 326
Barschall, Heinz, 388 n. 4
basic vs. applied research, 27
beam separators, 406 n. 18
Beers, Norman, 98
Bell Telephone Laboratories, 13, 189
Bendix, 191
Berkeley Radiation Laboratory. See Lawrence Berkeley National Laboratory
Berkner, Lloyd, 149, 185
Berlin, Irving, 375 n. 16
Berman, Helen, 341–42
Bernardini, Gilberto, 289
Bethlehem Steel, 94, 130
Bethe, Hans, 174
BeV Day (Cosmotron), 141
Bhabha, Homi, 165
big discovery complex, 283–89
biology department, 63–66
Birgeneau, Robert, 338
Blachman, Nelson, 129, 134
Blewett, Hildred, 139, 221, 223–24, 377 n. 5
Blewett, John, 128–29, 133–37, 141, 145, 147, 219, 221, 223–24, 226, 255, 365, 377 n. 5
Blume, Martin, 332, 350
Boffey, Phillip, 352
Bohm, David, 233
Bohr, Aage, 169
Bohr, Niels, 148, 165
Bohr-Mottelson model, 169, 325
Bonestell, Chesley, 186
boron neutron capture therapy, 154, 182–89
Borst, Lyle, 26, 69–83, 87–92, 170–71, 192, 317, 319
Breit, Gregory, 160
Breit-Wigner formula, 160, 173–74, 212
Brobeck, William, 123, 126
Brockhouse, Bertram, 331, 390 nn. 13, 14
Bronk, Detlev W., 150
Brookhaven Council, 262, 363

Brookhaven Concept, 57, 64, 210, 239
Brookhaven Employee Recreation Association (BERA), 58
Brookhaven/Facade (revue), 353–54
Brookhaven Free Press, 355–57
Brookhaven Graphite Research Reactor. See reactors
Brookhaven Medical Center, 188
Brookhaven Medical Research Reactor. See reactors
Brookhaven, naming of, 33
Brown, Donald P., 309
Brueckner, Keith, 212, 243
bubble chambers, 233
  6-inch (BNL), 234–35, 237
  10-inch (Berkeley), 241
  10-inch (BNL), 237
  12-inch (BNL), 238
  14-inch (BNL), 235, 294
  15-inch (Berkeley), 241
  20-inch (BNL), 239–41, 282, 289, 302–4
  30-inch (BNL), 241, 282, 287
  72-inch (Berkeley), 241
  80-inch (BNL), 241, 272, 282, 301–9
  7-foot (BNL), 272–73, 313
  12-foot (Argonne), 272
  14-foot (BNL), 270–72, 367
  Gargamelle (CERN), 270
Burns, Mary (scanner), 396 n. 11
Bush, Vannevar, 9, 349
Business Week, 277
Butterworth, Ian, 304

Cabibbo, Nicola, 300
Cabibbo angle, 300–301
Cambridge Crystallographic Data Centre, 341
Cavendish Laboratory, 165, 329
CERN, 141, 143, 145, 147, 223–24, 241, 251–52, 280, 282–83, 286–93, 304–12
Chadwick, James, 165, 178, 183
chamber days, 207–8
Chandrasekhar, Subrahmanyan, 165
charge conjugation (C), 398 n. 41
chemistry department, 116, 213, 215–17
Chernick, Jack, 172, 195–97, 198, 317
China Syndrome (film), 404 n. 45
Chinowsky, Willy, 218, 231–32
choppers, 158–59, 163, 170–71

fast, 160–64, 326, 334, 347, 389 n. 10
slow, 160, 162, 334, 344–45, 390 n. 13
Chrien, Robert, 326–27
Christofilos, Nicolas, 219–20, 396–97 n. 21
classification of research, 47, 54–55
Clinton National Laboratories. *See* Oak Ridge National Laboratory
cloud chambers, 205–7, 231–33
cloudy crystal ball model, 160
Cocconi, Giuseppe, 289
Cockroft, John, 165
Cohen, Michael D., 357
cold neutron source, 343–45
Collier's, 185–87
Collins, Arthur, 120
Collins, George, 128, 143, 200–201, 218, 220, 226, 228, 232, 242, 247, 253–55
Collins Radio Company, 119–20
Compton, Arthur H., 9–10, 14–15, 24
Compton, Karl, 14, 23
computers, 171–73, 246, 294, 334, 340–42, 391–92 n. 28
Cosmotron department, 200, 254
Condon, Edward U., 23
Conrad, Robert D., 63
contract, AEC-AUI, 27–29, 41, 43–44, 47, 55–58
Cool, Rodney, 212–13, 245, 283, 289, 298, 313
Corliss, Lester, 179, 181, 339–40
Corning Glass Works, 134
Cotzias, George, 277–78
Coulson, Charles, 105, 340
counter groups, 212–13, 282–83, 287
Cox, George, 161
Cowley, Roger, 338
CP violation, 294–98, 301, 314, 361
Crick, Francis, 341
critical facilities, 182
critical point, 390 n. 14
Cronin, James, 287, 295–98
Cronkite, Eugene, 354, 357–58, 364
cross-sections
neutron, 158–60, 170, 254–55, 324–26, 388 n. 6
particle, 212, 282, 284–85, 298–301, 304
Crystallographic Computing Network (CRYSNET), 341–42

crystallography, 329–31, 339–43
Cumming, Jim, 217
Curtis, Howard J., 65

Daddario, Emilio Q., 349–50
Daddario-Kennedy bill, 350
Dahl, Louis, 277, 279
Dahl, Odd, 143
Dalitz, Richard, 242
David, Edward, 350
Davis, Raymond Jr., 280
Dean, Gordon E., 148–49
Deficiency Appropriations Act, 83, 86, 119
Delner Corporation, 76, 79
Department of Energy (U.S.), 364, 327
der Mateosian, Ed, 167–68, 327
de Solla Price, Derek, 263
Devons, Samuel, 267, 269
Dexter, Ed, 141
diamonds, irradiating, 191
Dienes, George, 154, 177, 195, 332, 335, 337
Dirac, P. A. M., 165
Dodson, Richard, 179, 260
Donora (Pa.), 100
Doppler broadening, 174
Downes, Kenneth, 197–98, 275, 277, 317
Drell-Yan process, 300
Dubos, René, 105
DuBridge, Lee, 16–17, 20–23, 28–29, 105

Eddington, Arthur, 165
Einstein, Albert, 13, 96, 165, 301
Eisenhower, Dwight D., 54, 160
Eisenhower, Julie Nixon, 359
electron analogue, 222–25, 397–98
electron loading, 117
electroweak force, 300–301
Eli Lilly Company, 65–66
Energy Research and Development Administration, 363–64
engineering department, 91
environmental concerns, 93–104, 194–95, 274–77, 348, 351–54, 358–60, 382 nn. 31, 32
Environmental Defense Fund (EDF), 358
environmental Maginot line, 352
Esso Laboratories, 189
Evening Star, 266

Faissner, Helmut, 289–92
Falk, Charles, 253–54
Farr, Lee, 65, 182–88
Federal Bureau of Investigation (FBI), 50
Federation of Atomic Scientists, 70, 105
Feinberg, Gerald, 265, 284–85, 300
Fermi, Enrico, 9, 69, 148, 157, 202, 205,
    212–13, 331
Fermilab, 269, 272, 315, 361
ferrite, 133–34, 179–81, 253
Feshbach, Hermann, 67, 160
Feynman, Richard, 233, 243, 286
Fisk, James, 125–27
Fitch, Val, 295–98, 315, 365
flux, 25, 74, 322, 325–26
Fort Devens, 32
Fort Hancock, 32, 34–35
Fort Slocom, 31
Fowler, Earle, 206, 210
Fowler, William, 206–7, 306, 309, 311
Fox, Marvin, 70, 72–73, 87–88, 153, 175,
    192, 194
Frazier, Chalmers, 179, 333, 335
Friedlander, Gerhart, 50–52, 213, 215–
    17, 226, 340
Fuchs, Klaus, 53, 389 n. 12

Gaillard, Jean-Marc, 287
Galison, Peter, 239, 363, 398 n. 38
Gell-Mann, Murray, 205, 228–31, 237,
    241, 284, 302–4, 307–8, 310–12,
    332
General Ceramics and Steatite Corpora-
    tion, 134–35
General Dynamics, 189
General Electric, 94, 112–18, 134, 189,
    213, 219
General Nuclear Engineering Company,
    198, 317
General Services Administration, 358–
    59
Gerlach, Walther, 166
Gibson, John, 103
Gingrich, John, 51
Glaser, Donald, 233–34
Glashow, Sheldon, 300
Glennan, T. Keith, 267, 269
Goco contracts, 56, 258
Goldhaber, Gertrude, 153, 165–70, 327
Goldhaber, Maurice
    as director of BNL, 260–62, 267, 269,

271–72, 274, 283, 289, 292, 298,
    300, 304, 311, 315, 323, 348, 354–
    55, 358, 360–61, 365
    neutrino helicity, 248–50
    nuclear structure research, 153, 165–
    70, 183, 218, 327
Goldwasser, Edwin, 269, 272
Goldsmith, Hy, 388 n. 4
Goudsmit, Samuel, 179, 227, 243, 254
Goulianos, Dino, 292
Goward, Frank, 143
Green, G. Kenneth, 124, 128–29, 137,
    147, 220, 225, 226, 252, 254–55,
    281–82, 287–88, 290, 294, 296,
    312
Gregory, Bernard, 241, 308
Grodzins, Lee, 249
group theory, 302
Groves, Leslie R., 9, 14–18, 20, 23–24,
    26, 28–29, 43–44, 47
Gurinsky, David, 50, 61, 71, 73, 80, 178,
    195
Gurney, Natalie and Ronald, 52–53,
    377 n. 5

Hafner, Everett, 113–17
Hahn, Otto, 165
Hale, William, 184
Halpern, O., 331
Hamilton, Alexander, 122
Hamilton, Walter, 339–43
Hartzell, Karl, 96, 383 n. 4
Harvard Medical School, 63, 191
Harvey, Paul, 93
Harwell, 178, 331, 344
Hastings, A. Baird, 63–64, 184
Hastings, Julius, 179, 181, 195, 197–98,
    339–40
Haworth, Leland
    and accelerator project, 113, 117, 119,
        120, 124–26, 128, 133–34, 141,
        146, 149
    as AEC commissioner, 281
    assuming directorship of BNL, 47, 67–
        68
    comparison with Goldhaber, 261–62
    and HFBR project, 195, 199, 317
    as lab consultant, 326
    last years as BNL director, 257–59
    and pile project, 63, 75, 81, 83, 88, 90
    and research programs, 153, 159, 172,

175, 178–79, 194–95, 199, 200, 220–27, 247
and WASH-740 report, 274–75
Heilbron, John, 109, 386 n. 40
Heisenberg, Werner, 141, 148
Hendrie, Joseph, 197, 198, 316–21, 346–47
Herb, Ray G., 111, 113, 117
High Energy Advisory Committee (HEAC), 283, 287–89, 296, 299, 303, 304
High Energy Study Committee, 365
High Voltage Engineering Company (HVEC), 111
Higinbotham, William, 78, 105, 351, 377 n. 8, 384 n. 22
Hill, Robert, 167–68
Hirs, C. H. W., 329
Hitler, Adolf, 165–66
H. K. Ferguson Company, 76–87, 90, 94
Hoey, George, 112
Hofstadter, Robert, 286
Hoffman, Roald, 60
Holifield, Chet, 276
Hollinger, David A., 262
Hough-Powell flying spot digitizer, 406 n. 18
Holton, Gerald, 208
House Committee on Un-American Activities (HUAC), 70
housing, availability of near Brookhaven, 60–61
Hughes, Donald, 153–64, 173–75, 195, 199, 244, 254, 323–27, 329, 335, 343
Hydrocarbon Research, Inc., 75–76, 79
hyperfine interactions, 174
hypertension, link with salt consumption, 277, 364

Initiatory University Group (IUG), 16–18, 20, 22, 25–33, 70
instrumentation and health physics department, 104
insulin, synthesis of human, 277
isomers, 167–69

Jameson, John D., 95, 98
Jacobus, David, 133–38, 147
Jensen, Jack B., 309
Johnson, M. H., 331

Johnson, Lyndon, 255, 349–50
Johnson, Thomas, 206, 222
Joint Committee on Atomic Energy (JCAE), 40, 84, 93, 258, 264, 274, 318, 367

K_L particle, discovery of, 230–33, 361
Kane, Walter, 327
Kapitza, Peter, 165
Kaplan, Irving, 44, 71, 74, 77, 89, 91–92, 159, 172, 380 n. 8
Kato, Walter, 163–64
Katsoyannis, Panayotis, 277
Kelley, William, 54
Kendrew, John C., 329, 331
Kenard, Olga, 341
Kennedy, Edward, 350
Kennedy, John F., 259
Kerst, Don, 129
Kevey, Andrew, 332–34, 343
Kevles, Daniel J., 9
Kistiakowsky, George, 23–24, 32, 63
Koerber, Kenneth, 63, 102
Kouts, Herbert, 71, 154, 181–82, 198–99, 257, 275, 316–17
Kruger, P. Gerald, 183
Kuper, Mariette, 18–20, 23–24, 31–33, 39, 44, 149–50
Kuper, J. B. H., 19, 104
Kusch, Polycarp, 316

Lagarrigue, André, 289
Lamb, Willis, 316
Landé, Kenneth, 210, 231
Landau, Lev, 390 n. 14
Langdon, Frank, 335
Langsdorf, Alexander, 206–7
Larsen, Richard, 294
Lawrence, Ernest O., 9, 21, 54, 108–9, 126, 129, 134, 143, 145, 233, 241, 267
Lawrence Berkeley National Laboratory, 108–11, 122–51, 219, 266, 280, 282, 303–4
layoffs, 358
leakers, 192–94
Lear, John, 185–86
Lederman, Leon, 210–12, 231–33, 244–46, 283, 286–96, 298–99, 315, 360
Lee, Tsung Dao, 242–44, 285–86

Leipuner, Lawrence, 235, 245–46, 254–55, 294
Leitner, Jack, 235–36, 238, 244
Lennon, John, 61
life sciences, 61–66
Lilienthal, David E., 43–44, 46, 94–95, 104
Limon, Peter, 299
Lindenbaum, Seymour, 201, 212
Lindgren, Robert, 113
Livingston, M. Stanley, 20, 25, 44, 108–11, 118–19, 122–29, 133, 141, 143–49, 219, 221, 223
Lloyd Harbor Study Group, 359
Long, Franklin, 98, 161
Long Island Lighting Company, 326, 359, 385 n. 24
Loomis, F. Wheeler, 22, 31, 166
Los Alamos National Laboratory (LANL), 9, 10, 12
Love, Robert A., 63
Loy, Myrna, 394 n. 47

Main Street Press, 358
Mallory, Donald, 39, 41, 47
Manhattan Engineer District (MED), 9, 16, 20, 28–31, 39–43, 47, 54, 74–76
Manhattan Project, 1, 9, 11–17, 25, 32, 65, 72, 90, 157, 160, 167, 198, 352
Manowitz, Bernard, 85, 89, 394 n. 47
March, James G., 357
Marshall, Walter, 331–32
Martin Marietta Corporation, 46
Massachusetts Institute of Technology Radiation Laboratory (MIT rad-lab), 9–10, 12, 19, 20, 22, 40, 68, 135
Matthews, P. T., 310
May-Johnson bill, 40
McCarthy, Joseph, 53
McClintock, Barbara, 277
McDaniel, Paul, 269
McKeown, Mike, 167
McMahon bill, 40, 70
McMillan, Edwin, 121, 123, 126, 267
McReynolds, Andy, 179
medical department, 61–66, 182–89, 364
Meitner, Lise, 165
Metropolitan Museum of Art, 189

Meyer, Edgar, 341
Mihelich, John, 167–68
Milband, Tweed, Hope, Hadley & Mc-Cloy, 33, 84, 117, 120, 150, 381 n. 27
Miller, Jack, 216
Mills, Frederick E., 313, 345, 365, 368
Mills, Robert, 242, 300
Monsanto Chemical, 46, 189
Moore, William, 130
moratorium activities, 354–55
Morris, Thomas, 210
Morse, Phillip, 22–25, 33, 40–41, 47, 49, 52–54, 63, 65–68, 70, 76, 97, 109, 112, 118, 123–26
Mottleson, Ben, 169

NASA, 348
Nathans, Robert, 333–36, 344–45
National Academy of Sciences, 266
National Defense Research Committee, 27
National Environmental Policy Act, 351
National Institute of Medicine, 349
National Science Foundation (NSF), 259, 339, 350
National Synchrotron Light Source (NSLS), 364
Nation's Business, 266
Naval Research Laboratory, 189
Naval Officers' Reserve Group, 357
Needell, Allan A., 90
Néel, Louis-Eugène-Félix, 180–81
Ne'eman, Yuval, 311
Nernst, Walther, 165
neutrino
    discovery of two types of, 283–94, 314, 361
    helicity of, 248–50
New Scientist, 310
New York Times, 103, 310–11, 337
New Yorker, 160
Newsday, 357
Nevis Laboratory, 122–23, 210, 213, 235, 237–38, 244–45, 290, 360
Nichols, Kenneth D., 14–15, 22, 26, 43
Nims, Leslie, 63, 65
Nixon, Julie. See Eisenhower, Julie Nixon
Nixon, Richard M., 350, 358, 359

north tract, 358–59
nuclear engineering department, 346, 350
Nuclear Regulatory Commission (NRC), 317

Oak Ridge National Laboratory (ORNL)
  anti-Vietnam War activity at, 354
  early history, 1, 9, 25–26, 45–46, 69, 72, 76–77
  tensions during BGRR construction, 83–84
  media sensationalism at, 94
  nuclear physics research at, 157–58, 162, 179, 196, 327, 331, 391 n. 14, 408 n. 26
O'Donnell, Eileen, 294
omega minus, 303–13, 361
open houses, 104–6
Ono, Yoko, 161
Oppenheimer, J. Robert, 28, 105, 243, 264, 368

Pais, Abraham, 205, 228–31, 237, 241, 265, 296
Pais-Piccioni regeneration effect, 231
Palevsky, Harry, 157, 172–73, 199, 254–55, 324, 326, 343–44, 390 n. 14
Palmer, Robert, 270–71, 305–6, 312
Panofsky, Wolfgang, 360
parity (P), 241–46, 248
Parkinson's disease, 277, 350, 361, 364
Passell, Lawrence, 175
Pauli, Wolfgang, 165
Pauling, Linus, 339
Pearson, Drew, 100–101
Pegram, George B., 12–17, 29, 33, 105
Pegram lectures, 105–6
Pendray & Liebert, 97
Pepinsky, Raymond, 179, 336
performance vs. political rivalries, 127, 224, 241, 283
Perlman, Morris, 50–52
Perutz, Max, 329
Peter, J. Georges, 40–41, 61
phase stability, 121–22
phase transition (accelerators), 221–22
phase transition (states of matter), 338–39
Philips Laboratories, 133

Phillips, Jack, 88
Phillips Petroleum, 189
phonons, 390 nn. 13, 14
Physics Today, 135
Piccioni, Oreste, 201, 212–13, 226, 231, 247, 298
Physical Review, 145, 232, 243, 249, 295
Physical Review Letters, 286, 292, 308
Placzek, George, 331
Planck, Max, 165
Planned Parenthood, 358
Plotkin, Martin, 122, 124, 127, 128, 133–34, 224
Polk, Irving, 135–36
Pontecorvo, Bruno, 285
Porter, Charles, 160, 173
Porter, Cole, 255, 354
Porter-Thomas distribution, 160, 173
Pough, Fred, 191
Powell, Jim, 346
Powell, Robert, 71, 78, 87–89, 103, 175, 178, 193–95, 323
Price, Melvin, 274–77
Primakoff, Henry, 265
Project Jason, 348
Protein Data Bank (PDB), 341–43
Purcell, Edward, 260

quarks, 312, 314
quadrupole, 145

Rabi, I. I., 7–22, 26–28, 32, 54, 58, 61, 87, 90, 96, 106, 123–24, 126, 128, 210, 224, 260, 267, 281, 316
Rahm, David, 234, 245
Rainwater, James, 316
Ramsey, Elinor, 32–33
Ramsey, Norman, 10–18, 20, 22, 29–33, 39, 47, 166–67, 206, 260, 262
Ramsey (PSAC/GSAC) Panel, 262–63
Rau, Ronald, 244, 246, 280, 283, 304–5, 308, 350
reactor department, 153, 175
reactor division, 175
Reactor Institut Laue-Langevin, 347
reactor science and engineering department, 91, 175
reactors, 11–12, 15, 23, 200
  Argonne Advanced Research Reactor (A²R²), 196, 346

reactors (*continued*)
Brookhaven Graphite Research Reactor, 69–92, 99–100, 152–99, 249, 263
power recovery proposed at, 74, 99
Brookhaven Medical Research Reactor, 188–89
CP-1 (Argonne), 69, 157, 379 n. 1
CP-2 (Argonne), 379 n. 1
CP-3, 167, 194, 383 n. 42
CP-5, 198
High Flux Beam Reactor, 195–99, 316–47
High Flux Isotope Reactor (Argonne), 195–98, 321
ORR (Oak Ridge Research Reactor), 196
project, 20, 25, 43, 45, 49, 61, 68, 69–92
pulsed, 346–47
X-10 (Oak Ridge), 25–26, 69, 72, 74, 79
Republic Aviation Corporation, 189
resonances
nuclear, 158, 325–26
particle, 212–14, 301
Reynolds, Edward, 41, 43–44, 55–57, 149
Rideout, Stuart, 173
Riste, Tormod, 336
Robinson, Denis, 111
Rockefeller, Nelson, 269, 359
Roosevelt, Franklin, 13
Ruby, Stanley, 248
Rorer, David, 326
Rustad, Brice, 248
Rose-Gorter method, 174
Rosenberg, Julius and Ethel, 53
rotons, 390 n. 14
Russell, C. R., 159
Rutherford, Ernest, 165–66

Sachs, Wilma, 63
Sailor, Vance, 83, 87–88, 152–53, 159, 170–77, 325–27, 343, 359, 381 nn. 25, 26
Sakitt, Mark, 358
Salam, Abdus, 300
Saland, Edward, 226, 312
Salisbury, W. W., 119–20
Samios, Nicholas P., 60, 235–38, 244, 280, 303–15

Sandweiss, Jack, 304, 360
Sayre, Ed, 189
scanning, 208, 232, 238, 294, 307
scattering, neutron, 162, 389–90 n. 13
elastic, 390 n. 13
inelastic, 162, 331–32, 389–90 n. 13, 390 n. 14
Schaefer, Vincent, 219
Schiff, Leonard, 249
Schoenborn, Benno, 329–31, 342
Schrödinger, Erwin, 165
Schwartz, Melvin, 235–38, 241, 244, 283, 285–94, 298
Schwinger, Julian, 265, 331
Scientific American, 163
Seaborg, Glenn, 259
security clearance, 41, 47, 49–54, 77–78, 237
Seidel, Robert, 109, 361, 392 n. 28
Serber, Robert, 129, 149, 166, 202, 204, 213, 216, 227, 231
Shanks, Camp, 32
Shaw, Milt, 346
Shirane, Gen, 179, 323, 335–39
Shoreham nuclear power plant, 326, 359
Shoup, Eldon C., 41, 55–58, 76, 90, 112
Shull, Clifford, 163, 174, 179, 181, 336, 390 n. 13
Shutt, Ralph, 201, 205–10, 226, 234, 238–41, 244, 271–72, 289–90, 306, 308–9, 311–13, 344–45
Sigma Center, 159–61
Slater, John C., 15, 44
Small, Max, 80
Smith, Bruce L. R., 348, 367
Smith, Lyle, 247–48, 254
Smithsonian Institution, 145
Snyder, Hartland, 124, 129, 144–45, 149, 202, 219–21
solid state physics, 177–81, 331–39
spark chambers, 287–88
spectrometers, 170–71, 321, 391 n. 23
crystal, 170–75, 179, 327, 329, 331
double-axis, 179, 331, 340
triple-axis, 331, 337–38, 340, 345
universal spectrometer 1 (US-1), 332–35, 343
sperm whale myoglobin, 329, 331, 342
Spelman, Joseph W., 102
Sputnik, 198, 258, 262, 348

sputtering, 117
Stassen, Harold, 148
Steinberger, Jack, 232, 235–38, 244, 283, 287–88, 290
Steiner, George, 97
Steinfield, Winton, 183–84
Sternglass, Ernest, 351–52, 359
Stone & Webster, 32, 34–35, 36–37, 251
strange particles, strangeness, 205–12, 228–33, 237–38, 241–43, 301–10
Strauss, Lewis L., 148, 188
strong focusing, 144–47, 219–20
strong interaction, 301–15
SU(2) × U(1), 301, 314
SU(3) × SU(2) × U(1), 313–14, 365
SU(3), 302, 310–14
Substitute Alloy Metals (SAM) laboratories, 70
Suffolk Scientists for Cleaner Power and Safer Environment, 326
summer student program, 60
Sunderman, F. William, 62–63
Sunyar, Andrew, 167–68, 249, 327
Sutton, O., 382 n. 32
Swartz, Clifford, 200, 218–19
Sweet, William, 184–85, 188–89

Tape, Gerald, 172, 259–60
Taylor, Hugh S., 14
technetium 99m, 191, 350
Telegdi, Valentine, 245–46
test shack, 131, 149, 224
Thomas, L. H., 145
Thomas, Lewis, 298, 367
Thomas, Robert, 160, 220
Thomson, J. J., 165
time reversal (T), 399 n. 57
Thorndike, Alan, 206–8, 210, 239
Three Mile Island reactor accident, 317
time of flight measurements, 158
Truman, Harry, 40
Tristan, 327, 347
Trump, John, 111
Tucker, S. M., 359
Tucker, Walter, 191
Turkevitch, John, 225
Turlay, René, 296
Turner, Clarence, 117–18
Tuttle, William, 207, 309

Union Carbide, 46
Upton, Camp, 30–41, 46, 84, 353, 358, 375–76 n. 16
Urey, Harold, 9

Van de Graaff, Robert, 111
Van Horn, Emery L., 29, 58, 76, 119, 149, 172, 267, 407 n. 11
Van Hove, Léon, 162–63, 331
van Kármán, Theodore, 219
Van Slyke, Donald D., 63–66, 183–85
Veksler, Vladimir I., 121, 123, 224–25
video game, first, 105
Vineyard, George H., 254, 337, 359, 360, 367, 393 n. 35
Vladimirsky, V., 224
von Dardel, Guy, 290
von Laue, Max, 165–66
von Neumann, John, 19, 23

Walton, Ernest, 165
Wang, Zuoyue, 270
Warren, Shields, 65, 184
WASH-740 report, 274–77
Watson, James, 341
Watson, William, 14, 20, 27–28
Webster, Medford S., 309
weak interaction, 205, 283–86, 293–301
Weart, Spencer, 94, 103, 360
Weil, George, 44
Weinberg, Alvin, 194, 197, 258
Weinberg, Steven, 251, 266, 301
Weisskopf, Victor, 160, 169, 266
Weneser, Joseph, 202, 345
Westinghouse Electric Corporation, 189
White, Milton G., 22
Whittemore, William, 206–7
Wideroë, Rolf R., 143
Wigner, Eugene, 25–26, 160, 177
Wigner effect, 177
Williams, Clarke, 20, 32, 33, 70, 198, 266, 268
Wilson, Carroll, 57
Wilson, Robert, 147, 267, 269, 361
Windscale reactor accident, 178
Winsche, Warren, 71
Winston, Harry, 191
Witczak, Ignatsi, 50–52
Wollen, Ernest, 179

Woodwell, George, 359
Wright Air Development Center, 189
Wu, Chien Shiung, 244–46, 248

Yang, C. N., 242–44, 246, 285–86, 300
Yang-Mills theory, 242–43, 300
Yosemite revolution, 258

Yuan, Luke, 212, 244, 263–66
Yukawa, Hideki, 316

Zacharias, Jerrold R., 8, 15, 20–25, 29,
    30–31, 96
Zimmerman, Robert, 326
Zinn, Walter, 26, 157, 198